The World Trade Organization and International Denim Trading

The World Trade Organization and International Denim Trading

Y. LI, Y. SHEN, L. YAO and E. NEWTON

The Textile Institute

WOODHEAD PUBLISHING LIMITED
Cambridge, England

Published by Woodhead Publishing Limited in association with The Textile Institute
Woodhead Publishing Limited
80 High Street, Sawston, Cambridge CB22 3HJ, UK
www.woodheadpublishing.com; www.woodheadpublishingonline.com

Woodhead Publishing, 1518 Walnut Street, Suite 1100, Philadelphia, PA 19102-3406, USA

Woodhead Publishing India Private Limited, G-2, Vardaan House, 7/28 Ansari Road, Daryaganj, New Delhi – 110002, India
www.woodheadpublishingindia.com

First published 2003, Woodhead Publishing Limited

British Library Cataloguing in Publication Data
A catalogue record for this book is available from the British Library.

ISBN 978-1-85573-693-1

Printed by Lightning Source

Contents

Acknowledgment

We would like to thank The Hong Kong Polytechnic University for funding this research through the Area of Strategic Development: "Fashion" Design and Technology Innovation".

1 International Textile and Apparel Trading

1.1 GENERAL TRADING ENVIRONMENT FOR TEXTILE-RELATED PRODUCTS

OVERVIEW

Since the textile and apparel industry is of special importance to almost every nation in the world, this sector has merited unique attention both in an economic and a political sense. While other manufacturing sectors have been progressively liberalized under the GATT and WTO rules by such means as "MFN status" and "tariff imposition rather than quantitative restrictions", the textile and apparel sector remains to be deviated from this mainstream. Furthermore, the increased sharper global competition in textile-related trade has led to a set of intricate and complicated multilateral and bilateral agreements regulating the behavior of all the participants.

The UK and US were once the leading textile and apparel exporters in the world, before the 1950s. However, Japan's quick recovery in textile and apparel production as well as consequent increased exports, followed by Hong Kong, South Korea, India, and Pakistan, soon aroused upset feelings among manufacturers in the developed nations. Confronted with the import surge of textiles and apparel from more and more developing countries, the developed nations, relying upon their economic strength and political power, began to manipulate the global trade through enforcement of a series of agreements. These measures, including the "Voluntary" Export Restraints Agreement (VER), and the short-term and long-term Arrangement Regarding International Trade in Textiles (STA/LTA), paved the way for later enforcement of the Multifiber Arrangement (MFA) in 1974, a multilateral commitment exerting great influence upon the present world textile and apparel trade pattern.

It is worth noting that there are no quantitative restrictions among the developed nations based on the so-called "gentlemen's agreement". With the developing countries gaining more economic and political power in the world arena, they strive harder for the elimination of unfair quota restrictions in the textile and apparel trade. The successful completion of the Uruguay Round in 1994 marked another milestone in the textile and apparel trade history. The ATC agreement, which replaced the MFA since 1995, finally set a clear date for the full integration of textile and apparel trade within the GATT regime.

1.1.1 MFA & ATC

1.1.1.1 MFA

The MFA was elaborately worked out to manage the exports from the developing countries into the developed countries. It is an instrument conceived to safeguard the domestic interests of the importing countries, including the US, EU, Canada and Norway. It became effective in January 1974 and underwent five successive negotiations until the year 1993 [1]. It provides for the application of selective quantitative restrictions when surges in imports of specific products caused, or threatened to cause, serious damage to an importing country whose domestic textile industry was not in a position to compete with the imports featuring low cost and cheap prices. The MFA umbrella only embraces very general principles, under which the bilateral agreements between individual importing and exporting countries are the basis for MFA operation [1]. Generally speaking, these bilateral agreements usually comprise product-specific quotas upon textile and apparel exports from a particular country, control of the annual growth rate of the base quota level, required documents upon the entry date, and possible steps the importing country can take in face of an import surge during a certain period.

A significant change meriting special attention is the "reasonable departure" clause concluded in 1977 [1]. It allows participating countries to negotiate bilateral agreements freely, regardless of the former provisions of the original MFA. Some experts referred to this clause as a "departure from departure"—a way of waiving the provisions of an agreement which was itself a derogation from GATT principles. Therefore, in each negotiation round, there were more restrictive measures added to the renewed agreements in the light of changing circumstances. For example, the covering scope extended from cotton to wool and manufactured fibers in the first round. But it was further expanded in 1986 to include other natural fibers such as ramie, silk and flax. Another example is the departure from the 6 percent growth rate of quotas set in the original agreement. The importing countries, based on domestic concerns and strategic considerations, are entitled to discuss various growth rates with individual exporting countries.

During the MFA enforcement period, both the number of countries and the number of categories of products subject to quota restraints have increased. By 1994, MFA had 44 members (EU countries are counted as one member), covering approximately 57 percent of world textile trade and 65 percent of world apparel trade [1]. One study assessed that the global welfare loss resulting from the MFA was US$ 7.3 billion annually [2]. Though it seems that few countries, both in the developed and developing groups, were completely satisfied with the MFA, it did exist for more than 25 years and its main principles will continue to function until the year 2005. It has played a remarkable role in shaping current trade patterns and building the worldwide development track in the textile and apparel sector.

First, the MFA provided a certain "stable and predictable climate" for textile and apparel trading. The exporting countries, despite their size and strength, secured certain market access into the importing nations according to the quota levels stipulated in the bilateral agreement. And at the same time, the manufacturers in the importing countries could evaluate the approximate scale of imports based on the aggregate/group/category ceiling agreed with each exporting country.

Second, some traditional exporting countries and regions, such as Japan and Hong Kong, shifted their textile and apparel production base to less developed neighbors who were not subject to the MFA at that time or who had relatively large quotas. The technology and expertise input, on the other hand, helped these later comers to develop their domestic textile and apparel industries, further intensifying world competition and adding new members to the MFA list as well. For example, the US expanded its restriction countries from 17 in 1976 to 31 in 1986 [3]. In order to completely avoid the existent and potential limitations on exports, some countries, including Japan and South Korea, made a stride forward to establish plants in the importing nations either through direct investment or acquisition.

Third, though technology innovation and product diversification are the natural course when an industry has developed to a certain level, the existence of MFA certainly played a role here. Restricting imports on some items, simply provided room and incentives for exporters to shift to uncontrolled areas or develop new types of materials. As a result, the fiber coverage under MFA was continuously extended to new categories during the various stages. The rapid development of manmade fibers can also be partly attributed to strict restrictions upon natural ones.

Fourth, the ability and possibility to export is sometimes beyond the exporter's control. It is not only dependent upon product features such as the design and quality, but also upon the availability of quotas and the required documents. The same is true for the importers. They have to consider the quota issues and related extra costs when planning to source from foreign countries. Therefore, the MFA complicated the traditional decision-making process and trade procedures, involving both parties in time-consuming document application and cost computation on the premise of available quotas.

1.1.1.2 ATC

The greatest achievement of the Uruguay Round is the establishment of the WTO, which brings all agreements, either in the former GATT regime or newly reached ones during the Uruguay Round such as the Agreement on Textiles and Clothing (ATC), under one institutional umbrella. It is a multilateral organization supervising not only the trade on goods, but also the trade on services and trade-related issues as well. Up to now, there are 137 member countries as of June 14, 2000, accounting for about 95 percent of world trade volume (WTO). The organization is trying to create an ideal environment in which all the participants can compete with each other on an equal footing.

The ATC is a landmark accord in the textile and apparel sector, and it will gradually bring the MFA back to the normal track of the GATT (or WTO at present) regime within a 10-year period. By the year 2005, trade in textile and apparel will have been completely free from any quota restraints or other barriers not compliant with the WTO spirit. The effects of the ten-year phase-out program, though uncertain and hard to assess at present, are bound to be profound upon the restructuring process going on in the textile and apparel sector in most economies.

Table 1.1 shows the liberalization and integration process from 1995 to 2005. The scope of the products is extended to those that have never been subject to the MFA. As regards the base annual growth rate, it rages from less than 1% up to 10% with 3 to 6% as the focus. If the base growth rate is 1% annually, the expected growth rates during stages 1, 2 and 3 will be 3.48%, 4.35%, and 5.52% respectively. Therefore, the real liberalization pace is insignificant due to the low base rate.

The importing nation determines at which products it will integrate at each stage to reach these percentage thresholds, with the only condition that the list of products at each integration stage must consist of each of the four categories: tops and yarns, fabrics, made-up textile products and clothing. Therefore, most importing countries make elaborate arrangements so that the quotas in the most sensitive categories are removed at the last minute. For example, the US leaves 89% of the quotas on apparel to the last date of the ten-year transition [4].

Table 1.1 Liberalization and Integration Process

Stage	*Percentage of products to be brought under GATT*	*Annual growth rate*
Stage 1: Jan 1, 1995 to Dec 31, 1997	16% (minimum, taking 1990 imports as base)	16%
Stage 2: Jan 1, 1998 to Dec 31, 2001	17%	25%
Stage 3: Jan 1, 2002 to Dec 31, 2004	18%	27%
Stage 4: Jan 1, 2005	49% (maximum)	No quotas left

(Source: based on relevant clauses in the ATC)

In order to prevent import surges during the ten-year transitional period, the US and EU signed more restrictive bilateral agreements with the so-called "dominant" suppliers, such as China and Korea. In light of the EU-China agreement covering the period from 1996 to 1998, most products are subject to growth rates lower than 1% because it was expected at the time that China would soon join the WTO and would then be subject to the ATC. As regards Hong Kong, exports of 23 clothing products are under quota regulation, of which 12 are confined to an annual growth of no more than 2% [5].

ATC contains other rules pertaining to a special transitional safeguard mechanism[1], quota circumvention, administration of restrictions, and commitments undertaken in all areas of the Uruguay Round as they relate to textiles and clothing. The Textiles Monitoring Body (TMB) has been set up to oversee the implementation of the ATC and to examine whether the measures taken under it are in conformity with the relevant rules.

It is hard to predict the future scenario in textile and apparel trade with the self-destruction of ATC at present. But one thing is certain: by the year 2005, those that are not ready for the competition in a more open environment will have encountered tremendous problems and difficulties, some may be deadly to survival. Under the MFA arrangement, most textile and apparel trade is conducted within the allowed scope of individual bilateral agreements, which keeps a relatively clear development track for every participating country. However, with the gradual liberalization in this sector, it is expected that the established market order and the accompanying game rules will undergo remarkable changes. The more or less guaranteed market access for exporting countries will no longer exist. What's more, it is important to note that ATC introduces the concept of balance of rights and obligations, or the principle of "Reciprocity". The traditional exporting countries are required to take drastic steps to improve their market access while the importing countries are gradually liberalizing their quantitative restrictions. Any reduction on the mandatory growth rates is possible if a member is found to be discriminating against imports without due reason,

[1] It is intended to protect importers against damaging import surges during the transition period from products which have not been integrated into GATT. It can be applied on a selective, country-by-country basis through mutual agreement, or if bilateral consultation comes to nil, by unilateral action for at most three years. The quotas raised shall grow at the rate of no less than 6% after they have been in place for more than one year.

or not in conformity with its commitments, such as tariff bindings, elimination of non-tariff barriers, and facilitation of administrative formalities.

1.1.2 Tariff

Tariff is the main instrument employed to protect a nation's domestic industries as well as to reap fiscal revenue. Sometimes, it also functions as a punitive weapon to remedy trade distortion, as in anti-dumping cases. One of the major contributions of GATT during the past 50 years is the significant cut of trade-weighted MFN tariff rates on industrial goods, from as high as 40% at the end of WWII to around 6% at the end of the Tokyo Round (OECD, 1997). According to a report prepared by Japanese Ministry of International Trade and Investment, the import-weighted average applied rates in Japan, the US and EU in 1999 were 1.5%, 3.6% and 3.6% respectively, much lower than the pre-Tokyo Round level [6, p40].

Tariff is one of the few legitimate measures allowed under the WTO regime as a means of protections domestic industries. Therefore, in view of the unique position of the textile and apparel sector in domestic industries, most nations set a higher tariff rate for textile-related imports in order to erode their potential price advantage in the domestic market. For example, the US imposed an import rate as high as 25% upon woolen fabrics [6]. Table 1.2 is a comparison of import tariffs imposed upon the textile-related sector and the manufacturing sector. It clearly shows that the textile and apparel sector still faces high import tariff rates despite the continuous drop in other manufacturing sectors.

Since HK is a free port, no import duties are imposed upon textile and apparel products. The zero-tariff is one of the key factors accounting for the large re-export volume every year. There are no tariffs upon imports between EU members, whereas towards those from non-EU suppliers, the EU exercises average applied rates of 5.3% for yarns, 9.1% for fabrics and 11.9% for clothing (Dehousse et al., 1999). The EU WTO bound rates upon HS chapters 50-63 by January 1, 2005 will be no more than 12%, with most fiber and fabrics below 8% [7]. The US imposes no tariff upon Canadian exports and there are also special NAFTA regulations concerning trade with Mexican exports. As regards countries outside NAFTA, the US has relatively high tariff rates upon textile products, 2.5-12.2% for yarns, 0-31.5% for fabrics, and 33% or more for most clothing items (Dehousse et al., 1999 p165). WTO consolidated rates for textile and clothing products in Japan reach up to 14.2%, with most yarns and fabrics below 10%. The tariff rates for yarn range from 2.4-8.4%, for fabric 8.0-12.0% and for clothes 4.5-13.9% (Dehousse et al., 1999 p92). China is still a non-WTO member, though the bilateral talks with the US and the EU have reached successful conclusions. Compared with the former four markets, China imposes the highest tariff rates upon textile products, 9.8-22.0% for yarn, 15.0-34.0% for fabrics (with 22 and 34% as the focus) and 33% for most apparel items[2]. With China's entry into WTO, it will conform to the relevant WTO regulations and reduce the rates progressively.

1.1.3 Non-tariff barriers

As mentioned above, the overall tariff rate has reached a relatively low level after 8 rounds of GATT negotiations. But the global trade is facing a more restrictive environment despite free trade prospects drafted on paper. Most nations resort to

---

[2] Import tariff adjustment in 1999, China Customs

Table 1.2 Tariff rates in the US, EU, Japan and Canada

	Production weighted average applied MFN tariff rates					
	1988		1993		1996	
Country	Manufacturing	T & A	Manufacturing	T & A	Manufacturing	T & A
US	4.7	11.6	5	11.8	5.4	11.3
EU	8.4	10	8.6	9.9	7.7	9.8
Japan	4.1	10.4	3.5	11.7	3.3	10.1
Canada	10	20.2	9.7	19.6	17.8	14.4

(Source: OECD 1997, p49)

much more complicated and obscure non-tariff barriers (NTBs) whose protection effects are hard to evaluate.

Since the textile and apparel sector is still one of the most sensitive sectors in most nations, NTBs have been widely used in addition to high tariff rates. They take various forms. The quantitative restriction mentioned previously is the most common NTB. There are other types of non-tariff barriers, such as technical and administrative barriers, green barriers and the current prevalent anti-dumping investigations. In view of the ten-year quota phase-out program, some developed nations have actually intensified the application of non-quota NTBs to protect their domestic textile and apparel industry. The following is a brief introduction of some commonly used trade policies designed to distort the world textile and apparel trade.

1.1.3.1 Quantitative restrictions

Generally speaking, quantitative restrictions are prohibited under Article XI[3] of the GATT, mainly due to their more protective nature than tariff ones. However, the import quota is the most frequently used non-tariff barrier in the textile and apparel trade, a clear "departure from the GATT". It solely aims at the vast developing countries whose strength in this labor-intensive sector is very obvious and whose textile-related exports have posed a serious threat to the domestic industry in developed nations. The enforcement of the MFA over the past three decades well explains the "vitality" of this approach in protecting domestic textile industries suffering rooted problems such as high labor costs and labor shortage.

Table 1.3 is an example of quotas imposed upon blue denim fabric imports into the US market from specific countries. China has the most restrictive base level, which is only 15.9% of the Malaysian level. And the annual growth rates for HK, Taiwan and the Chinese mainland are no more than 2.5% while those for Malaysia, Egypt, Brazil and Indonesia are over 8.7%. These figures indicate that imports from traditional "big suppliers" face more severe quota restrictions than those from relatively insignificant exporters. That is why some sources point out that "big suppliers" actually gain little from the ATC arrangement. Imports from other countries[4], though not restricted by specific limits, are subject to other restrictive measures such as designated consultation levels.

[3] It states that no prohibitions or restrictions other than duties, taxes or other charges shall be instituted or maintained by any Member unless they are justified under the exception rules provided in Article XI of the GATT.

[4] They include India, South Korea, Pakistan, Philippines, Singapore, Sri Lanka and Thailand

Table 1.3 Import quotas upon blue denim fabrics by the US[5]

Quotas upon blue denim fabric imports into the US			
Country/Region	Level Coverage	Base Level (sq.m.)	Growth (%)
Hong Kong	218/225/317/326	71,437,169	2.175
Malaysia	225	40,636,560	8.700
Taiwan	225/317/326	39,408,907	2.500
Egypt	225	30,766,900	10.001
Brazil	225	12,863,788	8.700
Macau	225	11,696,601	6.350
Indonesia	225	7,721,621	8.700
China	225	6,461,344	2.500

(Source: http://otexa.ita.doc.gov/SOFA/sofa225.htm, retrieved on May 2000)

1.1.3.2 Anti-dumping and countervailing measures

Anti-dumping and countervailing measures are another two policies allowed under the WTO regime. If the importing country can prove, based on WTO substantive and procedural requirements, that there are dumped imports, material injury to a domestic industry, and a causal link between the two, it can take unilateral remedies by imposing punitively high import tariff rates upon the alleged "dumped" products [8]. As regards the countervailing measures, they can also be carried out unilaterally to offset injury caused by subsidized imports. Subsidy here is defined as a financial contribution by a government or any public body within the territory of a Member which confers a benefit [9]. The justification underlying these two measures is to restore "fairness" and normal market order. However, recent practices in the developed nations, especially in the US and EU, indicate that they are abused as an effective weapon against the imports mainly from the developing countries.

Table 1.4 shows the number of AD investigations between 1969 and June 1998. The US and EU together account for 40.9% of the total AD investigations during the 30-year period, Canada and Australia for another 37.3%. It is clear that the developed

Table 1.4 AD investigations by country

No. of Anti-dumping investigations by country between 1969 and 1998							
	1969-74	1975-79	1980-84	1985-89	1990-94	1995-98	Total
US	125	140	146	219	249	75	954
EU	19	55	138	101	147	110	570
Japan	0	0	0	0	4	0	4
Canada	42	74	176	115	90	31	528
Australia	n.a.	120	242	180	252	71	865
Others	39	64	10	74	227	395	809
Total	225	453	712	689	969	682	3730

(Source: WTO, from [6]; Period is from 1969 to June 30, 1998)

[5] Categories 218/225/317/326 refer to cotton or MMF denim, cotton twills and cotton sateen, other cotton or MMF fabrics of yarns of different colors, except jacquard woven fabrics, among which Category 225 covers blue denim fabric. Categories 218/225/317/326 from HK are covered by one level while Categories 225/317/326 from Taiwan are also covered by one level.

Table 1.5 Anti-dumping investigations in Brazil, India, Mexico and South Africa

No. of Anti-dumping investigations in some LDCs						
	1980-90	1990-92	1992-94	1994-96	1996-98	Total
Brazil	0	11	34	13	31	89
India	0	5	4	14	31	54
Mexico	11	38	47	21	13	119
South Africa	2	0	0	23	34	57

Note: Column 1: from 1980 to June 30, 1990; Column 2-5: the year is defined as from 1/7-30/6

(Source: [10])

countries play a dominant role in AD investigations. Another trend worth mentioning is the increased action carried out in some developing countries during recent years, such as Brazil, India, South Africa and Mexico. (Table 1.5) Anti-dumping investigations are mainly focused upon several specific sectors, such as the textile and apparel sector, the steel sector and the chemical sector. The less developed nations are the major targets in these anti-dumping cases, especially during the 1990s. During the 1990/98 period, about 64.7% of the total anti-dumping measures were against LDCs, among which about 40% were aimed at East Asian countries [10].

As regards countervailing duties, they are also widely employed as a means to increase the import price of the alleged subsidized imports to a level comparable with or even higher than the domestic price. Because of the ambiguity of the GATT's wording about subsidies, it is relatively easy for the importing countries to determine the existence of unfair subsidies in an arbitrary manner. That's why there are so many trade disputes concerning the imposition of countervailing duties, which sometimes have a negative effects upon the trade flow. The US is the most frequent user of this measure, initiating 235 investigations from 1985 to June 1999, among which 53 countervailing duties are still in place. The EU resorts less to subsidy investigations with only 28 cases. But there is an increasing trend in the EU market since July 1997. 89.3% of all the investigations were initiated between July 1997 and June 1999 [6, p91].

Frequent usage of anti-dumping and countervailing duty measures has certainly brought about market disruption to both importers and exporters, frustrating the normal trade flow. The developing nations are the major targets among these investigations. There is no denying that some imports from the developing nations did constitute "dumping" in the destination market. However, in terms of the striking differences between the developing and developed markets, especially concerning the resource input cost, the lower-priced imports from the developing nations sometimes are justified, requiring more careful and comprehensive examination. The unfair practice under the fair excuses clauses cast shadows upon the global liberalization pace.

1.1.3.3 Technical barriers

Though today's world is woven into a more integrated picture, each nation still retains its own legal and administrative system, leading to different rules and regulations concerning Customs procedures, labeling, inspection, product standards, sanitary requirements, and so on. Therefore, when products of Country A are exported to Country B, they have to meet the importing country's relevant requirements upon entry into its territory. A study carried out by OECD has found that different standards and technical regulations in different markets, along with the costs of testing and certifying compliance with those requirements, can account for 2-10 percent of a firm's overall production cost. The most widely used NTBs in the area of standards are conformity assessment procedures or the requirement to prove compliance with standards or technical regulations in another market, which may be sometimes unnecessary and difficult to conform with [11].

Today, technical barriers have triggered more international disputes than ever before. Because there are various kinds of technical barriers, it will take a long time frame to truly realize a unified system of international standards and procedures. In section 1.2, some specific examples are given to illustrate the negative effects of these technical barriers.

1.1.3.4 Green barriers[6]

With the worldwide enforcement of ISO 14000 and other environment-related regulations in the 1990s, "green" products are no longer strange to consumers and manufacturers. Enterprises with long-term vision focus their attention upon the development of environmental-friendly production processes and products, which enable them to gain a competitive edge over competitors.

However, though it is of great importance to maintain sustainable development, the over-use of environment-related regulations constitute de facto green barriers towards global trade. In fact, some regulations have little or nothing to do with environmental protection. For example, there are about 70 environmental seals in the German textile market, of which only 10 are for ecological purposes [12]. The Multilateral Environmental Agreements (MEA) allow trade restrictions necessary to protect human, animal or plant life or health, or relating to the conservation of exhaustible natural resources. The target is desirable, but some GATT panel reports point out that measures taken outside the jurisdiction of a regulatory country or taken to force other countries to change their policies are not justified under Article XX.

The jungle of environmental regulations not only complicates the protection task but also hampers the healthy development of future global trade. Some countries deliberately use "eco-excuses" to achieve their ultimate trade distortion aim and there are more and more green barriers in today's global trade. (Detailed examples will be given in 1.2.2)

1.2 NON-TARIFF BARRIERS IN SPECIFIC COUNTRIES

1.2.1 US[7]

1.2.1.1 Technical barriers

A. Standard and labeling barriers

- Different system of units: while most countries use the international system of units (SI) based on the metric system, the US still uses yards and

[6] Strictly speaking, a green barrier is a kind of technical barrier. But because of its rising significance in recent years, it is listed separately.

[7] These information is mainly from the following sources: [49]; 4. Hughes, J.K., *The ATC sky clouded over.* Textile Asia, 1998?p.52-61.; 13. Commission, E., *1999 report on United States barriers to trade and investment.* 1999?EU.

pounds in many circumstances and doesn't make any substantive effort to bring its domestic system into line with the international one.

- Complicated technical requirements on the product composition such as the type of dye used and the type of finishing technique employed: For example, for clothes with an outer shell constructed of more than one material, the relative weight, percentage values and surface area of each component should be marked clearly. And if these components are blends of different materials, it is required to include the relative weights of each component material. Furthermore, the label must be placed in the collar of the imported clothes.

B. Administrative barriers

Since the beginning of the ATC, the US government has carried out a number of new administrative actions that form de facto trade restrictions. They includes:

- The 592A List: The list has been issued by the US Customs since September 1995 in accordance with Section 333 of the Uruguay Round Agreement Act on a biannual basis. It names the foreign entities accused by the Customs under Section 592 of the Tariff Act for illegal transshipment of textile products. These names will stay on the list for three years and goods imported from these entities will undergo stricter procedures to ensure all the information presented is precise.
- Special import measures for Macao and Hong Kong: the US Customs raised new entry requirements on imports from these two regions, including cotton and manmade products. It also published a list of Hong Kong and Macao companies which had been prosecuted for illegal transshipments. Due to the unclear information given in the list, such as the possibility of same or similar names for more than one company, imports from the innocent company will no doubt suffer unnecessary losses.
- Increased bond requirements for importers: The present minimum bond amount is 2% of the total value of annual cotton, manmade fiber and wool imports by that importer, instead of 10% of the import duties and taxes paid to Customs every year, which increases the cost to importers. And if the importer has previously broken Federal rules, the bond can be increased up to 5%.
- Mutilation of samples: Samples that are not correctly marked as a sample or mutilated to be unsaleable will have to be re-exported or destroyed, which may cause serious losses to the importer due to the delay.
- Quotas cutback: In September 1997, the US informed WTO Member nations, without offering any appropriate adjustment to the affected countries, that the implementation of the Integration process under the ATC will include cutbacks in those quotas which are only partially freed on January 1, 1998.
- Extension of the liquidation period up to 210 days: For example, as regards the fiber content of textiles, the maximum difference between the invoice declaration and Customs control analysis should be no more than 3%. Non-conformity may trigger penalties reaching 100% of the goods value. In view of the short life span of fashion items, this regulation certainly causes trouble and extra burdens for traders.

Most of the above measures used by the US Customs are to monitor and control imports in the hope of eliminating illegal transshipment of textiles and apparel. However, it will hurt both importers and exporters who are honest and acting in good faith.

1.2.1.2 Rules of origin

In 1996, the US changed its rules of origin for finished products, establishing that the origin was conferred no longer by the operations of dyeing and printing plus two finishing operations, but by the origin of the fabrics from which the products were made.

In many ways these rules are even more protectionism because of their intricacy. Countries such as China will be negatively influenced under this new rule of origin. For example, China often exports gray fabrics to developed nations such as Italy for dyeing and finishing and then the final products are exported to the US with origin in Italy rather than in China. However, with the enforcement of the new rules, China will be the originating country and may have to be subject to quota limitation.

In addition to the general rule of origin, there are also specific NAFTA rules of origin featuring the "yarn forward" principle. This stricter rule places outsider nations in a more disadvantageous position.

1.2.1.3 Anti-dumping and countervailing measures

As regards anti-dumping measures, the US initiated 28 cases between July 1, 1997 and June 30, 1998, accounting for about 11.9% of the total cases in the world. Figure 1.1 shows the number of anti-dumping measures initiated in the US from July 1, 1989 to June 30, 1998. It indicates that the US resorted frequently to this measure in the early 1990s and that it has taken an upward trend since 1996.

Chinese exports to the US are subject to 37 anti-dumping cases up to date, 27 of which were initiated in the 1990s [14]. As regards textile and apparel exports, cotton shop towels and polyester cotton print cloth suffered anti-dumping investigations in the early 1980s.

In the US, a large number of countervailing duty cases was brought by the US industry in the mid-1980s against countries which were not signatories to the Subsidy Code. As non-signatories, these countries were not entitled to an "injury test" under US law, so the domestic industry could resort to imposition of punitive tariffs without having to prove actual injury or threat caused by the subject merchandises.

Table 1.6 US new rules of origin

US new origin rules	
Textile products	**Country of Origin classified by**
Yarn products	Spinning place
Fabric products	Weaving place
Knitted products	Knitting place
Finished products (apparel)	Sewing place
Apparel sewn in multi-country	The most important and/or final sewing place

1.1 Anti-dumping measures initiated in the US between 1989 and 1998

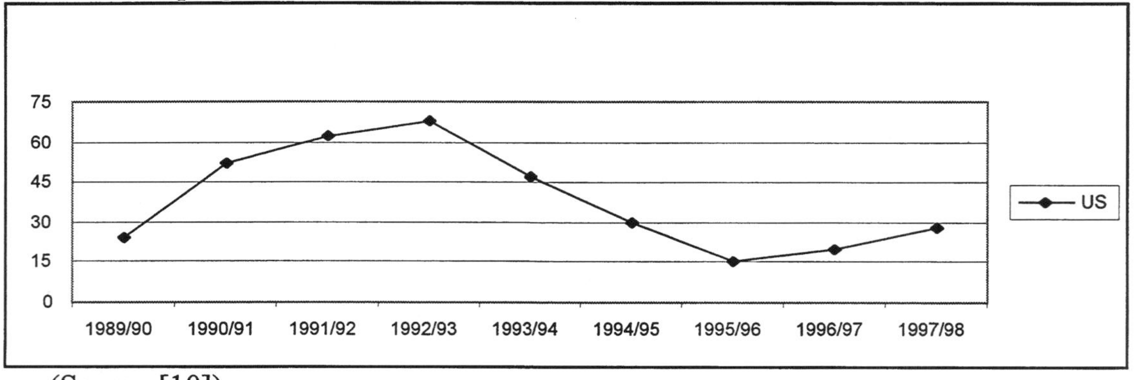

(Source:[10])

1.2.2 EU

1.2.2.1 Quota restrictions

The EU still keeps 209 quotas on textiles and clothing imports covering a total of 21 countries, of which 14 are WTO members [15]. About one fourth of EU textile and apparel imports were subject to quota restrictions in 1998. There is no quota restriction concerning imports from least developed countries.

While the main aim of the US's integration process is to prevent further import surges by signing more restrictive bilateral agreements with the so-called "dominant" suppliers during the transition period, the EU adopted a product-focused approach without publishing a full product list for each integration stage in advance [5]. In addition, the integration process will be linked to actual progress in access to third country markets.

Just as the US did, the EU established more restrictive and tougher terms when entering into bilateral agreements with supplying countries, particularly with the leading ones, including China, Hong Kong and South Korea. In the light of the EU-China agreement covering the period from 1996 to 1998, most products are subject to growth rates lower than 1% because it was expected at the time that China would soon join the WTO and would then be subject to the ATC. As regards Hong Kong, exports of 23 clothing products are under quota regulation, of which 12 are confined to an annual growth of no more than 2% [5].

1.2.2.2 Eco-labeling requirement

The EU has taken the lead in implementing Eco-friendly policies. In 1992, the EU adopted an EU-wide eco-labeling scheme, a voluntary program that permits a manufacturer to obtain an eco-label if its product meets certain criteria for 11 consumer product categories, including bed linens and T-shirts. As early as 1995, Germany prohibited the usage of certain dyestuffs containing harmful particles, such as aromatic amines and carcinogenics, in the manufacturing process. Many European dealers also require that textile imports have attached the OKO-TEX STANDARD 100, an environmental label issued by the Swiss Textile Testing Co., Ltd. What's more, the EU will pass a formal law stipulating the ban of more than 300 kinds of harmful dyestuffs from applications in textiles and apparel. Such a law may be essential for the sake of man's health. However, if this legislation is abused, it will become a green barrier hampering export activities.

1.2.2.3 Labeling

EU regulations require that textile products be labeled to show fiber content, using generic names instead of brand names [16].

1.2.2.4 Anti-dumping investigations

The Europeans have already revealed their determination to resort to "unfair trade actions" after alleging that fabrics produced in a variety of countries are being "dumped" in their market. The main product sectors covered by these investigations were those of textiles, electronics, and iron and steel (See Table 1.7).

According to the annual report released by the EU [48], the Community had 141 measures in force covering 63 products and 33 countries by the end of 1997, of which China involved 32 (or 22.7%) involved China. In the five-year period from 1993 to 1997, 167 investigations were initiated concerning imports from 38 different countries. China was one of the major countries with 25 investigations (Table 1.8). In 1997 alone, China exporters faced 5 investigations including products of gray cotton fabrics, certain footwear with textile uppers, handbags, and certain ring binder mechanisms. While investigation on cotton gray fabrics was initiated shortly after the first failed attempt to impose definitive anti-dumping duties, the latter three products were subject to high definitive anti-dumping duties ranging from 7.7% to 49.2%.

The EU used to treat China as "Non-market economy" so that "normal value" was established based on the prices or the constructed value in a third country such as

Table 1.7 Anti-dumping investigations in the EU

A breakdown of the product sectors						
Product	**1993**	**1994**		**1996**	**1997**	**Total**
Textiles and allied	1	17	4	10	8	41
Electronics	7	3	7	/	14	31
Iron and steel	/	7	2	9	4	22
Chemical and allied	5	3	4	/	8	20
Other metal	5	3	5	1	1	15
Other mechanical engineering	2	4	3	/	1	10
Wood and paper	/	/	1	/	7	8
Other	1	6	7	5	2	21

Table 1.8 EU anti-dumping investigation by country

A breakdown of the country by export						
Country of Origin	**1993**	**1994**	**1995**	**1996**	**1997**	**Total**
China	4	5	5	6	5	25
India	/	4	1	4	6	15
Thailand	2	5	4	/	3	14
Indonesia	/	4	4	1	1	10
South Korea	2	/	4	1	3	10

(Source: [48])

India and Indonesia. Such practice was unfair and brought about many difficulties and uncertainties to the investigations. In April 1998, the EU removed the label "non-market economy" from China and began to conduct anti-dumping investigations on a case-by-case approach. However, according to the Sino-US agreement, the US will still use the current "non-market economy" approach to determine whether imports from China are being dumped for another 15 years after its accession to the WTO.

1.2.3 Japan

Japan places few formal barriers on imported textiles and apparel. In fact, its import regulations on textile items have been among the most liberal within the industrial world since 1970 [17]. In 1980, its trade-weighted average tariff rate on textiles was 11%, while those in the EU and US were 14% and over 23% respectively [17]. However, according to a report prepared by the US Embassy in Tokyo [18], Japan is well known for tactics that can effectively keep non-Japanese products out of, or delay entry into its domestic market. Some commonly used NTBs are identified, including a requirement for prior experience in Japan, interconnection of business interests such as cross stockholding, powerful industry associations with arbitrary licensing powers, market influence, information control and/or limited membership, and discriminatory official regulations favoring domestic suppliers.

1.2.3.1 Quota restrictions

Japan used to be a textile and apparel exporter subject to various bilateral/multilateral agreements. After WWII, Japan recovered quickly from its nightmare and its textile and apparel exports soon aroused concern in the US and UK. Growing pressures in the domestic market forced the US government to "persuade" Japan to sign the VER on certain cotton textile products in 1955. This was the first post-WWII textile-specific restriction [1].[8] Later, Japan was the member in STA, LTA and MFA, And, by Japan becoming a developed nation, it lost its comparative advantages over less developed ones and its domestic textile and apparel industry shrank a lot. However, it still managed to lift quotas on imported textile products in 1970, hoping that the industry could move to higher value-added sectors under strong import pressure [17]. As a result, textile and apparel imports from Asian countries increased very rapidly, which aroused great domestic concerns and forced the government to negotiate some "voluntary" measures with major exporters. An agreement was reached between the Chinese and Japanese governments about the "voluntary" restraint of poplin broadcloth exports from China to Japan within the limit of 300 million meters per year [19]. There is also a Sino-Japan agreement in 1998, upon which China should "voluntarily" administer its textile exports to Japan. Before exporting the "two kinds of yarns and two kinds of cloth" to Japan, the exporter should consult and sign a contract with the Zhongda Corporation, the General Agency in Japan.[9]

1.2.3.2 Marking/labeling/packing requirements

Standard textile and apparel products are highly regulated in Japan, subject to both

[8] Textiles and Apparel in the Global Economy, third edition, Kitty G. Dickerson

[9] Almanac of China's Foreign Economic Relations and Trade, 1999/2000; the two kinds of yarns/cloth refer to cotton yarn, cotton/polyester yarn, grey cotton fabric and grey polyester/cotton fabric.

official rules and voluntary private standards, such as Japanese Industrial Standards. Textile and clothing labeling is currently subject to the "Household Goods Quality Labeling Law" and the "Act Against Unjustifiable Premiums and Misleading Representation", which will be replaced by a new law by September 30, 2000. There require that clothing labels should include a description of fabric types and textile yarn content with percentage figures for lining, thread, materials, etc, home washing instructions and other handling methods, size in metric measurements, the water resistance, and relevant information concerning the manufacturer/supplier. As regards sizing, there are specific JIS (Japanese Industrial Standards) regulations for ready-made items based on different categories and Men/Boys/Women/Girls/Infants groups. Straw packing materials are prohibited. The Japanese Measurement law requires that all imported products and shipping documents show metric weights and measures [18, 20].

1.2.3.3 Complicated distribution system

Japan's highly fragmented and complicated distribution system is widely recognized as a significant trade and investment barrier, responsible for Japan's relatively low level of manufactured goods imports from the US and Western Europe (Chen, 1995; Larke, 1994; [21]. Though a de-regulation process has been carried out since 1998, it still remains as the major concern for foreign counterparts.

Distribution channels in Japan are multi-layered with numerous wholesale and retail outlets. Wholesalers hold a dominant position in the distribution system, a unique characteristic in Japan. As a result, wholesale turnover is almost 4 times that of retailing, making consumer prices extremely high [22]. Moreover, manufacturers have dominant powers in the distribution system, around which "keiretsus" are formed. Compared with other leading industrialized nations, Japan has the highest wholesale to retail sales ratio. As regards the retail network, it is much denser than those of the US and Western Europe [23]. The Department Store Law of 1956 and the Large Retail Store Law of 1974[10], aiming to protect small and medium sized retailers, effectively restrict entry of foreign competitors with strength and economies of scale. The distribution system features strong personal ties and interdependence as well as long-term partnerships with local suppliers, making foreign entities hard to penetrate into this closely intertwined market. Some long-established business customs in the system, such as the return of unsold goods, long credit periods for payment on goods, and financial help during slump periods, are somewhat anti-competitive in nature and result in unclear accountability of risk sharing.

In recent years, there have been some significant changes in the Japanese distribution system, such as the much shorter approval time for opening large stores from 7 or 10 years to about one and a half years, application of IT in retailing and emergence of non-store retailing. But the unique characteristics in the system prevalent for so many years are hard to change in the short run and will continue to be a trade barrier.

[10] The 1956 Law restricts the establishment or expansion of department stores with a sales space of over 3000 m^2 in the seven largest cities and over 1500 m^2 in all other cities. The 1974 Law broaden this restriction to cover al large retails stores [Chen, 1995; Larke, 1994; 21. HKTDC, R.D., *A General View of Japan's Distribution System for Consumer goods*.1987?HKTDC?HK..

1.2.3.4 Anti-dumping and countervailing measures

Compared with the US and EU, AD and AC measures are not popular in Japan. There were only 4 AD cases from 1985 to June 30, 1998 and no AC cases during the same period [6, 10].

1.2.4 HK

One of the most important reasons for HK being as the entrepot port is its zero import tariff rate upon most items and relatively clear and efficient Customs procedures. There are few trade barriers to textile and apparel trade. One of the US concerns is IPR protection.

1.2.5 China

China exercises quotas, licensing, special tenders and other non-tariff measures on imports of 385 product categories' [24]. According to the World Bank, the coverage of NTBs on China's imports was reduced from about 50% in 1992 to one-third in 1996 and the tariff equivalent of China's NTBs is calculated as 9.3% in 1996 (World Bank, 1997). As regards textile and apparel imports, there were no NTBs[11] imposed upon apparel imports, while about 12.7% of textile imports were subject to NTBs, of which 0.3% were affected by state trading, 5.7% by designated trading, 12.7% by import licensing and 12.7% by quotas. China has promised to raise these restrictions within a five-year time frame after its entry into the WTO.

1.2.5.1 Quota restrictions

There are 28 categories subject to import quotas [25]. China's textile and apparel exports are subject to MFA-type restrictions and, at the same time, there are general import quotas imposed upon more than 40 tariff lines, covering wool products (9), cotton products (2) and synthetic yarns (30).

1.2.5.2 Import and export licensing

About 35 categories of import commodities (by the end of 1999) and 54 of export commodities (from January 2000) were subject to licensing control [25]. Most textile products are subject to compulsory prior import licensing, a program managed by MOFTEC. The application is first reviewed by different state agencies and MOFTEC has the final voice in the license issue.

As regards export licensing, it is not automatic and is applied to the exportation of raw materials such as cotton, silk, ramie and wool.

1.2.5.3 Highly regulated imports and export rights[12]

State-owned foreign trade companies enjoy preferential treatment over other enterprises engaging in import and export activities, especially if they are concerned with raw materials such as silk and cotton exportation. Though the cotton supply

[11] NTBs here refer to state trading, designated trading, import licensing and quotas.
[12] Some information is from 20. Franklin Dehousse, K.G., Philippe Vincent, *Market access study to identify trade barriers affecting the EU textiles industry in certain third country market.*1999?EU.

system is undergoing significant changes, including the introduction of market mechanisms and relaxation of distribution channels, cotton producers still cannot conclude transactions directly with foreign partners, who have no alternative but to purchase from a state agency. As regards silk, the state agencies enjoy a monopoly position.

Not all manufacturing enterprises enjoy export and import rights, though the relevant regulations have been progressively relaxed during recent years. The right to import and export is granted by MOFTEC based on certain criteria, including past manufacturing performance. As regards foreign investment enterprises, they face strict regulation of domestic sale ratios and are mainly subject to exportation of their own manufacturing products and importation of their manufacturing-related ones.

Individuals in China are still forbidden by law from engaging in direct import or export trade, which is a violation of Article XI of the GATT. The right for individuals to trade is a basic feature of the WTO system.

1.2.5.4 Lack of transparency and uniformity in government regulations and information release

One of the most serious problems in China is the difficult access to relevant trade information and regulation, including Customs procedures, government agencies' responsibilities and the appropriate channels for data access. Most foreign importers and exporters also complain about the arbitrary practice by Customs and other government agencies despite the existence of clear wording in the regulations.

1.2.5.5 Technical barriers

In addition to the different regulations concerning standards, labeling and certification, a lack of transparency still dominates and complicates the problem. China has identified over one hundred tariff-line items that require a safety license and about 780 imported goods are subject to statutory inspection conducted by the State Commodity Inspection Bureau. Foreign importers and exporters also complain that there is obvious discrimination against foreign entities concerning safety and inspection procedures (USITC, 1999).

Lack of intellectual property protection is another serious concern for foreign enterprises that are seeking access to the Chinese market. Meanwhile, trademark and copyright violations prevail. China was once placed on the "Priority Watch List" under the US Custom's Special 301 trade law which empowers the US to impose immediate trade sanctions on China at any time. Though the threat was removed after China's efforts to combat IPR transgressions, there is still much room for China to improve. Many world-famous fashion brands once encountered similar miserable experiences in China. The more successful the brand is, the more fakes there will be in the domestic market. If the situation is not improved, it will become an "effective" trade barrier for foreign enterprises.

1.2.5.6 Others

Anti-dumping and countervailing issues are still new in China, and the relevant law was not published until the year 1997. The first anti-dumping case in China was filed on October 1, 1997 by domestic newsprint suppliers. Domestic chemical fiber suppliers also claim that some foreign producers dumped their products in the Chinese market, causing material injury to the domestic industry. However, no further steps have been taken except oral warnings of possible anti-dumping activities. Since these two defensive measures are WTO-allowed, it is expected that Chinese enterprises will resort more to these approaches after China enters WTO.

Strict foreign exchange control in China is another major trade protection, but steps are being taken to progressively relax the regulation. In December 1996, China announced full convertibility in the current account, with the capital account remaining controlled. In October 1997, "Provisions on the Administration of Foreign Exchange Accounts in China" was issued, setting up new procedures for the opening and maintenance of foreign exchange accounts by FIEs and foreign individuals. Moreover, selected foreign banks in the Shanghai Pudong area (since 1996) and Shenzhen (since August 1998) are allowed to do RMB business for foreign companies and individuals.

1.3 INTERNATIONAL DENIM TRADING PATTERNS[13]

OVERVIEW[14]

Born in Europe and gaining its popularity in the US, denim is probably one of the most universally recognized fabrics with a history of more than three hundred years. The original Levi's button-fly shrink-to-fit jeans were introduced in 1873 when Levi Strauss and Jacob Davis patented the use of copper rivets. Since the 1960s denim, together with its related apparel items, has stepped out of its original "workwear" image, and has become a well-established category in the fashion world.

The US is the biggest denim market in the world, consuming 41.7% of the global denim supply. The denim consumption per capita reached 5.0 linear meters in 1996, compared with 2.1 in Europe, 2.4 in Japan and 0.2 in the rest of the world. An American owns 6 items of denim apparel on average. The annual consumption for denim bottoms is 3.1 units per capita, of which 2.2 units are denim jeans alone.

The colonization of Europe by jeans started after WWII, a matter associated with the glamorous heroes of the American armed forces. Today, Europe is the second largest market for denim products, accounting for 23.3% of the total world consumption of denim fabrics. The EU as a whole is the biggest consumer in the European continent.

Japan is another important market for denim products, holding 10% of the market share. Compared with the US and European countries, Japan only started denim manufacturing in the 1970s. However, based on its special focus upon technical innovation, Japanese denim products are now very popular, featuring superior quality and a high technology content.

Hong Kong, as an important fashion manufacturer and trader, especially the taking into account its unique role in the transit trade, is worth specific study as well. China is now the world No. 1 textile and apparel exporter. In view of its rapid economic development and huge domestic market, it can be anticipated that China will become

---

[13] The following analysis concerning the US and EU markets divides the world into 16 groups based on geographic proximity and regional preferential arrangements with the two most important markets--the US and EU. There are 3 in America, 5 in Europe, 4 in Asia. Africa is viewed as a whole. Due to their special location, Turkey and CIS are separately listed. Though the last group is categorized as "Oceania", it in fact includes Australia, New Zealand and those not belonging to any other group. As regards HK and Japan denim trading, the division is much broader, mainly focusing upon Asia itself. (The division uses [26] for reference.

[14] The data is mainly from [27]

another important market for denim products in the near future. Furthermore, China's WTO entry will no doubt have an impact upon the world textile and apparel trade patterns. Therefore, detailed analysis will be conducted in the above five markets.

There are comprehensive databases concerning US, EU and HK denim trade. As regards the Japanese and Chinese markets, the analysis is somewhat fragmented due to the lack of systematic trade data. Therefore, databases in the US, EU and HK are used to get a partial picture of the two denim markets. What's more, their overall textile and apparel trade performance is also useful to identify specific denim trading patterns.

1.3.1 US denim trading pattern

1.3.1.1 Background of US textile and apparel trading environment

Before going into an in-depth analysis, several key events will be introduced first since they help shape the current trading pattern.

- In January 1989, Canada and US concluded an agreement concerning a ten-year phase-out plan of all tariffs and quotas. Tariffs imposed on textile and apparel are being dropped by one-tenth each year. Because of this, and the non-existent quota restraints, trade in textile and apparel has already realized liberalization between the two countries.
- The US-Canada agreement set the stage for the launch of NAFTA on January 1, 1994, covering America, Canada and Mexico. Under the NAFTA agreement on textiles and apparel, textile products meeting specific NAFTA rules of origin are immediately exempt from quota restraints and related duties. Quotas and tariffs for other non-originating goods will be phased out in ten-year time frame.
- Since 1986, the US has carried out certain preferential trade policies to encourage economic pickup in the Caribbean region. The 9802 (formerly 807) production arrangement permits cut garments to be exported for assembly and re-imported into the US, with import duties only imposed on the value-added part. If fabrics are both made and cut in the US, then a more liberal quota system for access to the US domestic market will be secured for the finished products assembled in these countries (807A or Super 807). Consequently, almost all Caribbean-made apparel is destined for the US market. More US-based manufacturers, viewing the low cost of labor and abundant natural resources as a premium for a competitive edge, are moving to build plants or production lines there.
- The US also has preferential access arrangements with Israel, so that, for examples, their textile and apparel exports have no quota restrictions. Since January 1, 2000, a new Outward Processing Program with Romania and Macedonia has been carried out concerning wool apparel categories [28].

These preferential arrangements greatly push forward regional trade development.

Take NAFTA for example. The fiber trade volume within NAFTA jumped from US$ 6 billion in 1993 to US$ 18.4 billion in 1998 [29], with an annual growth rate of 41.3%. Mexico has replaced China as the leading textile and apparel exporter to America. Its market share concerning MFA products rose from 11.6% in 1997 to 13.7% in 1998, while that of China dropped from 8.6% in 1997 to 7.5% in 1998 [30].

1.2 US Denim fabric import pattern

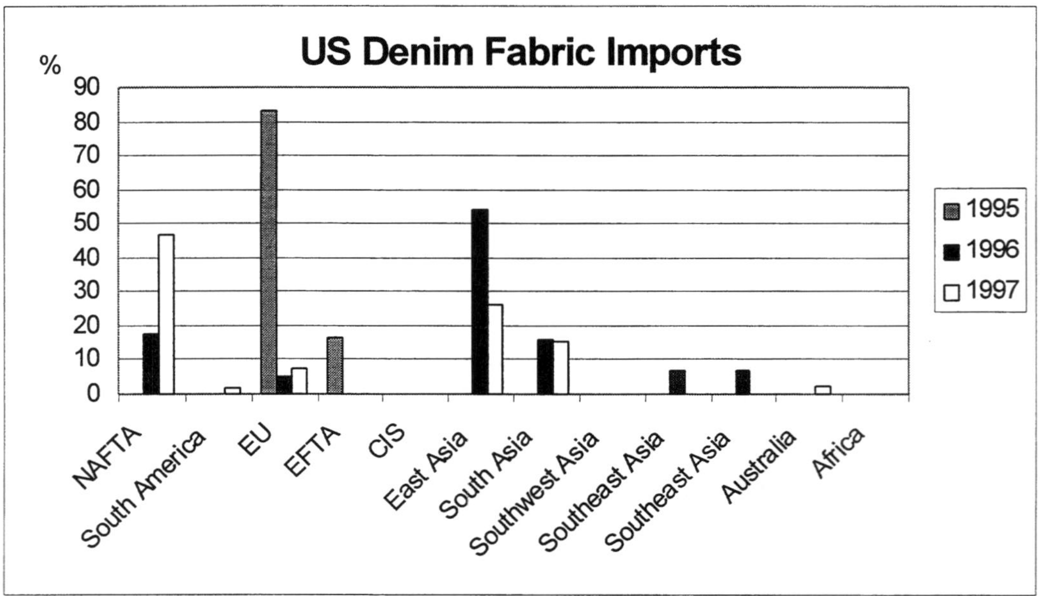

1.3.1.2 US denim fabric and apparel trading pattern

A. Denim fabric import pattern

In 1995, as Fig. 1.2 shows, the EU was the single significant denim fabric supplier to the US market, accounting for 83.3% of US total denim fabric imports. EFTA held another 16.7%. Thus, there were no imports from other regions. However, in 1997, Canada and Mexico (NAFTA) became the major suppliers with 46.9% of the market share, followed by East Asia (26.1%) and South Asia (15.3%). These three groups together accounted for 88.3% of US total denim fabric imports. The EU was reduced to the fourth supplier group with only 7.2% of the share in 1997.

B. Denim fabric export pattern

The US denim fabric export pattern is different from the import pattern, covering more export destinations as seen in Fig1.3. Canada and Mexico again accounted for almost half of US total denim fabric exports in 1997. In 1993, a year before the formation of NAFTA, only 27.2% for US denim fabrics were destined to Canada and Mexico, of which Mexico captured 24.7%. With the launch of NAFTA in 1994, exports to the two countries increased to 35.1% immediately and in 1997, the ratio jumped to 46.2%.n Exports to Mexico experienced the fastest growth, absorbing 53.3% of US total exports to NAFTA in 1997. Exports to Caribbean/Central America increased from 4.3% in 1993 to 5.8% in 1997 thanks to the elaborate OPT arrangement. The intra-NAFTA trade grew rapidly at the expense of trade diversion away from non-member nations. From 1993 to 1997, exports to South America, EU and East Asia dropped from 12.0% to 5.3%, from 38.2% to 32.2% and from 9.9% to 2.5% respectively. Exports to Africa and Southeast Asia experienced a health growth, increasing by 0.7 and 2.3 percentage points respectively.

Though US denim fabrics were exported to a large number of countries, the market concentration was high. NAFTA and the EU together absorbed 78.4% of US

1.3 US denim fabric export pattern

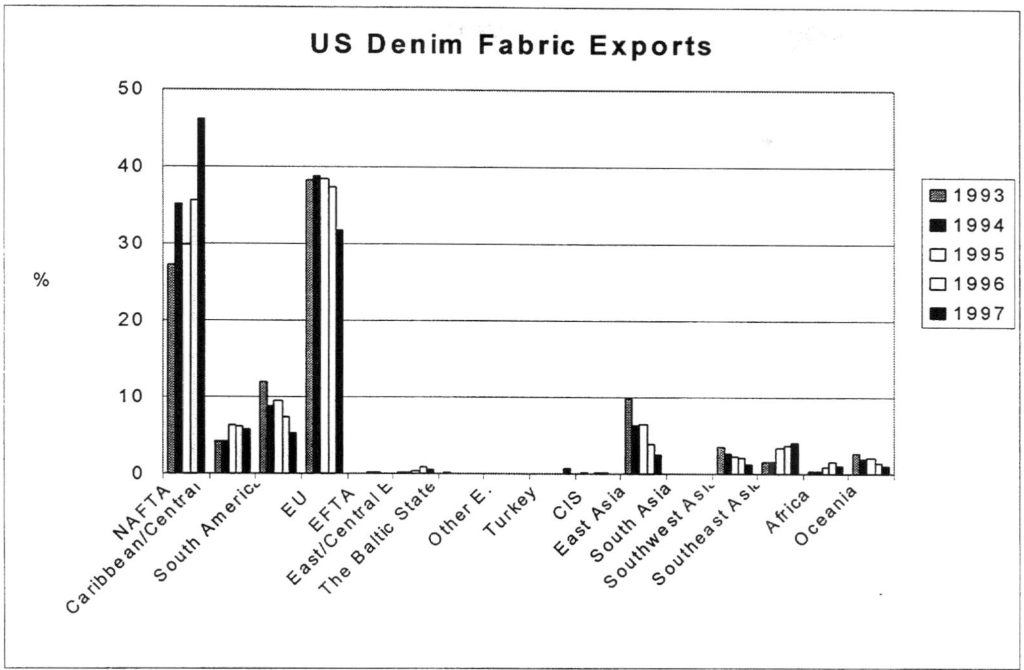

1.4 US MB denim apparel import pattern

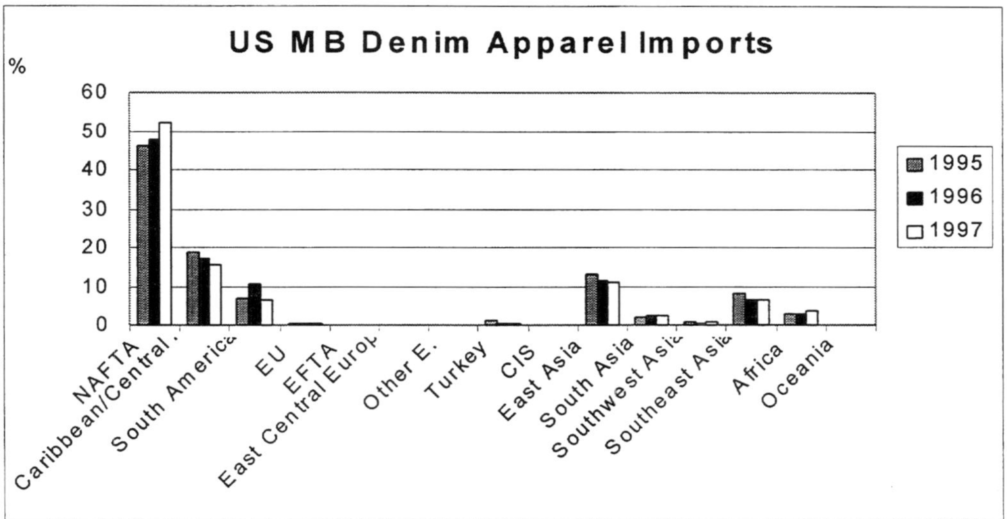

total denim fabric exports in 1997. If Caribbean/Central/South America are included, the figure rose to 89.5%.

C. Denim apparel import pattern

- Men's or boy's (MB) denim apparel

As regards US MB denim apparel imports, Mexico was its most important supplier, accounting for 93.0% of total NAFTA MB denim apparel exports to the US (see Fig. 1.4). Imports from NAFTA continuously increased from 46.0% in 1995 to 52.2% in 1997. The Caribbean/Central America was the second largest supplier with 15.5% of the market share, followed by East Asia (11.2%), Southeast Asia (6.6%) and South America (6.4%).

About 74.1% of US MB denim apparel imports were from suppliers in the American continent and 21.1% from those in Asia. Though the contribution from Africa is still small at present, its share increased a little from 2.7% in 1995 to 3.8% in 1997.

- Women's or girl's (WG) denim apparel

WG denim apparel imports from NAFTA members increased by 20 percentage points from 1995 to 1997, rising from 37.5% to 57.5% of total US imports (as see in Fig 1.5). Those from the Caribbean/Central America also increased a little from 12.3% to 13.8% in 1997. By sharp contrast, the share of Asian suppliers slumped from 40.3% to 23.8%, with East Asia suffering the most serious drop by almost 10 percentage points. Other regions, such as South America and Africa, also exported less to the US market during the three years. It is clear that Mexico and the Caribbean regions have become important WG denim apparel suppliers to the US market.

- MBWG denim apparel

Taking MB and WG denim apparel imports as a whole, Mexico has been the biggest gainer since the launch of NAFTA in 1994. Its strong export expansion to the US market pushed the overall share of NAFTA up from 42.7% in 1995 to 54.4% in 1997 (see Fig 1.6). The outstanding performance of NAFTA dwarfed that of other regions, especially the Asian nations. East Asian suppliers were the biggest losers in the hotter market competition. However, due to Asia's traditional strength in apparel manufacturing, it remained as one of the important players in the US import market and in 1997, it still held 22.2% of the market share. The Caribbean/Central/South America together accounted for another 18.6% in 1997. EU suppliers carried little weight in the US market, only holding 0.6% of US total denim apparel imports. Italy contributed over four-fifths of the total EU exports to the US.

1.5 US WG denim apparel import pattern

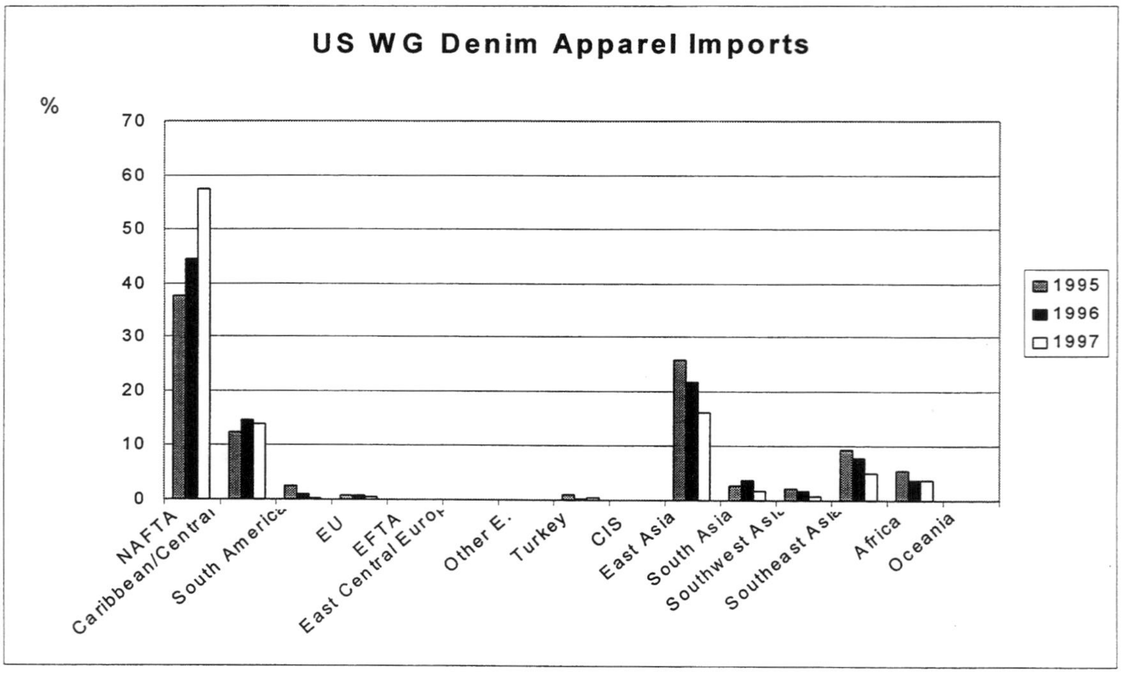

1.6 US MBWG denim apparel import pattern

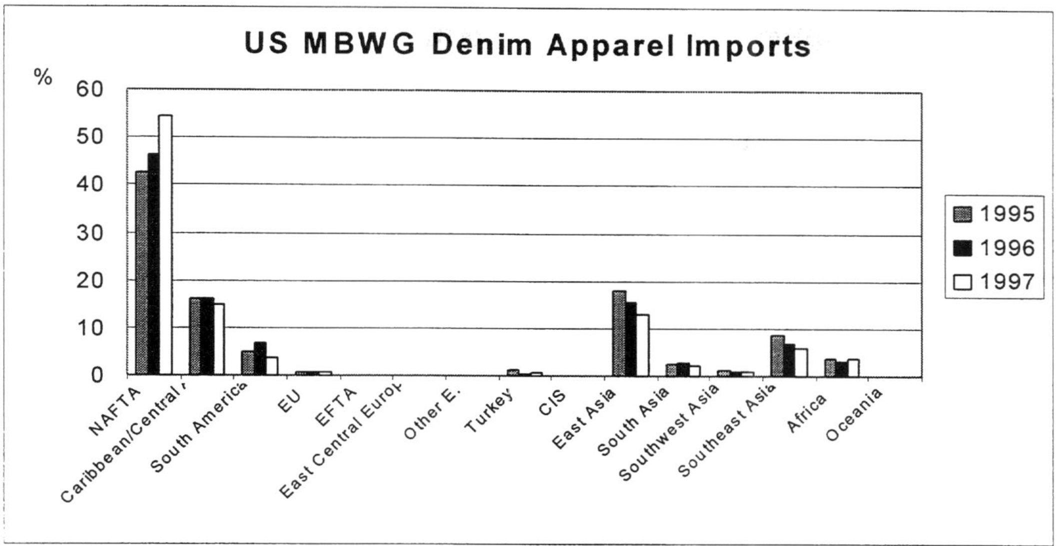

1.7 US MB denim apparel export pattern

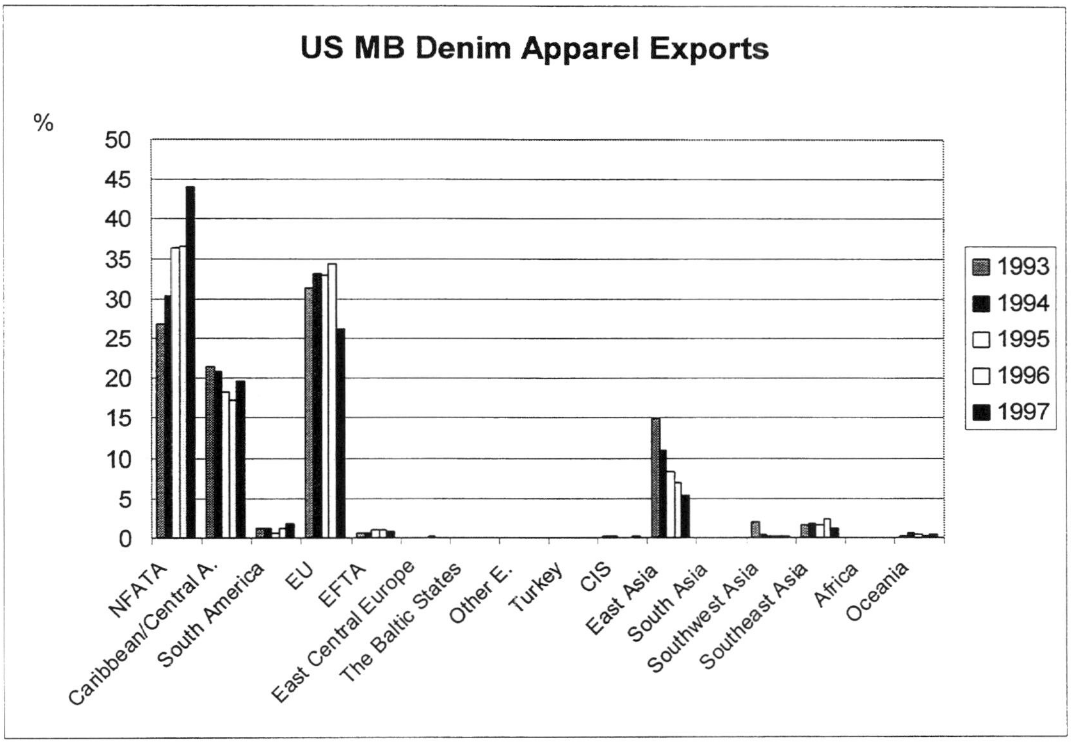

D. Denim apparel export pattern

- MB denim apparel

NAFTA and the EU are the top two leading importers of US MB denim apparel exports as can be seen in Fig 1.7. However, the two show a different development pattern. Exports headed to NAFTA enjoyed a very sharp growth, up from 26.8% of the US total MB denim apparel exports in 1993 to 44.1% in 1997; while those exported to the EU dropped to 26.3% in 1997 after the peak level of 34.2% in 1996. In addition to these two groups, the Caribbean/Central America was the third most

1.8 US WG denim apparel export pattern

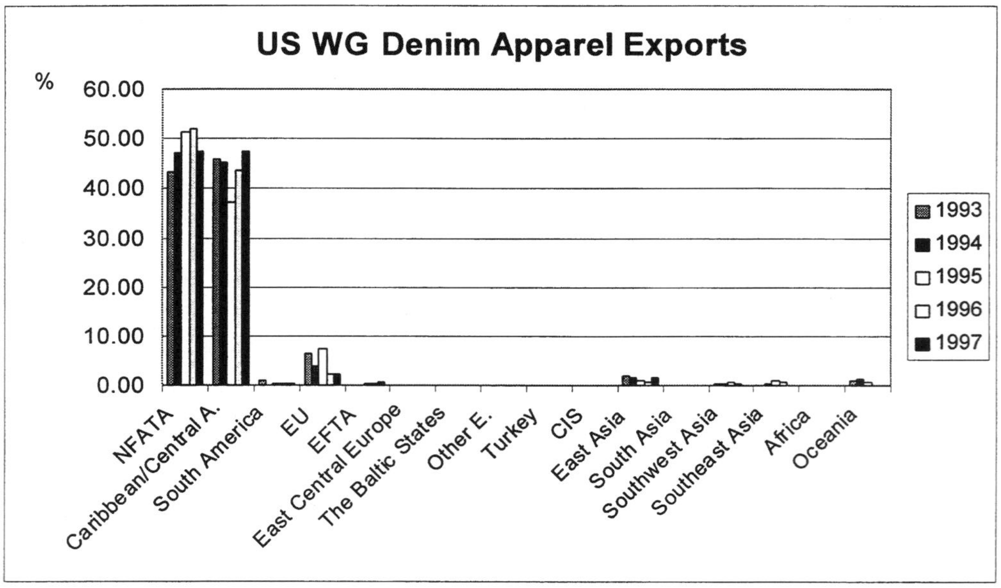

important destination for US MB denim apparel, accounting for 19.6% of US total exports. Japan was once one of the important export destinations for US MB denim apparel exports. It alone captured 95.8% of the US total MB denim apparel exports to the East Asia. However, it has imported less and less since 1993. As a result, the market share of East Asia was down from 14.8% to 5.3% in 1997. The Southeast Asia financial turmoil in 1997 caused tremendous troubles for economies in this region, negatively affecting denim apparel consumption as well. US MB denim apparel exports to this region have been continuously up until a year 1997 when it hit the bottom level of 1.1% during the five-year period.

- WG denim apparel

As regards US WG denim apparel exports, the picture was somewhat different from its MB exports. About 95.3% of WG exports were focused upon two regions: NAFTA and Caribbean/Central America, with equal market shares of 47.6% in 1997. In 1993, exports to Mexico and Canada already held 43.2% of the market share and rose to the peak level of 51.9% in 1996. Exports to the EU dropped from 6.4% in 1993 to 2.8% in 1997. The drop was not so large in East Asia, down from 2.0% to 1.5% in the same period. (See Fig. 1.8)

- MBWG denim apparel

Despite the unbalanced market concentration for US WG denim apparel exports, the overall US denim apparel was mainly destined to four regions: NAFTA, Caribbean/Central America, EU and East Asia. The four together captured 96.5% of US total WG denim apparel exports in 1997, almost the same as in 1993. The enlarged share of NAFTA (from 30.3% in 1993 to 45.2% in 1997) was achieved at the expense of shrinking EU and East Asian markets, which were 18.9% and 4.1% respectively in 1997. (See Fig. 1.9)

E. Summary

Based upon the above analysis, it is clear that the formation of NAFTA in 1994 contributed a lot to the rapid growth of US denim trading with its NAFTA partners, especially with Mexico. The Caribbean/Central American region was another important importer and exporter for US denim products, mainly due to the preferential arrangement between the region and the US. About 70% of US denim trading was conducted with these two regions. The traditional East Asian suppliers faced strong pressure from this regionalism move. Though they were still important suppliers to the US market, their market shares were continuously eroded by competitors on the American continent. The EU was an important consumer of US denim products rather than a significant supplier.

In short, US denim product imports were mainly from NAFTA, Caribbean/Central America and East Asia, together holding 82.3% of the market share. US denim product exports were mainly headed to NAFTA, EU and Caribbean/Central America, which together captured 89.5% of the market share.

1.9 US MBWG denim apparel export pattern

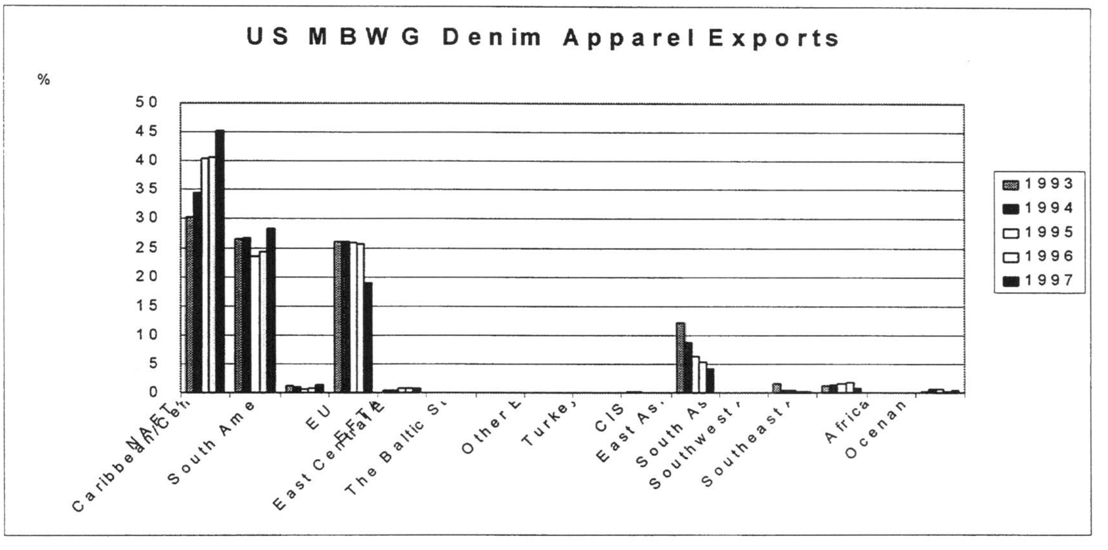

1.10 US denim products import pattern

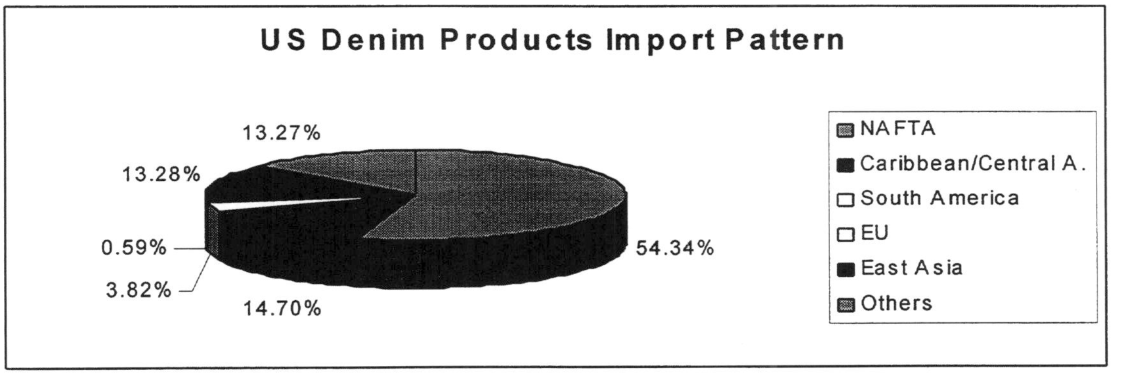

1.11 US denim products export pattern

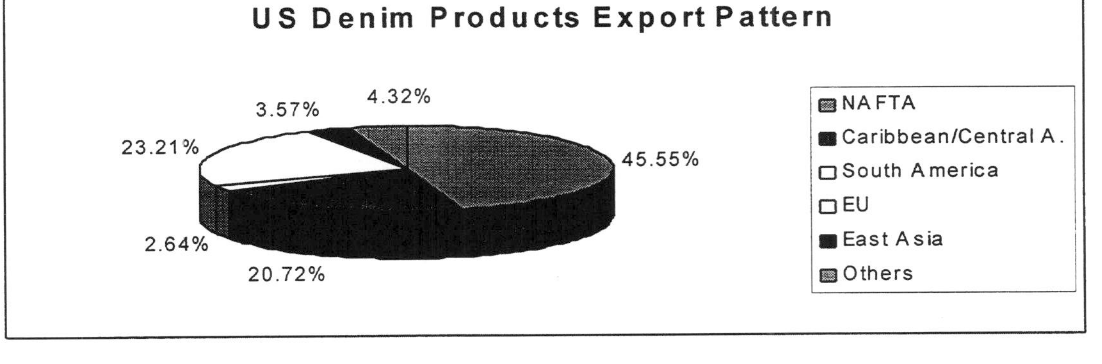

1.12 US overall denim products trading pattern

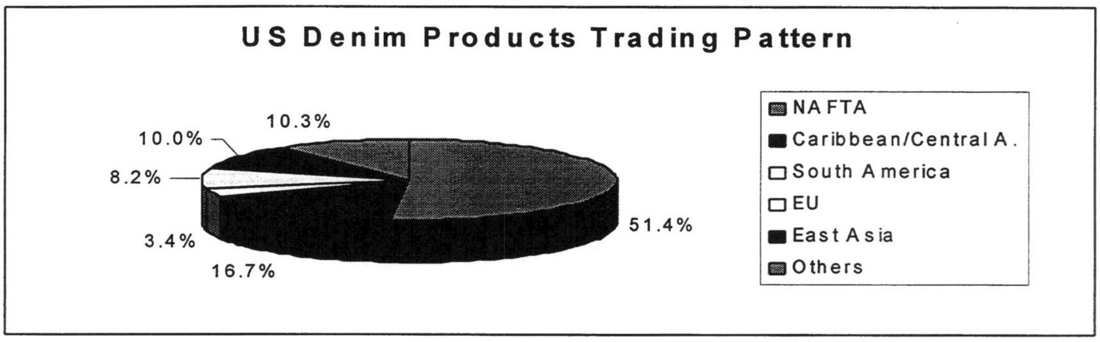

1.3.2 EU denim trading patterns

1.3.2.1 Background of EU textile and apparel trading environment

- Compared with NAFTA which was just a free trade area formed in 1994, EU goes far beyond that. The Treaty of Maastricht embraces a very ambitious goal, setting the stage for a political as well as economic and monetary union. The formal introduction of the Euro on January 1, 1999 marked the currency integrity among 11 EU Members[15], an effort to enhance the freer flow of goods, services and capital. The completion of a "single market" in 1992 has intensified intra-EU trade, which now represents about two thirds of the total EU trade [31]. About 60% of EU textile exports and 70% of EU apparel exports were conducted within EU in 1997 [32].
- The EU and the EFTA countries16 set up a "European Economic Area" (EEA) in January 1994, aiming to liberalize the movement of people, goods, capital and services and to promote cooperation in R & D, environment and other issues, a blueprint expanding the scope of a Free Trade Area.
- The implementation of the Customs Union between Turkey and the EU in January 1, 1996 has boosted the textile and apparel trade between the two sides, making Turkey one of the most important trading partners with the EU. The EU also established a Customs Union with Andorra, Malta, San Marino and Cyprus.

The closer political and economic relationship between EU and Central/Eastern European countries has strengthened the trade links between the two groups, making more EU companies shift their sourcing directions to these countries. The EU had phased out import duties on CEEC textiles and apparel by January 1, 1997 and removed quotas on January 1, 1998 [33]. The CEEC is now the second buyer worldwide of European textiles and the third apparel supplier.

- The Lome convention between the EU and 71 ACP (Africa/Caribbean/Pacific) countries provides a great preferential access to these former colonies. 92% of the products originating in the ACP enter the EU duty free. The preferential margin enjoyed in the textile sector reached 7.0% in 1996 and 6.8% in 2000. As a result, the volume growth rate between 1988 and 1997 hit 66.5% [34].

[15] Greece is now the 12th EU member who joined this monetary union.

[16] Sweden, Finland and Austria joined the EU in 1995. The rest three are Norway, Iceland and Liechtenstein

1.13 EU denim fabric import pattern

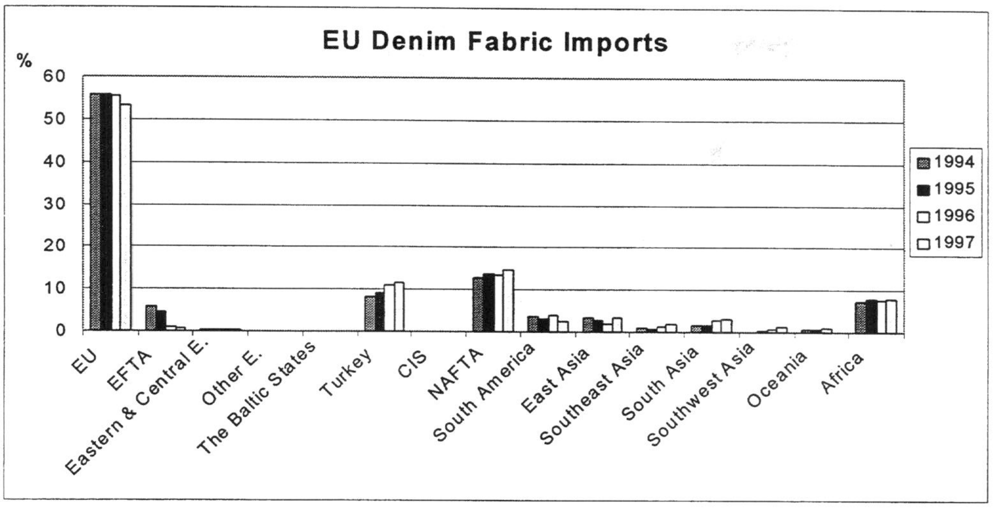

1.3.2.2 EU denim fabric and jeans trading patterns

A. Denim fabric import pattern

In 1997, about 53.4% of EU denim fabric imports were intra-EU trade(as shown in Fig 1.13). This ratio was 2.3 percentage points lower than that in 1994. NAFTA was another important supplier with 14.5% of the share, the US providing over 93.1% of NAFTA's total denim fabric exports to the EU. One should remember that the jeans industry in the EU, in fact, started with direct fabric imports from the US. With the establishment of a customs union with the EU, Turkey became the third largest supplier, with a continuously increasing market share from 8.2% in 1994 to 11.4% in 1997. Though the share of East Asian suppliers in 1997 was almost the same as that in 1994, the EU progressively increased its sourcing from other Asian regions, such as South Asia (from 1.6% to 3.0%), Southwest Asia (from 0.1% to 1.1%) and Southeast Asia (from 0.8% to 1.8%). As a result, the overall share of Asian suppliers rose from 5.8% in 1994 to 9.3% in 1997. Tunisia is a large denim fabric supplier to the EU market. It alone accounted for 66.8% of total African exports to the EU. Thus, imports from Africa held another 7.9% of total EU denim fabric imports in 1997. (See Fig. 1.13)

B. Denim fabric export pattern

The ratio of EU denim fabric exports destined to member countries reduced sharply from 57.0% in 1994 to 46.8% in 1997. At the same time, those to Eastern/Central European countries, as well as to Turkey, registered a rapid growth, from 7.7% to 10.3% and from 1.2% to 6.6% respectively. Africa was another important destination, accounting for about one quarter of EU total denim fabric exports. Tunisia and Morocco are the top two importers holding 97.1% of total Africa denim fabric imports from the EU. Other regions, excluding EFTA and NAFTA, also enjoyed growth. (See Fig. 1.14)

C. Jeans import pattern

EU members contributed about 54.1% of total EU MB jeans imports in 1997, almost the same as in 1994 (Fig 1.15). Eastern/central Europe and Turkey held 3.8% and 3.0% respectively. As regards non-European countries, Africa had a dominant position, capturing 16.0% of EU total MB jeans imports, 2.3 percentage points lower than in 1994. Those from NAFTA increased from 5.3% to 6.9%, among which the USA held 95% of total NAFTA exports to the EU. East Asia and South Asia encountered different experiences. The share of the former continuously dropped from 10.3% to 6.3% in 1997 while that of South Asia increased from 1.6% to 4.0% during the same period.

- WG jeans

As regards WG jeans imports, those from EU members fell from 54.7% in 1994 to 48.7% in 1997 while most of other regions more or less increased their market share. Africa contributed about 15.2%, followed by East Asia (13.3%), Turkey (5.4%), South Asia (4.7%) and NAFTA (3.5%). (See Fig. 1.16)

1.14 EU denim fabric export pattern

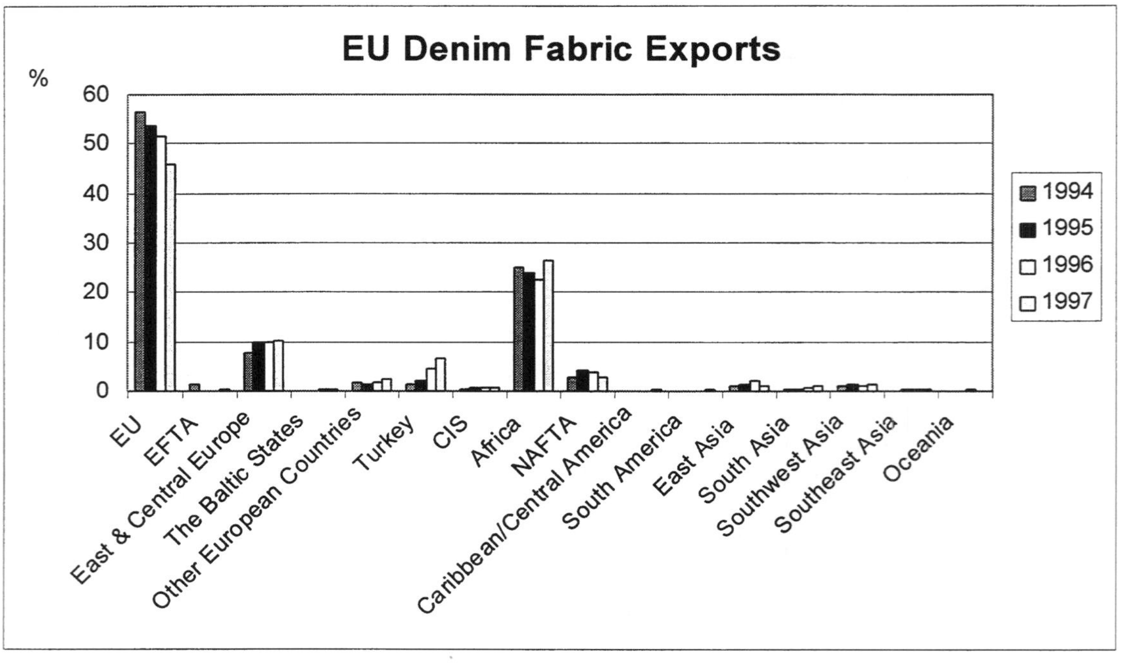

1.15 EU MB jeans import pattern

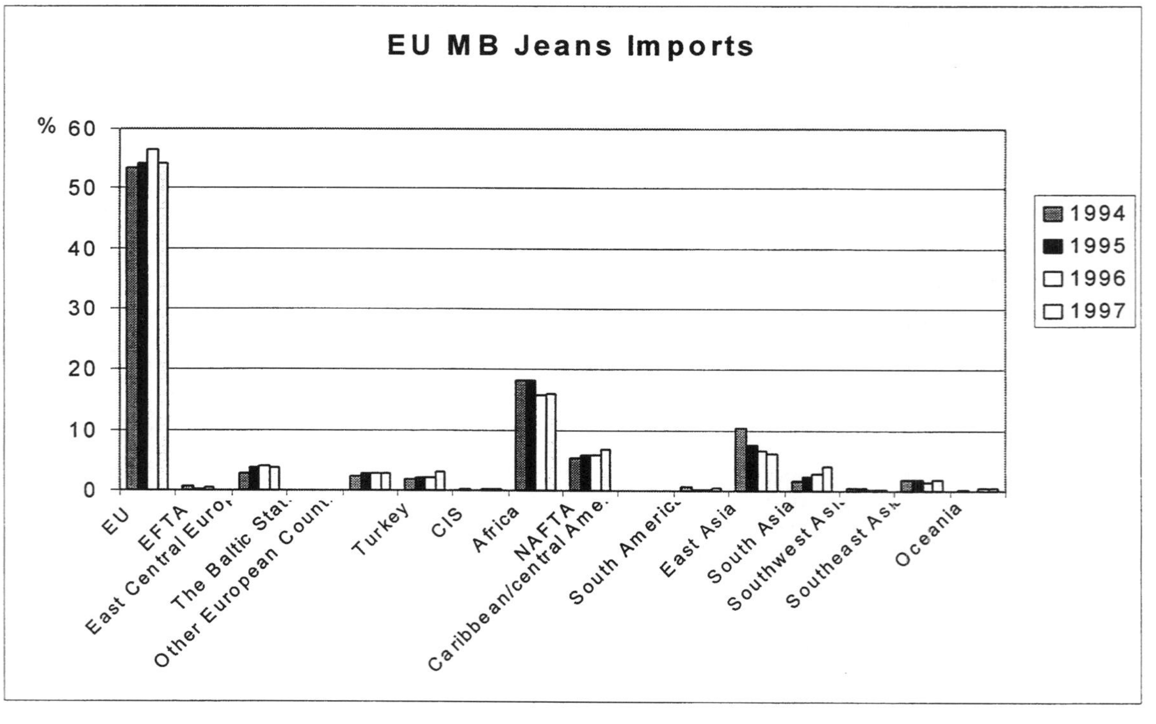

1.16 EU WG jeans import pattern

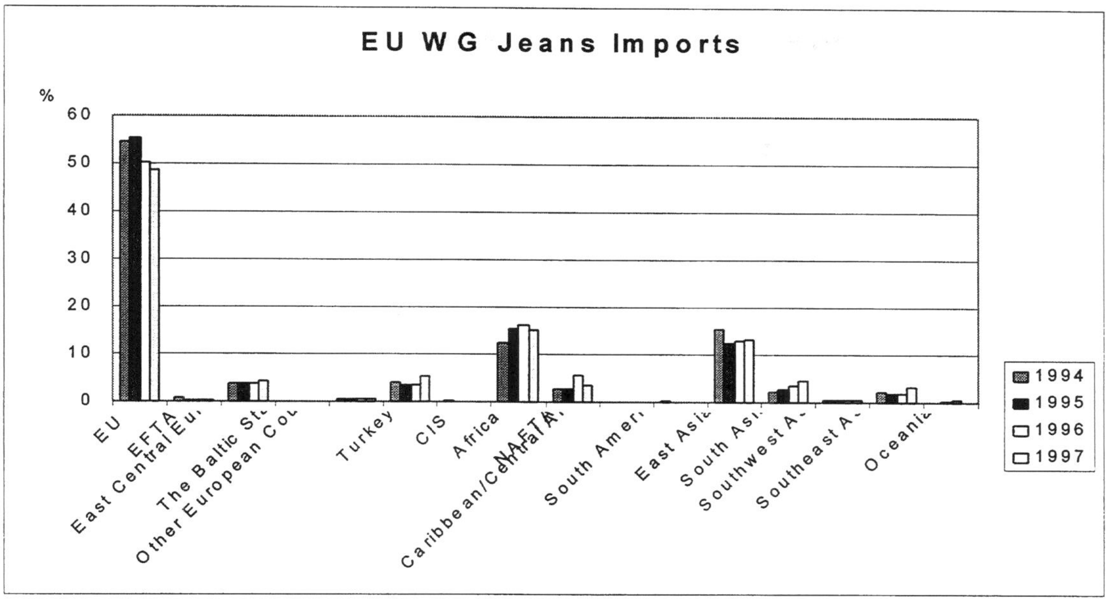

1.17 US MBWG jeans import pattern

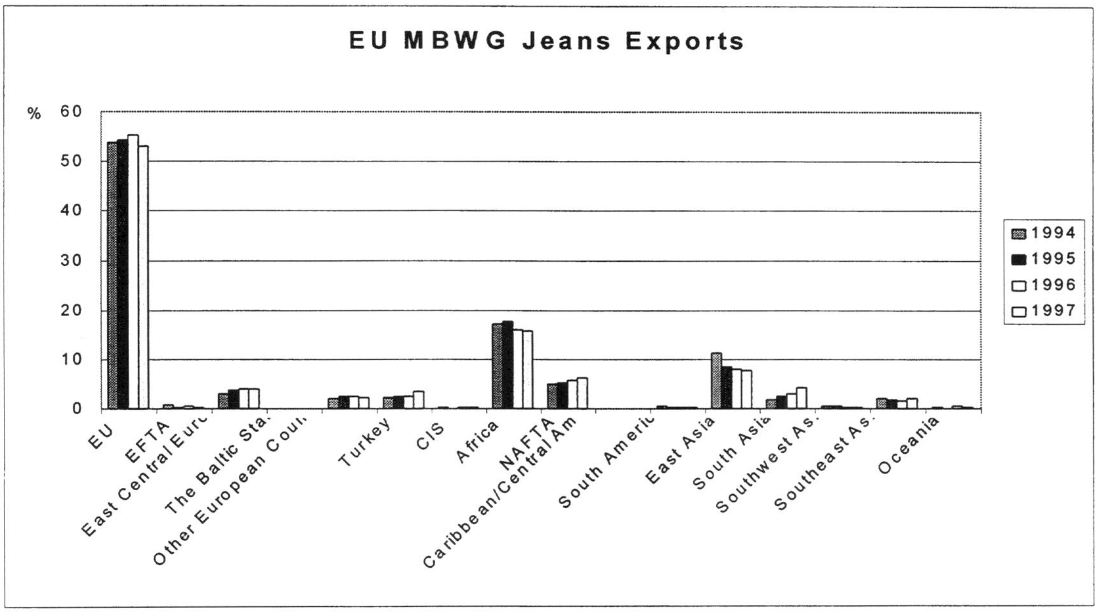

- MBWG jeans

More than half of EU jeans imports were generated within the EU itself. CEEC and Turkey were another two important sources. Though Africa and East Asia's share dropped during the investigating period, they still held a relatively significant position with 15.8% and 7.8% respectively. NAFTA, especially the US, increased its contribution to 6.2% in 1997. Exports from South Asia also enjoyed rapid growth, from 1.6% in 1994 to 4.1% in 1997. About 59.3% of EU jeans imports were sourced from the European continent and another 30.1% from Africa and Asia. (See Fig. 1.17)

A. Jeans export pattern

- MB jeans

The EU jeans export pattern is completely different from its import pattern as

1.18 EU MB jeans export pattern

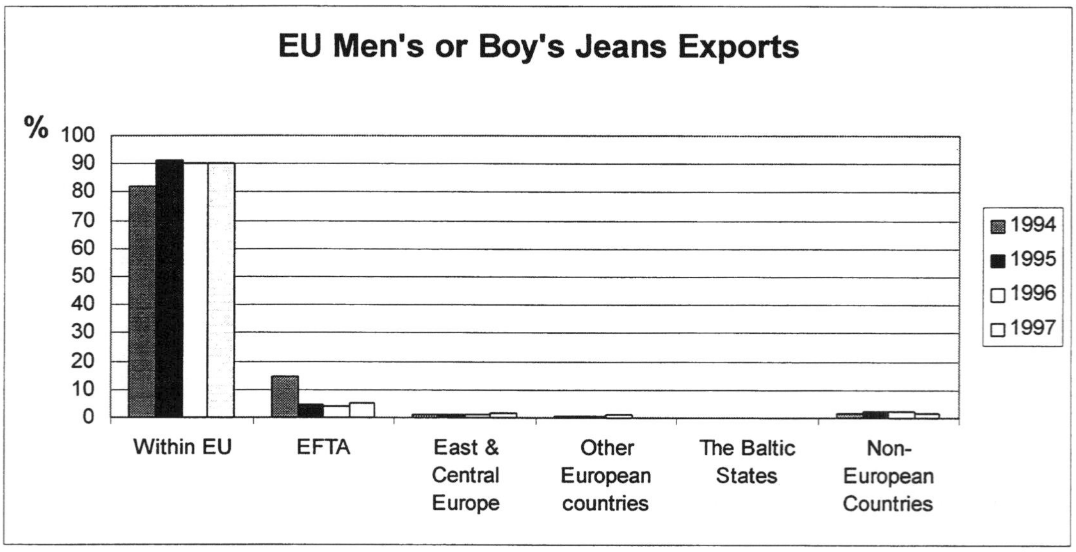

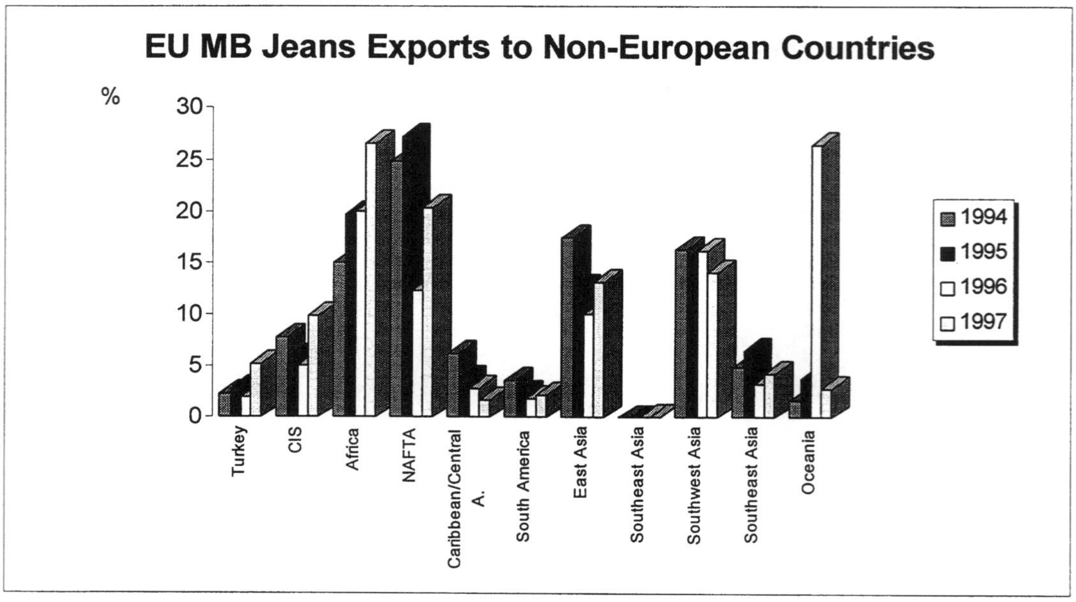

evidenced in Fig 1.18. EU members consumed around nine-tenths of the total EU MB jeans exports in 1997. The entry of Austria, Finland and Sweden into the EU in 1995 remarkably increased the EU's share from 81.9% in 1994 to 91.3%in 1995. EFTA was another important MB jeans consumption region, accounting for 5.2% in 1997. Together with CEEC and other European countries, about 97.3% of EU MB jeans exports were concentrated upon the European continent. As regards those to non-European regions, Africa absorbed 0.7%, followed by NAFTA (0.6%), Southwest Asia (0.4%) and East Asia (0.4%).

- WG jeans

The EU WG jeans export pattern is similar to its MB export pattern with some small differences (Fig 1.19). About 86.8% of EU WG jeans exports were absorbed by EU members themselves. EFTA and CEEC imported 3.6% and 3.0% respectively. Therefore, the European continent totally consumed 94.1% of total EU WG exports.

1.19 EU WG jeans export pattern

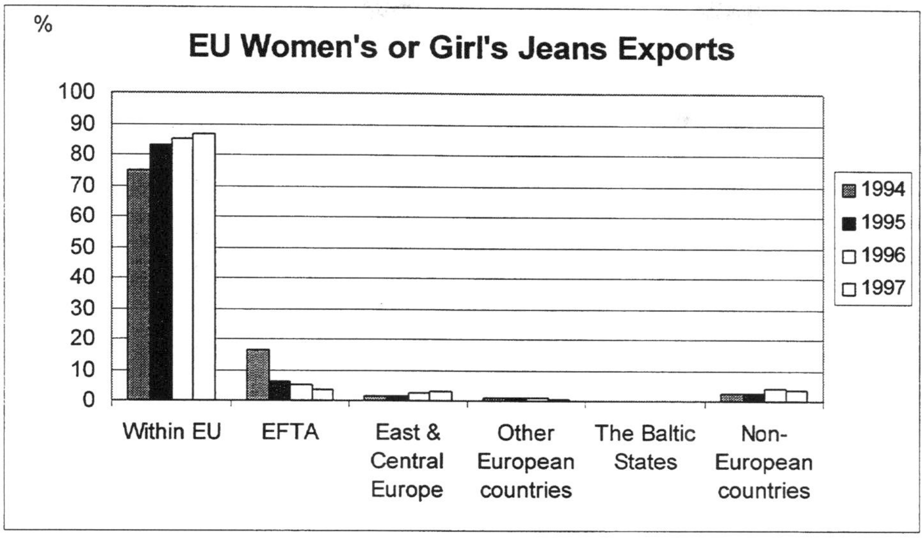

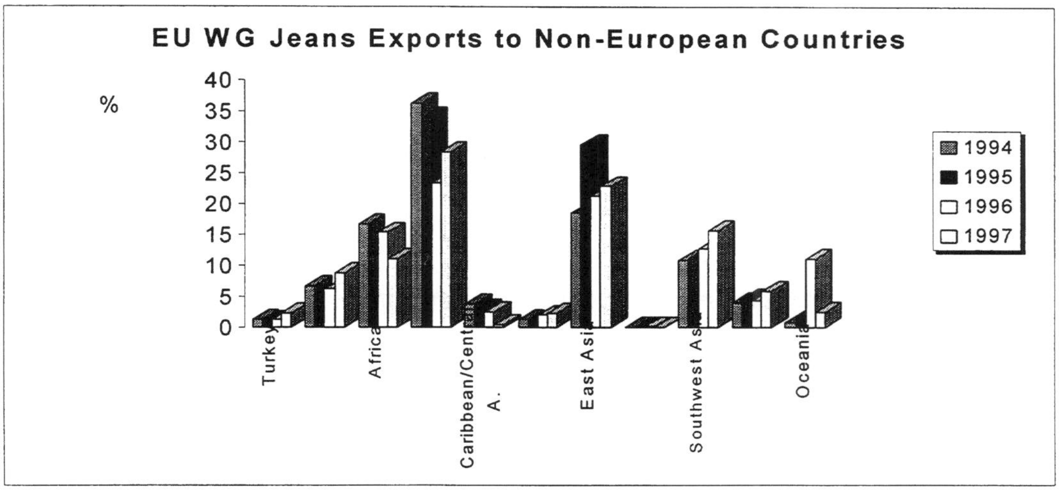

1.20 EU MBWG jeans export pattern

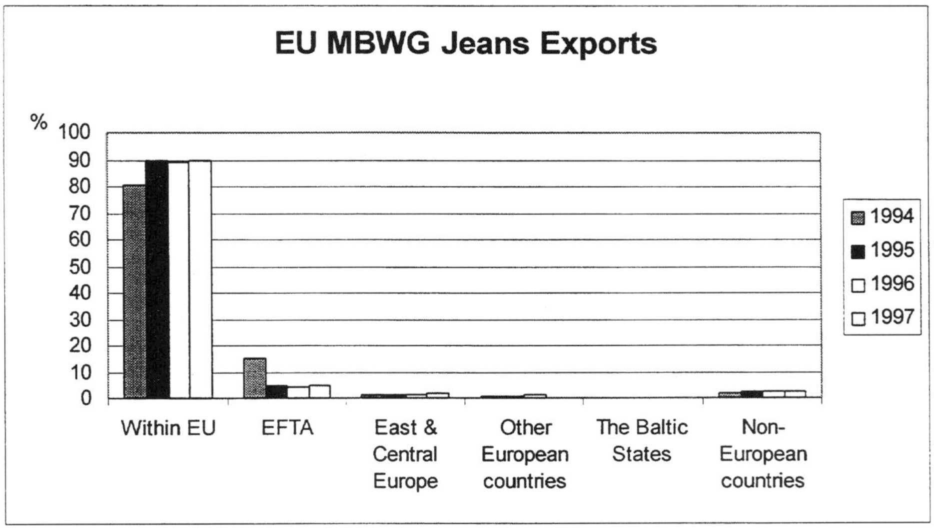

1.20 cont.

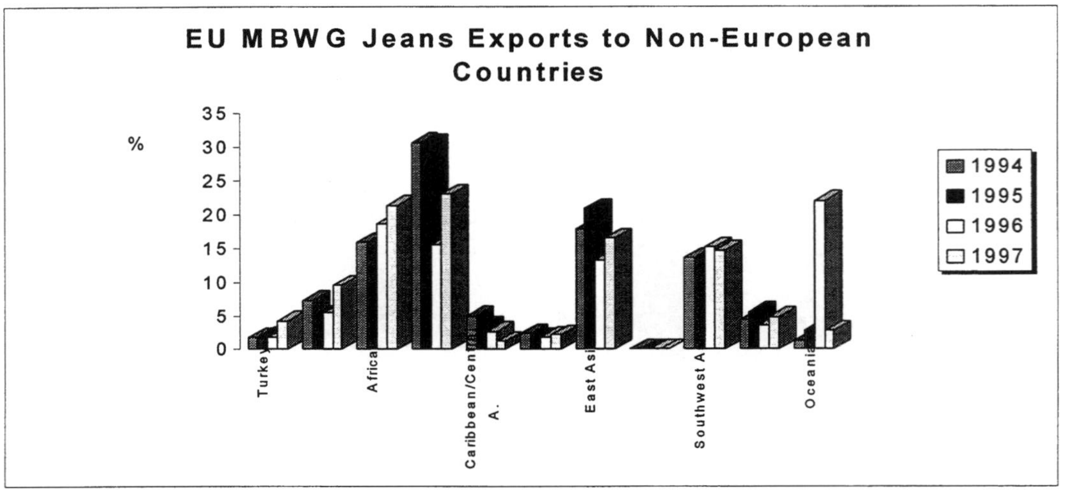

NAFTA was the largest non-European importer with 1.7% of the market share, followed by East Asia (1.3%), Southwest Asia (0.9%) and Africa (0.7%).

- MBWG jeans

EU jeans exports were mainly focused upon the EU itself. Non-EU markets only accounted for about one-tenth of total EU exports. EFTA and CEEC were the top two non-EU importers with 4.9% and 1.8% respectively. Asian importers together took 1.2% with East Asia 0.6% and Southwest Asia 0.5%. NAFTA and Africa held 0.8% and 0.7% respectively. Other regions' shares were negligible. (See Fig. 1.20)

B. Summary

EU members were both the largest suppliers and importers of EU denim product imports and exports, accounting for 53.0% and 79.4% respectively. Turkey has played a more and more important role in recent years, especially as a supplier with a 4.6% market share. Based upon geographic proximity and preferential access to the EU market, Africa is of much significance in supplying denim products to the EU, roviding 14.7% of total EU denim product imports. US, as the home of denim and jeans, was another important EU trading partner. Asian countries, backed by their comparative advantages in labor-intensive sectors, are a traditional sourcing base for the EU, and provided 13.6% of EU total denim product imports.

1.21 EU denim product import pattern

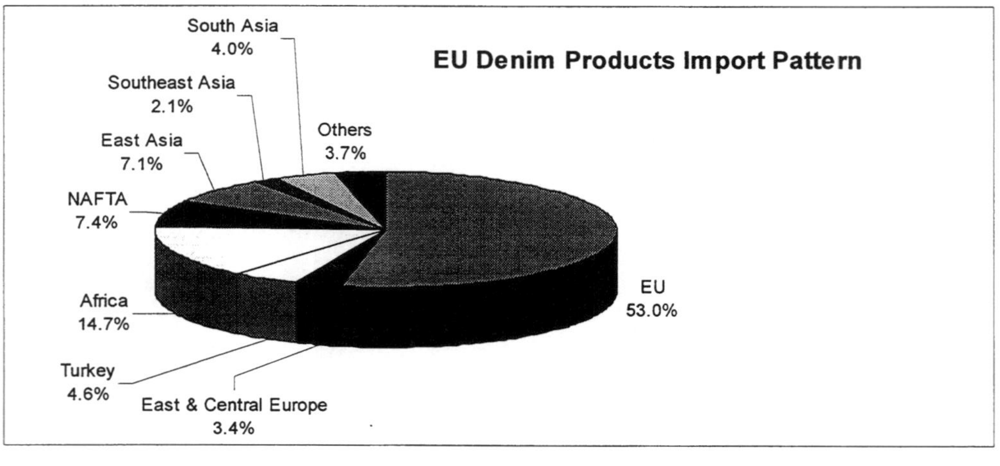

1.22 EU denim product export pattern

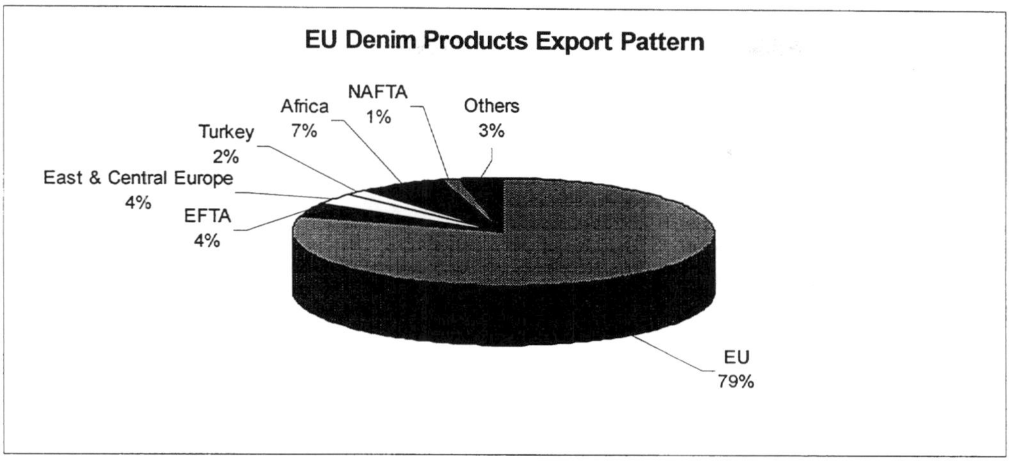

1.23 EU overall denim product trading pattern

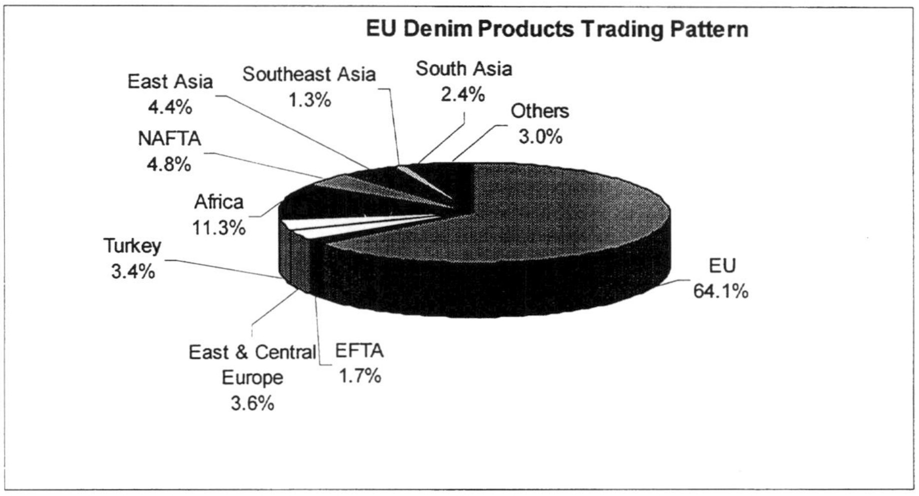

1.3.3 East Asia denim trading

After the US and the EU, Japan is the third largest denim consumer in the world and its denim-related industry is well known for the persistent pursuit of technological innovation and product development. HK, due to its unique geographic location and free port characteristics, is a world famous sourcing center and entrepot port for fashion items. The Chinese mainland is a newly emerged player in the denim world since the mid-1980s. Its traditional comparative advantages in labor-intensive sectors enabled it to develop the denim industry at a fast pace. HK's return to China in 1997 marked a new are of cooperation between the two sides. Since these three are all located in East Asia and have strength in different aspects, import and export trade among the three is very active, along with North/South Korea, Taiwan and Macao. The following analysis is centered on HK because it has a complete set of denim trading data. One point should be mentioned first, that is, there is no free trade area or regional integration in East Asia as there is in North America or Europe. Therefore, active trade links in this region, especially concerning textile and apparel items, cannot be explained with preferential access arrangements.

1.3.3.1 HK denim trading patterns

A. Denim fabric import pattern

As is evident from Fig 1.24, East Asia was the dominant denim fabric supplier to HK, accounting for more than nine-tenths of total HK denim fabric imports in 1998. Southeast Asia was the second most important region, holding another 2.8%. If the four Asian regions are taken as a whole, Asia totally contributed about 94.9%. NAFTA and the EU squeezed only 2.7% and 1.5% respectively.

B. Denim fabric export pattern[17]

Though HK denim fabric exports were not as concentrated as its imports, Asian countries still absorbed 69.6% of the market share, with East Asia taking 42.1%, Southeast Asia 14.7%, South Asia 11.5% and Southwest Asia 1.3%. NAFTA was a significant destination for over one-fifth of HK denim fabric exports. EU held about another 2.5%. (See Fig. 1.25)

1.24 HK denim fabric import pattern

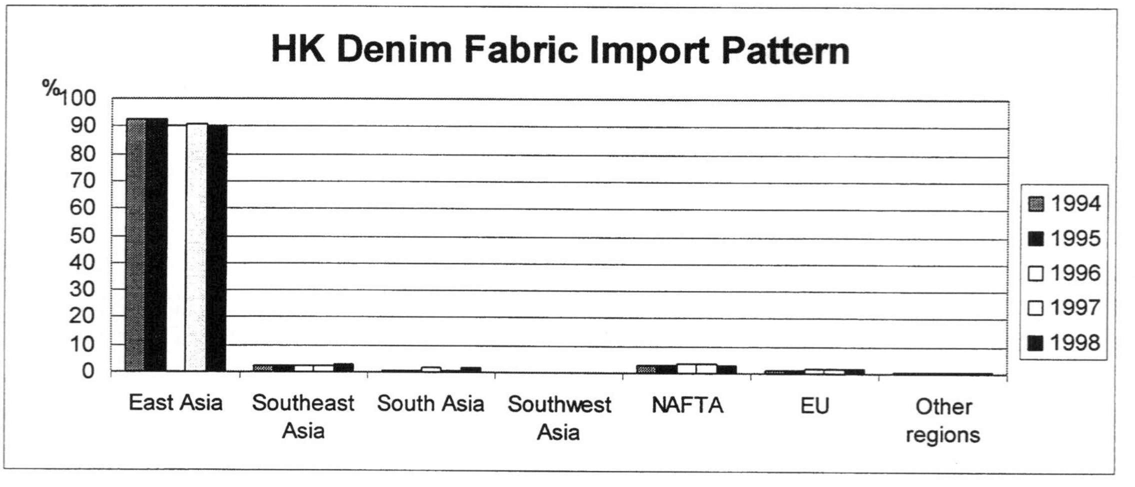

1.25 HK denim fabric export pattern

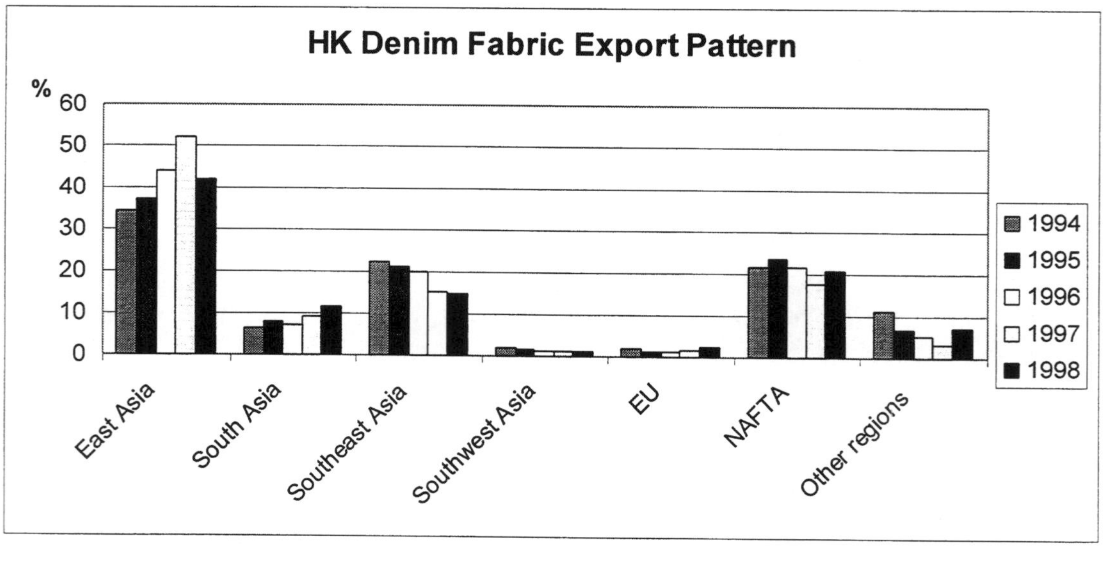

[17] HK denim fabric and apparel exports refer to domestic exports only.

1.26 HK denim apparel import pattern

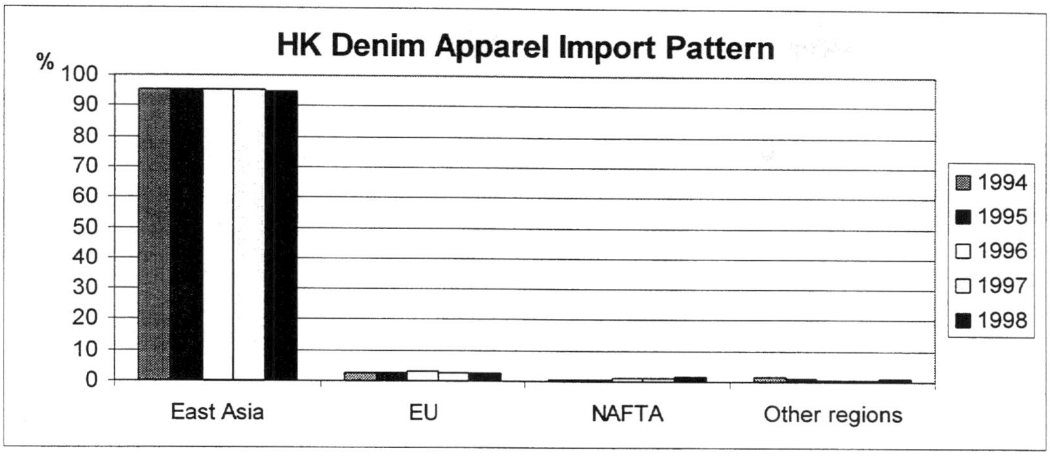

1.27 HK denim apparel export pattern

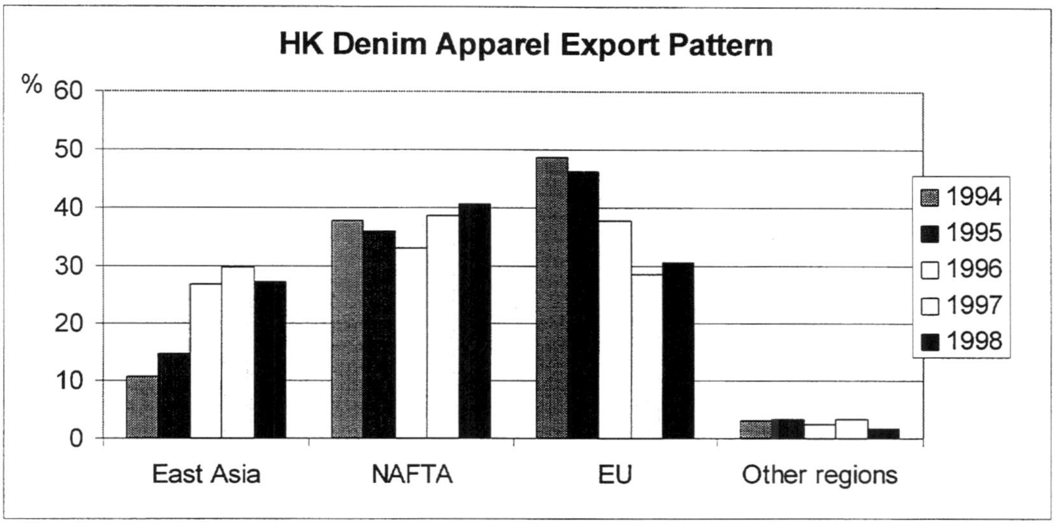

C. Denim apparel import pattern

HK denim apparel imports were highly centered upon East Asia, especially on China whose contribution alone was 94.5% of HK total denim apparel imports. In this sense, it can be said that China is the single denim apparel supplier of overwhelming importance to HK. EU and NAFTA accounted for 2.6% and 1.4% respectively. (See Fig. 1.26)

D. Denim apparel export pattern

As regards HK denim apparel exports, East Asia was no longer the key player though its share jumped from 10.8% in 1994 to 27.2% in 1998. The EU was once the largest importer with 48.2% of HK total denim apparel exports in 1994. However, its share slumped rapidly to 30.5% in 1998. The ratio of NAFTA imports from HK was relatively stable, and was 40.5% in 1998, the No.1 export destination for HK denim apparel. (See Fig. 1.27)

E. Summary

HK denim product imports were concentrated upon East Asia which had a more than nine-tenths market share. Together with other Asian regions, Asia was responsible for more than 95% of HK total denim product imports. NAFTA and EU, though still small suppliers to the HK market, increased their shares during the five-year period with 2.3% and 1.9% respectively. (See Fig. 1.28) The HK denim product export pattern was different. Though East Asia replaced NAFTA as the largest importer in 1996, its share was far less dominant than its export contribution. NAFTA was a very important market, accounting for 30.7% of HK total denim product exports. The EU's share was continuously down from 28.5% in 1994 to 16.5% in 1998. Southeast and South Asia absorbed 7.6% and 5.8% respectively. (See Fig. 1.29) All in all, though there isn't any form of FTA or special preferential arrangement among East Asia nations, HK denim product trading still focus as upon this region. NAFTA and EU were another two important partners. More than 83% of HK denim product trade was conducted within Asia, a percentage higher than NAFTA and EU's intra-regional denim product trade.

1.28 HK denim product import pattern

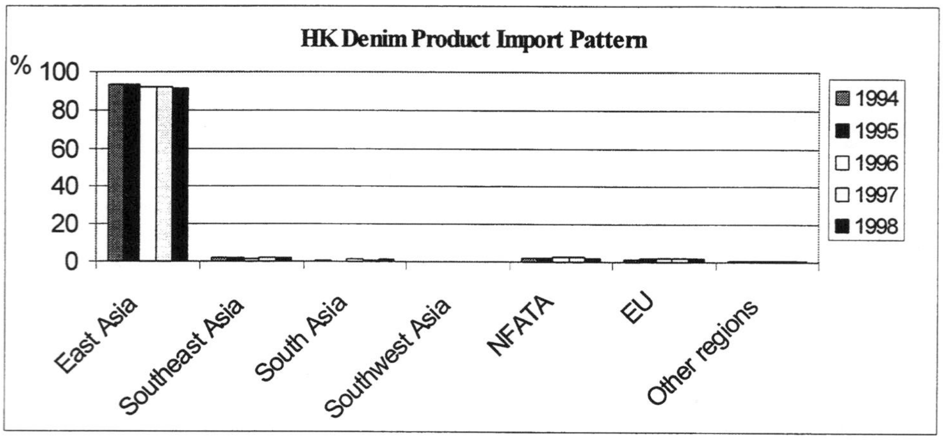

1.29 HK denim product export pattern

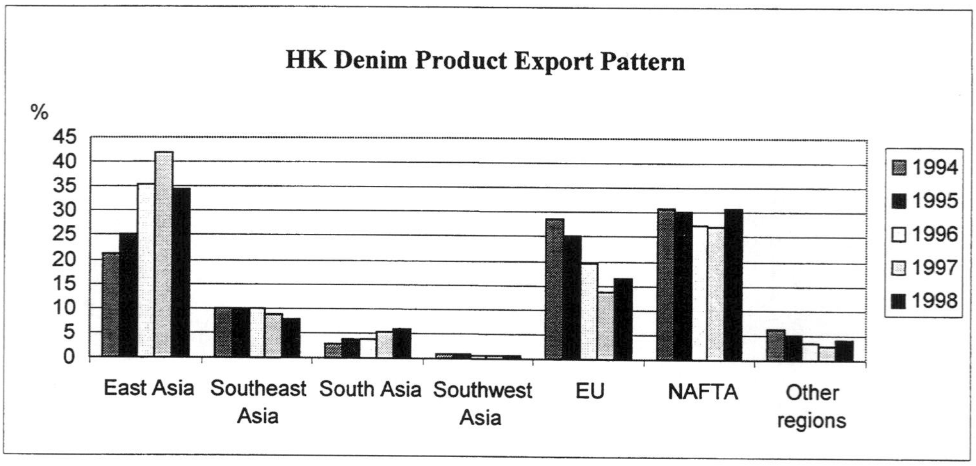

1.3.3.2 Japanese textile trading patterns

Since Japan denim product trade data is not available, a brief analysis of its overall textile trading pattern may be useful to get a partial picture of its denim trade. East Asia was the largest textile exporter with 63.1% of Japan total textile imports, followed by EU (13.0%), ASEAN (8.8%) and NAFTA (6.1%). The top four import regions for Japan textile exports were East Asia (47.3%), ASEAN (12.1%), EU (11.7%) and NAFTA (9.74%). (See Figures 1.31 and 1.32) Therefore, East Asia was both the largest supplier and importer for Japanese textile products. China held a

1.30 HK overall denim product trading pattern in 1997

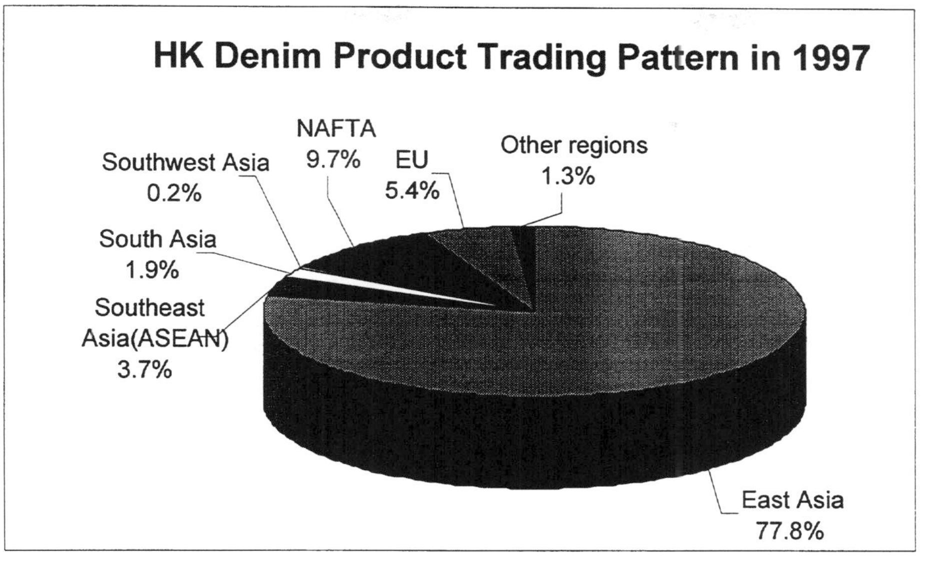

1.31 Japan textile import pattern

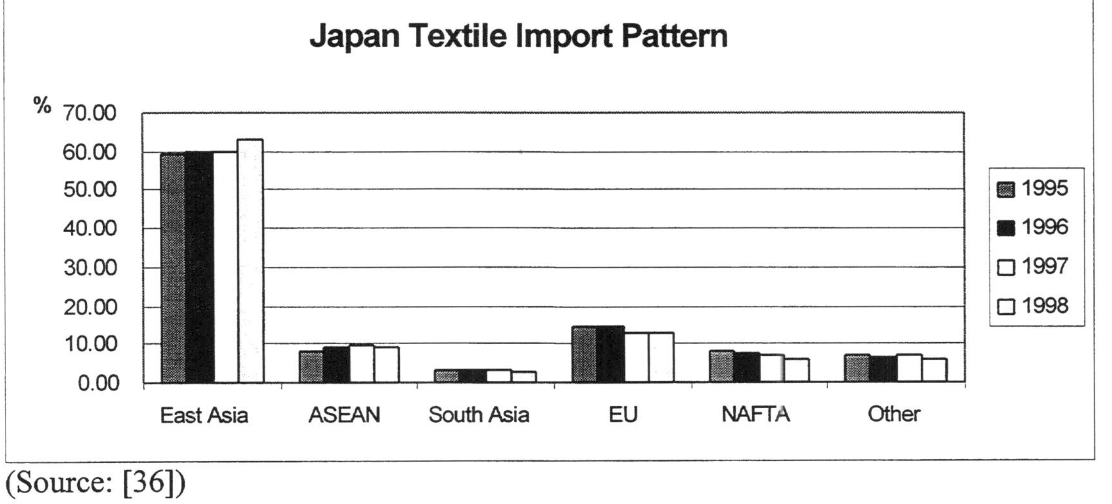

(Source: [36])

dominant position in Japan's textile and apparel trade. In 1997, apparel imports from China accounted for 63.7% of Japan total apparel imports, followed by Italy (8.0%), South Korea (5.1%) and the US (4.5%) [35]. The data are summarized in Figs 1.31 and 1.32.

Figure 1.33 lists the most important importers for Japan denim exports. In dividing these countries into various regions[18], it is clear that the pattern coincides with Japan's textile export pattern. The top three important regions were East Asia (54.2%), ASEAN (20.9%), and EU (10.1%). The US share was less than 1%.

[18] East Asia: China, HK, Macao, Taiwan, Korea; EU: UK, Belgium, Italy; Southeast Asia: Singapore, Philippines, Malaysia, Indonesia

1.32 Japan textile export pattern

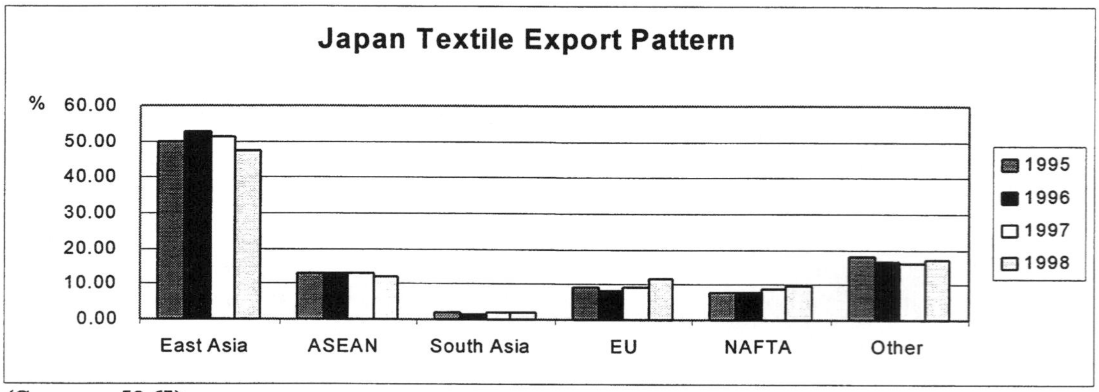

(Source: [36])

1.33 Japan denim export pattern

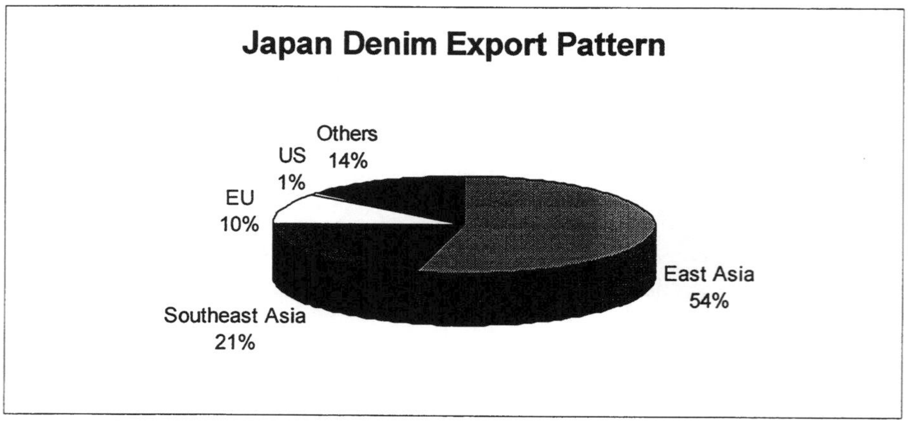

Table 1.9 US denim trade with NAFTA members and the East Asian region

US denim trade with its NAFTA partners and the East Asian region				
%	Denim fabric exports		Denim appparel exports	
	To NAFTA	To Asia	To NAFTA	To Asia
1993	27.2	15.0	30.3	14.9
1997	46.2	7.8	45.2	5.1
%	Denim fabric imports		Denim apparel imports	
	From NAFTA	From Asia	From NAFTA	From Asia
1996	17.6	76.8	46.5	26.4
1997	46.9	41.4	54.4	22.2

1.3.4 Major findings

1.3.4.1 Increased importance of trading blocs

As regards of the top 30 countries/regions that hold 90% of the world combined GDP, only Japan, China (including HK), Korea and Taiwan are not involved in some kind of regional integration or free trade agreement [37]. Some 90% of the WTO members are also parties to certain kinds of regional trade agreements at the same time [6, Chapter 15]. The EU and NAFTA are the two most outstanding examples of regionalism, though the former has a far more ambitious vision than the latter. Scholars have advanced numerous qualitative and quantitative explanations justifying this phenomenon, but no consensus has yet been reached concerning its ultimate effects upon the globalization path. However, one thing is certain--regionalism is now in the trend. The emergence of regional trading blocs has had a remarkable influence upon the countries both inside and outside the blocs.

An analysis of the US denim trade data clearly indicates the fast rising intra-regional trade volume with the formation of NAFTA in 1994. Table 1-9 is a comparison between US denim trade with NAFTA partners and that with the Asian region. The trade diversioneffects are very obvious for both denim fabric and apparel imports and exports. The NAFTA's share expanded at the expense of Asian suppliers and importers. US denim fabric importation is a concrete example. In just one year's time, imports from NAFTA members increased by 29.3 percentage points while those from Asia shrank by 35.4 percentage points. The very strict "yarn forward" rules of origin concerning textile and apparel trade greatly and effectively keep non-NAFTA originating goods out of the NAFTA door.

Denim trade within the EU covers more than a half of the EU total denim trading volume. At the same time, it is interesting to note that EU denim trade within the EU itself actually took a downward trend between 1994 and 1997 except for jeans exports whose increase is attributed to the entry of three former EFTA members. However, it cannot be concluded that all of the non-EU members have gained benefits from this trend. In fact, the drop of intra-EU trade is mainly due to the EU's expanded preferential arrangements with its neighboring nations such as CEEC and Turkey, which enjoyed the most rapid growth due to tariff removal or concession as well as quota elimination (Table 1.10). Though there are various voices concerning the concept of "an enlarged EU", there are signs indicating the possible emergence of a closely connected and integrated Greater Europe in the future. There are 12 countries applying for EU membership, among which 10 are CEE countries and the other two are Turkey and Cyprus. Six countries have undergoing the first screening, including Poland, the Czech Republic, Hungary, Slovenia, Estonia and Cyprus [38].[19] In this sense, the trading bloc concept in the European continent is no longer confined to the EU alone. It has been expanded to cover more countries with a more blurred national geographic boundary.

1.3.4.2 Importance of geographic proximity

As mentioned earlier, the denim trade pattern of HK and Japan cannot be explained by "regionalism", because there is no special preferential arrangement among East Asian nations and regions. However, the intra-East Asian denim trade is very intensified, particularly concerning denim product imports. The underlying reason is the geographic proximity which plays a more and more important role in today's quick response climate. The different economic development stages in this region also help to explain the active trading relationship in the region. HK's return to the Chinese mainland in 1997 further enhanced the bilateral trade flow between the two sides. Of course, such close cooperation far exceeds the "geographic proximity" argument.

Another example in afforded by the EU denim fabric and jeans imports from Africa for another example. In 1997, jeans imports from Africa accounted for 15.8%, among which Morocco and Tunisia contributed 77.7%. Denim fabric imports from the two countries were even more significant, accounting for 97.0% of EU total imports

[19] The other five CEE countries are Romania, Slovakia, Latvia, Lithuania and Bulgaria

Table 1.10 EU denim trade with EU members, CEEC and Turkey in 1994 and 1997

EU denim trade within EU, with CEEC and Turkey				
	Denim fabric exports			
%	To EU	To EFTA	To CEEC	To Turkey
1994	57.0	1.4	7.7	1.2
1997	46.8	0.2	10.3	6.6
	Jeans exports			
%	To EU	To EFTA	To CEEC	To Turkey
1994	80.4	14.9	1.2	0.1
1997	89.6	4.9	1.8	0.1
	Denim fabric imports			
%	From EU	From EFTA	From CEEC	From Turkey
1994	55.7	5.7	0.3	8.2
1997	53.4	0.5	0.4	11.4
	Jeans imports			
%	From EU	From EFTA	From CEEC	From Turkey
1994	53.7	0.7	3.1	2.2
1997	52.9	0.1	3.9	3.5

from Africa. Their geographic proximity to the EU may render a strong explanation for their good performance.

1.3.4.3 Trade manipulation with more subtle approaches

When developing countries may have a competitive edge in some products, such as labor-intensive items, the market rules are often changed to prevent free and open competition [26]. Since textile and apparel industries in developed nations have faced more serious challenges from their counterparts in developing countries during recent years, they turn to various policy instruments to manipulate trade. High import tariffs and quota imposition are the two traditional methods which have become less effective in the new trading environment. New approaches have been adopted, featuring subtle and clear policy orientations.

The US provides a very good example here. Previous sections have listed some measures, including change of general rules of origin, specific NAFTA rules of origin, and the 807/super 807 program with Caribbean nations. Recently, the US House of Representatives passed a bill designed to expand its apparel trade with 48 African countries/regions south of the Sahara, as well as the 25 Caribbean countries/regions. Though this measure, together with those mentioned above, seems to help these LDCs promote economic development, the underlining argument is possibly the obvious and direct benefits to the US textile manufacturers. For example, denim fabrics exports from the US to CBI members jumped from US$ 7.8 million in 1993 to US$ 16.8 million in 1997. Therefore, these specific trade policies give a development space for domestic textile mills. The largest losers are those low-cost suppliers in other regions such as China, rather than domestic apparel producers, because it just involves with re-distribution of import channels. With the complete quota abolition in 2005, it can be expected that world textile and apparel trade will be subject to more carefully designed trade policies in individual countries.

1.3.4.4 International labor divisions in denim-related products manufacturing

A. World trends in textile and apparel trade

- *Apparel trade*

Because of the labor-intensive nature of this sector, the developing countries are the major players in the world apparel export market. In 1997, the top ten apparel importers were all developed countries, except for HK. Though there were still five developed countries among the top ten apparel exporters, their share only accounted for 23.4% of the world apparel export volume, while that of China alone was 18.0%.

- *Textile trade*

Compared with the apparel sector, the textile sector is now capital- and technology-intensive rather than labor-intensive. Because of the high labor cost and failure to compete with low-priced products from the developing countries, manufacturers in developed countries are focusing upon high value-added aspects, such as the development of new chemical fabrics and new fashion designs, to gain new competitive edges. The textile industry in the developed countries is highly automatic in spinning, weaving, dyeing and finishing. Based on strong R&D capabilities and well-equipped facilities, the developed countries have further widened the technological gap with the developing countries. In 1997, there were six developed countries among the top ten leading textile exporters, accounting for 36.2% of the total world textile exports.

- Factors contributing to the international labor division in apparel/textile production as well as in the different links of the value chain

First, geographic relocation. The developed countries as well as the newly industrialized nations have moved their assembly and production bases to the less developed countries to take advantage of the low-cost labor and abundant resources. Examples include US apparel plants shifted to Mexico, EU to Turkey, Japan to China and Southeast Asian countries, and HK to China. At the same time, most of the core activities including design, branding, product development and technical innovation are retained in the developed nations.

Second, OPT arrangement. To further maximize their competitive strength in high-value added activities, developed nations have intensified their efforts in outward processing trade. Fabric is made and sometimes cut in the home country and then shipped to a developing nation for assembly. The final products, usually with the home country's brand name, will be shipped back to the home country or to a third country for consumption. Such arrangements are supported by both developed and developing countries' government policy, though the aims are different. The developed countries use OPT as an effective way to boost fabric exports, as well as to lower the total production cost; while the developing nations regard it as a way to promote employment and earn foreign exchanges. According to a report by the Japan Chemical Fibers Association, nearly 80% of the apparel made in Mexico and exported to the US uses materials manufactured in the US [39]. An OPT arrangement in the EU originated in Germany. It stipulates that fabrics exported to Poland, Hungary, the Czech Republic, Romania, Slovenia, Morocco and Tunisia for processing will be free of duty when the final products are imported later. Such "well-designed" OPT arrangements, though to some extent helping to promote the development of domestic industries in developing countries, ultimately turns them into simple assembly lines for the developed nations.

Third, lack of specialized factors. Lack of specialized factors is another key reason and maybe the fundamental one. Most developing countries, though having comparative advantages in labor cost and natural resource availability, fail to acquire advanced factors such as brain reserve, design, brand building and product development and innovation. Section 2 will give a detailed explanation of the current Chinese textile and apparel industry, pointing out that lack of specialized factors significantly influences its international competitiveness and limits its future development. It is no easy job to acquire these specialized factors because it is not equivalent to simple capital investment or global sourcing. It takes time, capital and doubled efforts to cultivate local designers, establish local brands and at last, gain strength in product development and innovation. In this aspect, such labor divisions and value-added patterns will last for a relatively long time.

B. International labor divisions in the denim product trade

The denim product trade pattern clearly reflects this world trend. For example, EU denim fabric imports from CEEC and Africa in 1997 only accounted for 0.4% and 7.9% respectively but exports to the two regions were 10.3% and 25.8%. About 15.8% of EU denim apparel imports were from Africa but only 0.7% of exprots went to this region. There were no denim fabric imports from the Caribbean region to the US in 1997 but the region supplied 14.8% of US total denim apparel imports. US denim apparel imports from the EU had a negligent share of 0.6% but the EU absorbed 18.9% of US total denim apparel exports.

Table 1.11 shows the top 10 suppliers/importers of the US denim product trade in 1997. While the ratio of developing to developed nations is 5:5 in the first column, that in the second column is 9:1, indicating that developing nations hold a dominant position in denim apparel supply to the US. In Columns 3 and 4, the ratio in both is 7:3. However, it should be noted that five of the seven developing nations/regions in Column 3 are close to the US market while in Column 4, only Mexico and Brazil are in the America and the remaining five nations/regions are all in the Asia. This comparison reflects the fact that US denim fabric exports are more concentrated upon the American Continent, and that most of the developing nations in the Caribbean and South American region mainly engage in denim apparel manufacturing.

Table 1.12 provides another strong argument for current fabric/apparel division patterns. As regards HK denim product exports[20] to the Chinese mainland from 1992 to 1998, more than 90% were denim fabrics. By sharp contrast, over 95% of HK denim product exports to Japan were denim apparel. As for HK denim product imports[21] from the Chinese mainland, less than a half were denim fabrics while 99% of denim products from Japan were made up of denim fabrics.

In addition to this labor division pattern, developing and developed nations also engage in different value-added activities. Figures 1-33 and 1-34 show the different unit prices for US denim fabric and apparel imports. Generally speaking, imports from developed nations such as Italy, France and Japan were priced at a higher level than those from developing countries such as India and Mexico. The price gap indicates that developing nations are still providing to low-/middle-range denim products while those from developed nations mainly serve the upmarket.

1.3.4.5 Emergence of more denim manufacturers in developing countries

Since the middle 1980s, more developing countries have entered the world denim market. China and India are the two newly emerged denim giants. The latter has a capacity of nearly 250 million meters [40]. In 1997, India was the third largest denim fabric supplier to the US market while China was its No. 9 denim apparel supplier. While there were no denim fabric imports from Brazil and Tunisia in 1995 and 1996, they began to export to the US in 1997 with US$ 175 thousand and US$ 4 thousand respectively. Cambodia began to export denim apparel to the US in 1997 with US$ 10.2 million. Other new entrants include British Virgin Islands, Venezuela, the Czech

[20] Exports here include domestic exports and re-exports by market

[21] Imports here include normal imports and re-exports by origin

Table 1.11 Top 10 US denim trading partners and its share

US Denim Product Trade							
Apparel exports		Apparel imports		Fabric exports		Fabric imports	
World	568562	World	1681097	World	287183	World	9137
Mexico	203721	Mexico	855772	Mexico	70671	Mexico	4250
Costa Rica	95208	Hong Kong	153524	Canada	62032	Hong Kong	2084
Canada	53353	Guatemala	64791	Belgium	59961	India	936
Belgium	50016	Canada	58436	UK	25511	Italy	476
Honduras	40677	Costa Rica	53578	Dominican R.	8985	Pakistan	459
France	24089	Dominican R.	37885	Philippines	7818	China (Taiwan)	295
Japan	19366	Nicaragua	37140	Colombia	5608	Australia	220
UK	17303	Colombia	37064	Hong Kong	4182	Brazil	175
Dominican R.	12138	China	36587	Honduras	3131	Germany	91
Spain	4950	Philippines	33914	Venezuela	2693	Spain	39
Top 10	520821	Top 10	1368691	Top 10	250592	Top 10	9025
%	91.6	%	81.4	%	87.3	%	98.8

Table 1.12 HK denim trade with China and Japan (1% of total trade)

Denim Fabric	1992	1993	1994	1995	1996	1997	1998
to China %	96.71	94.90	93.28	94.28	93.92	94.39	91.02
to Japan %	1.95	0.40	1.20	1.17	1.31	3.82	4.48
from China %	34.13	34.79	39.64	48.06	43.46	45.67	47.64
from Japan %	97.12	96.57	98.50	99.25	99.41	99.01	99.33

1.33 US denim fabric import price

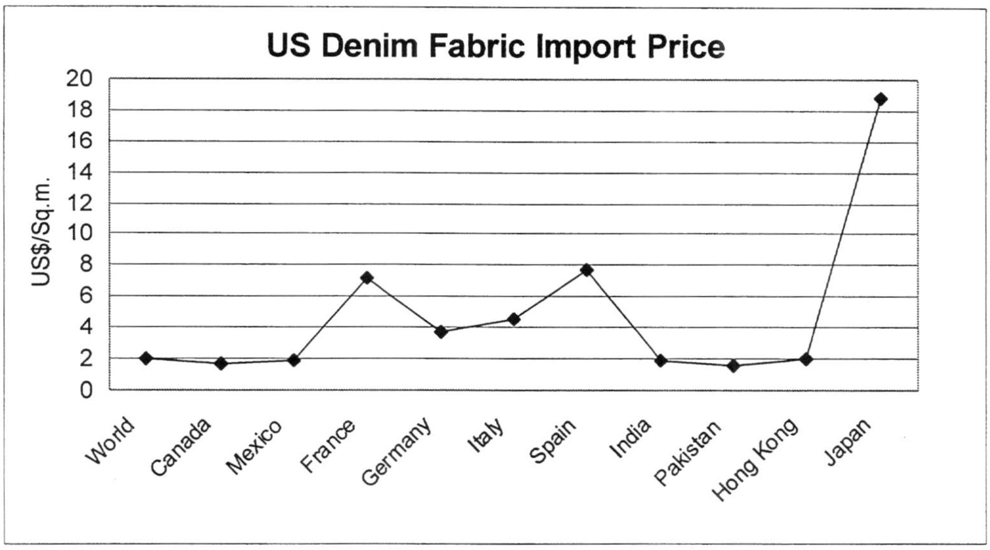

Republic, Slovenia, Romania, Saudi Arabia, and Botswana. New faces in the EU jeans import market include Bosnia Herzegovina, Serbia, Mauritania, Mali, Cameroon, and Trinidad and Tobago, which didn't start exports until 1996 or 1997. As regards denim fabric imports from the non-EU countries, there were 14 new suppliers, among which Pakistan enjoyed the quickest growth from zero in 1994 to 2.6 million Euro in 1997. The rest were Bulgaria, Albania, Estonia, Lithuania, Tadjikistan, Peru, Pakistan, Maldives, Malaysia, Sri Lanka, Bahrain, Qatar, Sierra Leone, and Madagascar, most of which are located in Europe or Asia.

1.34 US denim MB and WG denim apparel import prices

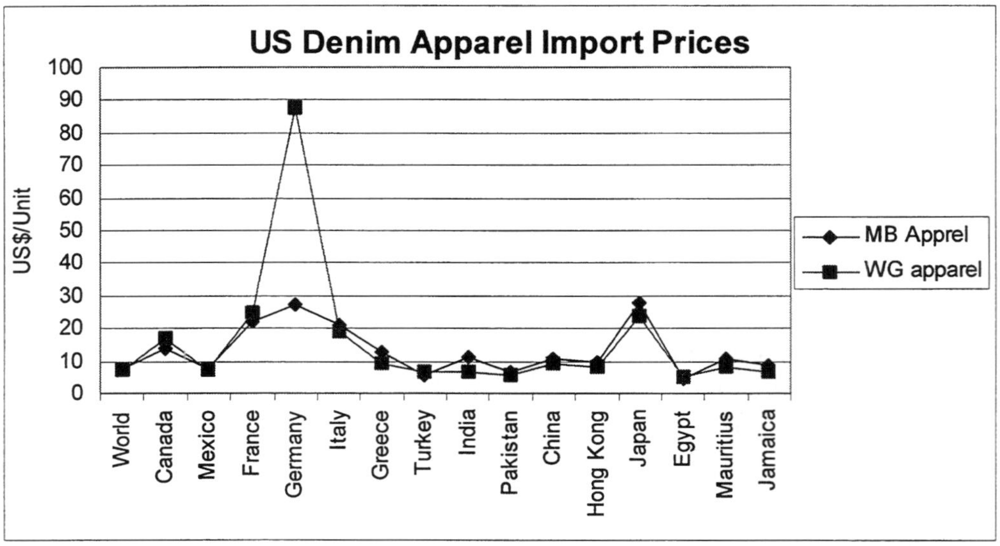

All of these new entrants hope to gain a share in the world market, greatly intensifying the already sharp competition in the commodity category where supply has far exceeded demand. The old and new players will fight hard to squeeze less and less profits in the low-end markets.

1.3.4.6 Key factors determining a nation's denim trade pattern

Based on all the above points and supported by the trade data in the US, EU and HK markets, several key factors can be identified in determining a nation's denim trade pattern.

A. Factors determining denim product exports

$P_E = F(C, G, A, D, E)$

P_E: Denim export pattern

C: Comparative and competitive advantages

G: Government policy

A: Multilateral/bilateral arrangements with other nations such as the MFA and preferential trade arrangements (PTA)

D: World market demand

E: Entry time

- Comparative and competitive advantages

There are various trade theories explaining the driving forces for a nation to engage in foreign trade (Appleyard & Field, Jr., 1998). The most classic one is the "comparative advantage" theory suggesting that nations can improve economic welfare when they employ resources to engage in activities with highest value. The Heckscher-Ohlin trade model points out that a nation's comparative advantages are determined by its factor endowments and each country will export their goods using the abundant factors most intensively. Porter further advances that a nation's international competitiveness is based on the capacity of its industry to innovate and upgrade rather than an inherited natural endowments [41].

The underlying reason for most developing countries to chose the textile and apparel sector as the engine for economic development and industrialization is their comparative advantages in basic factors such as land and labor. Less capital and technological requirements make entry easier at the same time. The industrial restructuring process in developed nations downsized the traditionally strong textile and apparel sector and gave a space for these new entrants to play in. The traditional textile and apparel manufacturers such as the US, Japan, France, Italy and Germany, when progressively reducing employment and relocating low-value added links to other nations, focus upon the upmarket with their strength in capital input, design, technical innovation and product development. Therefore, the different comparative and competitive advantages decide the current denim export pattern in individual countries.

- Government policy

The government's policy shift from import-substitution to export-orientation in most developing nations significantly promotes their textile and apparel exports where they have clear comparative advantages in apparel assembly and commodity goods production. Some measures may be argued to be unfair for foreign players in view of WTO-based rules, such as direct production subsidy and export credit, but such a shift does promote the export performance in these labor-intensive sectors. At the same time, governments in developed nations, in order to maintain the survival of domestic competitive textile and apparel enterprises, have tried hard to work out sector-specific policies, such as OPT arrangements and the imposition of high import tariffs.

- Multilateral/bilateral arrangements with other nations such as the MFA and preferential trade arrangements (PTA)

MFA (ATC) and PTA are actually the result of government trade policies, but they are far more influential power, as they have already gone beyond national boundaries. These "artificial" arrangements mean that world trade in carried out in a very strictly managed manner, greatly affecting the direction of trade flows. Exporters in a nation subject to the MFA have to acquire quotas first, an issue sometimes out of their control and having little to do with competitiveness. And even if they get the quota, they have to limit the export volume within the quota level and fail to realize economies of scale. An outsider of a particular PTA may find it hard to compete with insiders, especially if his/her products are similar to those of the insiders'. The loss of market share of China to Mexico is the best example of this.

- World market demand

Since the US, EU and Japan are the three largest denim consumers in the word, their domestic demand is very important for other countries' export performance. The economy in the US has developed healthily for a long time and as a result, demand for fashion items is strong. From 1995 to 1997, denim apparel imports into the US increased by 30.9%. On the contrary, the Japanese economy is sluggish after the collapse of the bubble economy. Its denim apparel import from the US dropped by 59.0% from 1993 to 1997 and those from HK slumped by 75.5% from 1992 to 1998.

- Entry time

Early movers enjoy special advantages such as the learning curve, a relatively larger market share and consumers' brand loyalty. The US is the home of "denim and jeans". It started denim manufacturing much earlier than other players. The well-established "Levi's" brand can be partly attributed to its earliest move in the denim world. After so many years' efforts, "Levi's" enjoys very high brand loyalty and brand awareness, to the extent that it is now denim's synonym. With the hotter market competition, it is much more difficult to achieve the same success as Levi's did 100 years ago when there were only a few players in the market and when consumers were not so demanding and sophisticated.

B. Factors determining a nation's denim import pattern[22]

P_I = F (I, SC, SD, A, C, E, M)

P_I: Denim apparel import pattern

I: Income

SC: Social and cultural context

SD: Domestic supply and demand situation

A: Availability of other causal items

C: Climate

E: Exchange rate

M: Market access

- Income

Denim apparel is less perceived as work wear than fashion items nowadays. Therefore, income per capita is a key variable in the above formula.

Table 1.13 shows the top 10 denim apparel (jeans) importers from the EU, US and HK in 1997. It is not surprising to see that nations with income per capita over US$ 25,870[23] dominate the scene. As regards EU exports, the top 10 are all EU members in the high-income category, accounting for 83.5% of EU total jeans exports. Among the top 20 importers, 18 are of high income per capita. The US export markets are a little different, with 13 of the 20 importers in the high-income group. Of the top 20 export destinations for HK denim apparel, 17 are high-income nations or regions. South Korea, Taiwan and Japan suffered economic setback in 1998. Their GDP per capita dropped from US$ 10590, US$ 13130, US$34534 in 1997 to US$ 6800, US$ 12009, and US$ 34092 respectively in 1998 [43]. As a result, their denim apparel imports from HK in 1998 were only 7.3%, 68.2% and 36.4% of the 1997 level. China is a developing country of low income per capita. But it is the second largest export market for HK denim apparel with an average annual growth rate of 50.6%. This can

Table 1.13[24] Top 10 denim apparel importers from the EU, US and HK in 1997

Top 10 Denim Apparel (Jeans) Importers from the EU, US and HK			
Ranking	EU jeans exports	US denim apparel exports	HK denim apparel exports
1	*Germany*	Mexico	*USA*
2	*France*	Costa Rica	**China**
3	*Netherlands*	*Canada*	*Germany*
4	*UK*	*Belgium*	*UK*
5	*Belg.-Luxbg*	**Honduras**	*South Korea*
6	*Italy*	*France*	*Canada*
7	*Austria*	*Japan*	*Netherlands*
8	*Ireland*	*United Kingdom*	*Italy*
9	*Spain*	Dominican R.	*Sweden*
10	*Sweden*	*Spain*	Venezuela
Top 10	83.5 %	91.6 %	94.0 %
Top 20	95.8 %	97.4 %	98.6 %

[22] This part will focus upon denim apparel.

[23]

[24] Italic countries are in the high-income group with income per capita over $25870. Bold countries are in the low-income group with income per capita below $490. Countries with single underline are in the upper-middle income group with income per capita between $1740-4600 and others are in the lower-middle income group with income per capita between $490-1740. This division is based on the World Bank standards 42. Sarathy, V.T.R., *International marketing*.Eighth Edition?ed.2000?The Dryden Press, Harcourt College Publishers.72-73..

be partly attributed to its rapid economic development which has significantly increased the income per capita, especially in the urban areas.

Though income is in direct proportion to denim import volume, it is not the only variable affecting denim apparel imports. Other factors play a role as well. These include:

- Cultural and social context

Thanks to great technological breakthroughs, the world is now becoming more integrated with blurred geographic boundaries. However, the different cultural and social contexts remain unique to individual nations, cultivating various tastes and preferences towards fashion items which are prone to the traditions, customs and value concept.

Denim apparel, such as jeans, has strong image implications. It is often associated with the pursuit of freedom, individuality and anti-establishment, the spirit championed by Americans for so many years. But this is not well accepted in some countries (such as most Arab-countries) whose cultural and social context is very different from that in the US. Therefore, denim apparel is not and will not be preferred in the near future in these nations. For example, Kuwait belongs to the high-income group. However, its denim apparel imports from the EU, US and HK were very small, only ranking the 59th, 53rd and 35th respectively.

At the same time, as the only super power in the world nowadays, the American influence is very strong in many aspects. Its economic penetration into foreign markets by large MNCs has brought American culture to these markets as well, directly or indirectly influencing the host country's view of American-labeled products, especially among the younger generation who are quick learning and ready to absorb new ideas and concepts. There is some debate about whether young consumers are becoming more homogeneous or more differentiated in their tastes. The debate result is relatively unimportant; what really matters is the increasing internationalization trend of domestic demand patterns in some consumer goods [44]. For example, Japanese young consumers favor American apparel embodying the American life style. Denim apparel imports from the US to Japan amounted to US$ 19.4 million in 1997, ranking the seventh in US export markets. Those from HK only reached US$ 0.9 million[25] despite closer transportation distance. Therefore, cultural differences, though impossible to be eliminated completely, can be reduced with time. Current globalization trend will no doubt accelerate this process.

- Domestic supply and demand conditions

Domestic supply and demand conditions also influence the ultimate import volume. Domestic production capacity, local industry's internationally competitive position and foreign investment are important in determining the available quantity and quality of final products. Demand conditions are closely associated with market size, based on population and purchasing power as well as consumers' preferences over fashion items. The gap between the two is crucial to the ultimate import volume.

The US is the home of denim products. Though US denim apparel manufacturers such as Levi's, Lee and Wrangler, are world famous, they can hardly meet the demand. Therefore, each year, the US imports a lot from the outside world, especially from Mexico.

China, on the other hand, imports much less denim apparel than fabrics. Though its denim apparel consumption has increased in recent years and market potential is huge, it is still a developing country with a slow urbanization process and wide

---

[25] US$ 1= HK$ 7.8

regional economic gaps. When most people are still troubled to provide basic living substances, how could they be expected to spend money on something "unnecessary" such as jeans and denim skirts? On the other hand, China itself is the world largest apparel manufacturer. Though there are no world famous brands, local brands are adequate enough to satisfy most market niches. Therefore, denim apparel imports to China are small in volume, and most of them are from HK, whose price is relatively cheaper than those from the US and EU. In 1997, denim apparel from HK to China was priced at around US$ 4.0 per unit with 12.3 million units, while those from the US were imported at US$17.3 per unit with a mere 1560 units.

In the US, women own almost twice as many pairs of denim jeans (7.8) and other denim apparel (18.8) as women around the world [45], followed by Colombian women (6.1 pairs of jeans) and Brazilian women (17.6 denim apparel). The international survey commissioned by Cotton Council International also reached the conclusion that Latin American women are the most fervent in clothes shopping. Keeping these facts in mind, it is not surprising to see different import patterns concerning the MB and the WG denim apparel trades. As regards US denim apparel exports in 1997, 65.5% of MB denim apparel were exported to other American countries while 95.5% of WG denim apparel were concentrated on the continent. At the same time, the American countries imported 0.7% and 1.8% of EU MB and WG jeans exports respectively. Since there is no special preferential arrangement between HK and other nations and HK is close neither to the EU nor to the US market, its trade data is perhaps more relevatn for analysis. In 1997, the US and Germany, of comparable economic level[26], were the second and third importers of HK MB and WG denim apparel. However, the former imported five times as much WG denim apparel than the latter, while the gap was narrowed to 1.6 times for MB denim apparel.

- Availability of other casual items

Consumers nowadays tend to prefer a casual approach both in the office and out of it. "Casual Friday" is the result of such a trend, leading to the increased sale of casual attire such as wrinkle-resistant cotton slacks. Then one question is raised: As a member of the whole casualwear family, will denim apparel take the lead again? A report from the Cotton Incorporated Lifestyle Monitor may give some hints. It seems that the biggest challenge comes from clothes made of khaki fabric whose history can be traced back to the Crimean War of 1853. The report cites one merchandising manager as saying: "khaki pants are the perfect clothing item to span the gap between weekend and business wear. Khakis fit that in-between place because they're versatile, they look polished but not uptight, and you don't look like you are trying too hard." It further reports that 70% of women aged 56 to 70 choose casual slacks over denim jeans, as do 43% of women aged 35 to 55. Almost half of the Fashion Innovators, the trendiest of shoppers, take the non-jeans option. Another survey conducted by the Cotton Inc. shows that the percentage of consumers between the ages of 16 and 19 saying they either like or love denim has dropped from 84% to 79% since 1994. Therefore, though denim apparel still enjoys a favorable market position, it will face more pressure from other casual items in the near future. Of course, these casual clothes are not perfect substitutes for denim apparel because the latter has its own unique image implications. But the rapid development of the former does present a potential threat for the latter's further growth, especially when the population in some developed nations, such as the US, is growing older. According to the Fairchild Fact

[26] In 1997, GNP per capita in the US and Germany were US$ 27738 and 28228 respectively.

File (1980), the median age of the male population in the US was 28.6 years in 1979, but it is expected to jump to 34.1 by 2000. The demographic change will lead to fashion preference shift as well, ultimately affecting the domestic demand for denim products and a nation's import volume.

- Climate

Climatic conditions also affect denim apparel consumption level. Countries having extremely hot or cold weather all around the year, such as those near the equator or the Arctic Ocean, are not suitable markets for denim apparel. Climate also plays a role in deciding the consumption patterns for the variety of denim apparel. For instance, soft jeans are much more popular in Japan than in the US or EU. Lightweight soft jeans made of rayon became the main stream instead of basic products made of 14 oz. denim in 1994 [46]. Around 60 to 80% of women's jeans from major jeans manufacturers are soft jeans [47]. Though one may explain that soft jeans fit Japanese better than basic ones both in the light of their smaller-than-the-west figures and reserved rather than straight dispositions, there is no denying that hot weather in the summer makes lightweight jeans more appealing than the basic or extra-heavy ones.

- Exchange rates

Frequently fluctuating exchange rates are always cited as one of the big risks associated with foreign trade. Therefore, the currency integration in the EU is expected to boost intra-regional trade in the near future. The exchange rate sometimes plays a key role in a nation's trade volume. Some nations in Asia (such as Japan, South Korea, Malaysia and the Philippines) witnessed a sharp currency devaluation during the devastating Southeast Asian financial turmoil in 1997. As a result, imports into these nations slumped significantly. Denim apparel imports in the above four nations in 1998 dropped by 53.6%, 92.7%, 36.0% and 91.3% respectively over 1997. Though such big reductions had other contributing factors such as less disposable incomes and a sluggish domestic market, the weaker domestic currency against the US dollar also played a role.

- Market access

Last but not least, market access in a nation's domestic market sometimes may exercise more influential power on its import volume and pattern because it goes far beyond economic implications and is the result of various trade and administrative policies.

Previous analysis has already shown that easier market access in specific countries resulting from regional preferential arrangements has led to clear sourcing bias against "outsiders". In addition to the "artificial" trade barriers, transportation is a natural one, which is particularly important for fashion items requiring quick availability to consumers and short turn around times. Among the top 20 export destinations for EU denim apparel, only the US and Japan, which ranked 17th and 20th, are not located in or close to the EU. As regards the top 20 importers for US denim apparel, Mexico and Canada ranked the first and third respectively. The other six Latin American countries were in the 2nd, 5th, 9th, 11th, 12th, and 13th position respectively. CBI countries totally imported 28.2% of US total denim apparel exports. As for HK denim apparel exports, four East Asian countries/regions (China, South Korea, Taiwan and Japan) and one Southeast Asian country (Singapore) are among its top 20 destinations, while there were only two Latin American importers. Since HK doesn't have any formal preferential arrangement with other nations, geographic proximity is a very important factor here. If it is difficult to access a nation's domestic market in view of various artificial and natural barriers, an apparel manufacturer in the exporting country may choose franchising or direct involvement in local production instead of direct export.

If this is the case, it will no doubt affect the target nation's final import volume and pattern.

REFERENCES

1. Dickerson, K.G., Textile and apparel in the global economy. Third Edition ed.1999 Upper Saddle River, New Jersey; Columbus, Ohio Merrill, an imprint of Prentice Hall. 336-418.
2. Yang, Y., The impact of MFA phasing out on world clothing and textile markets. Journal of Development Studies, 1994? p.892-915.
3. MITI, J., The Future of Japanese Textile Industry.1998 Shanghai China Textile University Press.
4. Hughes, J.K., The ATC sky clouded over. Textile Asia, 1998 p.52-61.
5. Majmudar, M., The MFA phase-out and EU clothing sourcing: Forecasts to 2005. Textile Outlook International, 1996 p.33.
6. MITI, J., 2000 report on the WTO consistency of trade policies by major trading partners. 2000 MITI.
7. OTEXA, Export Advantage: EU.2000 OTEXA, US.
8. WTO, Agreement on implementation of article VI of the general agreement on tariffs and trade.1994 WTO.
9. WTO, Agreement on subsidies and countervailing measures. 1994 WTO.
10. Spinanger, D., Liberalization good, but open markets... Textile Asia, 1999 p.50-57.
11. Stephenson, S.M., Standards and technical barriers to trade in the free trade area of the Americas, in Trade rules in the making: challenges in regional and multilateral negotiations P.L. Miguel R. Mendoza, and Barbara Kotschwar Editor.1999 Organization of American States, Brookings Institute Press Washington DC.p.280-301.
12. Galli, A., Germany--new goals in halting downturn. Textile Asia, 1999 p.103.
13. Commission, E., 1999 report on United States barriers to trade and investment.1999 EU.
14. HKTDC, Antidumping likely to remain US weapon of choice even after China joins WTO.2000 HKTDC.
15. Anon, The textile sector: overview.2000 EU.
16. OTEXA, Export advantage: EU.2000 OTEXA, US.
17. Torre, J.d.l., Clothing-industry adjustment in developed countries.1986 London Macmillan.
18. US Embassy Tokyo, J., Country commercial guide Japan, FY 2000.2000 US Embassy Tokyo, Japan.
19. HKTDC, Market profiles: Japan.2000 HKTDC.
20. Franklin Dehousse, K.G., Philippe Vincent, Market access study to identify trade barriers affecting the EU textiles industry in certain third country market.1999 EU.
21. HKTDC, R.D., A General View of Japan's Distribution System for Consumer goods. 987 HKTDC HK.
22. Larke, R., Japanese retailing.1994 London and New York Routledge.56-126, 226-236.
23. Chen, M., Asian Management Systems: Chinese, Japanese and Korean Styles of Business.1995 London and New York Routledge.263-278.
24. HKTDC, R.D., China to adjust trade policies.2000 HKTDC.

25. HKTDC, Market profile on Chinese mainland.2000 HKTDC.
26. Cole, J., Geography of the World's major regions.1996 London and New York Routledge.
27. Zimmermann, R., Denim: what might the future hold for the world's hottest textile sector ITS Textile Leader, 1998 p.34-48.
28. Anon, Outward processing program.2000 OTEXA.
29. Anon, The textile industry is facing a new international competitive environment.2000 www.ctei.gov.cn.
30. Robin Anson, P.S., World textile and apparel trade and production trends. Textile Outlook International, 1999 p.20.
31. Anon, EU world trade & WTO: overview.2000 europa.eu.int.
32. Coster, J.d., Where quality tells. Textile Asia, 2000 p.9.
33. Coster, J.d., CEC threat to Asia. Textile Asia, 2000 p.14-15.
34. Anon, Analysis of trends in the Lome Trade Regime and the consequences of retaining it.2000 EU.
35. Anon, Apparel imports in a period of change. JTN Monthly, 1998 p.16-18.
36. Anon, Comparison of EU, US and Japanese trade--textile sector.1999 europa.eu.int.
37. Anon, Why a free trade agreement (FTA) now .2000 www.jetro.go.jp (Japan External Trade Organization).
38. Anon, EU world trade & WTO: legislation, reports and texts.2000 europa.eu.int.
39. Anon, NAFTA's influence on textile & apparel industries in the region and in Asia. JTN Monthly, 1999 p.28-30.
40. Anon, Denim in distress. JTN Monthly, 1998 p.68.
41. Porter, M.E., On competition.1998 Boston Harvard Business School Publishing.1-271.
42. Sarathy, V.T.R., International marketing. Eighth Edition .2000 The Dryden Press, Harcourt College Publishers.72-73.
43. Anon, OTEXA export advantage: overseas market information.2000 OTEXA.
44. Singleton, J., The world textile industry.1997 London and New York Routledge.
45. Anon, A Global: US leads the world in casual mindset.2000 www.cottoninc.com.
46. Anon, Japan: soft jeans become the main-stream. JTN Monthly, 1995 p.72.
47. Anon, Great changes in Japan's jeans industry. JTN Monthly, 1996 p.69-70.
48. European Commission., Sixteenth annual report from the Commission to the European Parliament on the Community's anti-dumping and anti-subsidy activities 1999 http://europa.eu.int/comm/trade/pdf/16report.pdf
49. Franklin Dehousse, Katelyne Ghemar, Philippe Vincent., Market access study to identify trade barriers affecting the EU textiles industry in certain third country market. 1999 http://europa.eu.int/comm/trade/pdf/mastb1.pdf

2 China's WTO Entry and its Impact upon the International Textile and Apparel Trade

2.1 COMPETITIVE POSITION OF CHINA'S TEXTILE AND APPAREL INDUSTRY

2.1.1 Historic review of the textile and apparel industry

China's textile industry emerged in the 1870s. However, continuous wars in the following years disrupted its normal development. Though the founding of the People's Republic of China in 1949 sent a positive signal to this labor-intensive industry, its fastest development was not achieved until the 1980s when government attention was shifted to economic construction and restructuring. In 1985, the industry was eventually included in the seventh "five-year plan" and, one-year later, it was selected as a priority industry by the State Council, hence entering a golden development period.

It is natural for China to choose the textile and apparel industry as the engine for economic takeoff due to its large labor force, relatively rich raw materials, and relating small capital and technology requirements. Table 2.1 shows the tremendous growth of the textile and apparel industry between 1978 and 1998.

With an average annual growth rate of 13.7%, textiles & apparel soon became a most important pillar industry generating almost one-fourth of foreign reserves from exportation. It has been the No. 1 textile and apparel exporter in the world since 1994. Between 1970 and 1998, China's textile and apparel exports grew at a rate five times faster than the world average level [39], amounting to US$ 43.1 billion in 1999.

However, the industrial performance itself is far less happy than its brilliant export achievement. The absence of appropriate planning, randomly repeated construction, a high percentage of outdated equipment and low efficiency have all contributed to today's loss-making situation. In 1996, the loss amounted to RMB10.6 billion with over 40% of the state-owned enterprises on the verge of bankruptcy [61]. Though great efforts have been made through strategic restructuring that started in 1998 and some positive results have been obtained, it still takes more time and inputs to completely solve these deep-rooted problems.

2.1.2 Competitiveness analysis of the Chinese textile and apparel industry

In view of China's huge output and export figures, no one will deny that China is a very important international player. However, if one considers the world's top fashion brands, none of the popular names originated in China, though many of them are actually manufactured in China nowadays. In world fashion circles, it is very hard to find either a top Chinese designer or a world-famous Chinese textile/apparel enterprise. China-labeled products are usually associated with a "cheap price" and

Table 2.1 Some facts about China's textile and apparel industry

China textile and apparel industry between 1978 and 1998		
	1978	1998
Cotton yarn (Million ton)	2.38	5.42
Wool knitting yarn (Thousand ton)	37.8	442
Cotton fabric (Million meters)	11030	24100
Wool fabric (Million meters)	89	268
Silk fabric (Million meters)	611	6390
Total textile and apparel exports (US$ bn)	2.43	42.8
Total textile and apparel imports (US$ bn)	0.2	12.15

(Source: China Customs; http://www.ctei.gov.cn/cteinew/english/fz.htm "China Textile Industry")

"street market" image. Thus, two questions must be asked: What is the competitive strength of China's textile and apparel industry that makes China the No. 1 exporter? And what is its competitive position compared with counterparts in the rest of world, especially those in the US, Italy and France that are the world fashion centers?

According to the Leeds Model, advanced by Peter Kilduff, who used seven factors to identify the different development stages in the textile and apparel industry, China's textile and apparel industry is now entering the third stage. [1] It features rapid growth in apparel exports and chemical fiber imports, increased vertical integration, a relatively large trade surplus and competitive advantages in the production of commodity products. In the view of Michael Porter's "Diamond Theory" [2], an industry's competitiveness is mainly determined by four aspects, including factor conditions, demand conditions, firm strategy and rivalry, and related/supporting industries. Analysis of these four aspects can give some insight into China's textile and apparel industry and help explain its dominant world position but "cheap label" image.

2.1.2.1 Factor conditions

Production factors refer to the inputs required during the production process, including basic factors such as natural resources, land, labor (unskilled and semi-skilled), and debt capital as well as advanced factors such as high-tech, talented personnel, strong R & D capabilities and a technology-based infrastructure. [2]

A. Labor input

The quick emergence of the Chinese textile and apparel industry has largely been attributed to its low labor cost. According to a Werner International Inc. survey on labor costs in 1998, China ranked 52nd of the total 58 investigated countries.

Figure 2.1 shows that the labor cost in China is only 4.8% of that of the the US, 4.4% of France, 4% of Italy and 2.9% of Germany. Since the textile and apparel industry, especially the latter, is still labor-intensive in nature, this huge difference is by no means an insignificant factor for China's presence as a significant textile and apparel supplier in the world, especially concerning the basic products of low added value.

2.1 Comparison of labor cost in selected countries in 1998

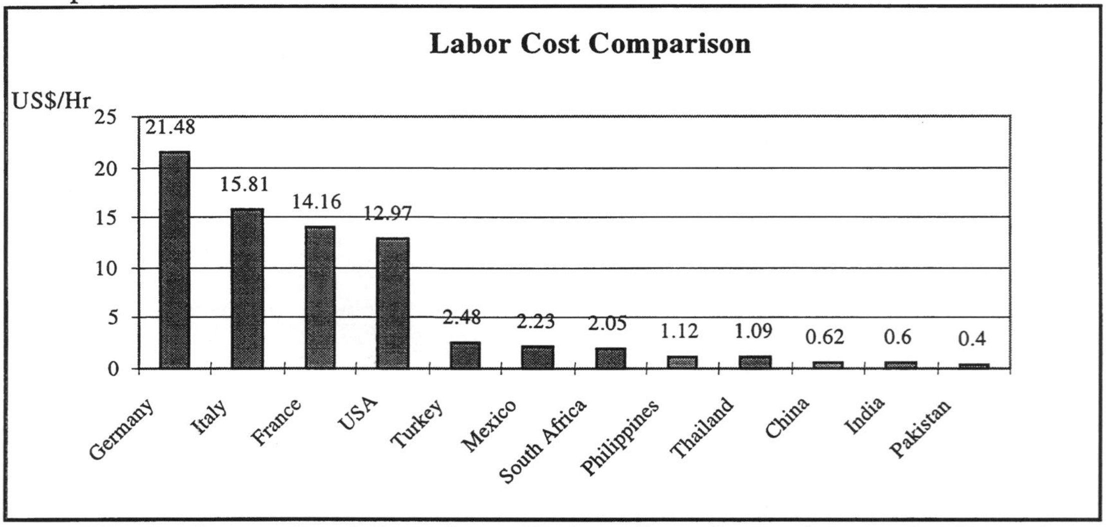

(Source: Werner International Inc., from "Labor Cost Comparison", Textile Asia, December 1998, p77-82)

However, if one takes growth rate into consideration, the future prospect is not so optimistic. The labor cost in China increased by 230% from 1988 to 1998, surpassing India, Indonesia, and Pakistan in 1998. This increasing pace is much quicker than that of Mexico, the Philippines and Thailand. Therefore, though China has a distinct comparative advantage in labor cost compared with developed countries and most developing countries, the gap is being bridged very rapidly.

By sharp contrast to its abundant availability of unskilled and semi-skilled labor, China lacks qualified personnel with a higher educational background. Technicians in the textile enterprises only account for 2.7% of the sector's total work force on a nationwide scale, of whom those with master degrees account for 0.1%, with bachelor degrees 46.5% and with three-year college experience 53.4%. [3] Shanghai is a well-developed metropolis in China. Shanghai is a well-developed metropolis in China. The ratio of technicians to the total textile employment in Shanghai is 9.48%; however, it is still 5.41 percentage points lower than Shanghai average technician ratio.[4] The poor performance of the whole textile and apparel industry in the 1990s cast a grave shadow over the human resource situation. Take the former China Textile University for example. It is a multi-disciplined university well known for its strength in the textile field. But it has faced a dilemma during recent years. Top students are reluctant to apply for the university, partly due to its "textile" associations. As a result, the university had to give up the established name and drop the "textile" indication completely. Name change is easy but the implications are not desirable: with fewer students studying in this field, it is hard to build a strong and solid base of human capital to help sustainable development. This would be the largest loss for the domestic textile and apparel industry.

Lack of qualified managers is another problem. Managers used to think or act in a central-planning frame, and their decisions were not responsible for profits and losses. Promotion was based on seniority rather than qualification and capability. Though the situation has been improved, it is far less than satisfactory by market standards. The complex and unclear division of ownership and management further complicates the problems.

B. Resource input

In addition to the labor cost advantage, China is also relatively abundant in some natural resources if viewed in absolute figures. Cotton production in China during 1999/2000 amounted to 4.1 million tons, accounting for more than one-fifth of the world total output. [5] As regards cocoon and silk production, China holds a dominant share of about 70%. [6] There is also a rich supply of other natural resources such as wool, ramie and flax. At the same time, limited availability of arable land[1] has led to a rapid growth in the proudction of chemical fiber. China is now the No. 1 chemical fiber producer in the world. Though there appears to be no correlation between a nation's competitive advantage in natural fiber production and the success of its textile and apparel industry [7], the availability of some textile raw materials does lay a foundation for China to develop a comprehensive textile sector starting from fiber production.

However, if the above figures are viewed against population, China is actually a resource-scarce country. The insufficient market mechanism also constrains efficient distribution flows. The much higher domestic cotton price compared with the international level well explains this point. Though corresponding reform has been carried out since 1999, it takes time to establish a well-functioning market. On the other hand, large production capacity is not bound to bring about desirable results if it fails to realize high value-adding processes and/or leads to a reduction in quality. Thirty percent of chemical fibers consumed by the textile and apparel industry are imported from other countries every year despite the large domestic output. As regards domestic fabrics, a survey conducted by China Garment Association indicated that those used in apparel exports were only 45.18% in spite of the fact that the annual fabric output in China is 16 billion meters. [8] Table 2.2 illustrates the self-sufficiency rate of domestic fabrics based upon material type. The self-sufficiency rate for chemical-fiber fabrics is the lowest, indicating the poorest performance of this sub-sector. The main problems for domestic fabrics include high washing shrinkage, serious color differences and faults, poor hand feel, dull pattern styles and unsatisfactory development of new-type materials. Long delivery time has placed domestic suppliers in a more disadvantageous position when quick response has become more and more crucial in today's competition.

Table 2.2[2] Self-sufficiency rate of domestic fabrics in apparel production

Self-sufficiency rate of domestic fabrics in apparel production				
Fabric Type	Pure Cotton	Cotton Blends	Pure Wool	Wool Blends
Self-sufficiency rate %	70.47	74.40	77.59	83.14
Fabric Type	Silk	Flax & Flax blends	Chemical Fibers	Others
Self-sufficiency rate %	79.03	98.46	42.56	66.02

(Source: http://www.chinagarment.net.cn/m104.htm, [9])

[1] China has 22% of the world population but only 7% of the world arable land.

[2] The investigated enterprises totally used 13.8-million-meter fabrics.

Table 2.3 Comparison of textile machinery in China, Germany, Italy and US

Textile Machinery Comparison, 1996				
Nation	Spindles %		Looms %	
	Ring	Open-ended	Shuttleless	Shuttle
China	5.81	94.19	5.38	94.62
Germany	0.85	99.15	94.02	5.98
Italy	1.60	98.40	88.30	11.70
US	0.55	99.45	87.15	12.85

(Source: Taiwan Textile Industries Newsletter, Volume 6, Issue 9)

C. Capital, technology and equipment

Many countries once selected the textile and apparel industry as the priority industry to develop partly because of its relatively low capital and technology requirements. Nevertheless, Hi-tech, along with strong capital support, is now playing a more and more important role in product innovation and industrial upgrading.

In the ninth "five-year plan", the industry plans to invest RMB200 billion in fixed assets, of which RMB80 billion is for infrastructure and RMB120 billion for technological innovation. Considering that China is a capital scarcity country, such investment is by no means small. However, the weak technological foundation and a high proportion of obsolete textile machinery make these investments insufficient to upgrade the industry. The technology-content in pure fiber products is 38.8% in China; while it is 86.0% in Japan, 71.5% in the US and 61.2% in South Korea. [10] As regards present equipment, 25% of cotton spindles should be dismantled, with only one-third of the rest up to the level of the 1990s. [11] Only about 15% of wool spindles and 6% of dyeing and finishing machinery are up to the international advanced level. [12] Table 2.3 clearly shows that the ratio of shuttleless looms in China is much lower than that in other three countries. The overall technological level of equipment in China is 10 to 15 years lag behind the world advanced level. [63]

The heavy losses of the industry since 1993 make technological investment much more difficult. As a result, the technical upgrading in SOEs was only RMB1.75 billion in 1998, an 81.62% drop over 1997. [13] Low-level technology and outdated machinery can both be partly attributed to insufficient capital, a problem requiring fundamental reform in the capital market. For example, enterprises of different ownership still receive very different treatment as regards bank loans, which have usually favor state-owned enterprises regardless of market-based consideration. By contrast, while SOEs are enjoying "soft budget", private ones find it difficult to access capital through normal financing channels. Therefore, capital availability in China not only refers to absolute volume but also equal opportunity for enterprises to access.

D. Other important factors

Traditional comparative advantages such as labor cost and raw material availability have progressively lost their historic luster in today's global sourcing activities. According to Michael Porter, these generalized factors are easily imitated and acquired. The labor cost in some developing countries, including Indonesia and Pakistan, has been lower than that in China. Therefore, factors that used to help China achieve great success will promote others' rapid development as well. It is a location-specific advantage easily accessed and acquired nowadays.

Compared with their counterparts in well-developed markets, China's textile and apparel industry lacks strength in specialized factors such as productivity, intangible assets and product development, which greatly hinders the industry from moving upmarket.

In China, the advantage of low labor costs is partly offset by low productivity, which is only a fourteenth of the American, Japanese, Italian and German levels and a seventh of HK, Korea and Taiwan. [14]

While low labor costs have effects upon the final costs, they can be outweighed by other factors such as raw materials, capital costs, and other auxiliary expenses. Considering the production cost of 30-count cotton fiber in the US and China, it was US$ 3.60/kg in the US and US$ 3.47/kg in China [15], a difference which is not as significant as the labor cost differences.

China is a large textile and apparel supplier, but it is by no means a fashion center featuring historic-inherited fashion culture and sophisticated fashion consciousness, as is the case in Italy and France. Italy is considered a "Mecca" in the world fashion industry, well known for its design creativity, famous brands and superior quality. France is another world fashion center, having a great influence upon fashion and fabric trends. Among the world's top 30 fashion brands, Italy owns 19 and France has 7, while it is still a blank in China, no are any of the world's top fashion designers based in China.

2.1.2.2 Demand conditions

A. Urge domestic market

China's huge domestic market itself is a magnet. Its population is one-fifth of the world total population. The EU as a whole has only 30% of the Chinese population, not to mention the US whose area is just a little smaller than China. Figure 2.2 is a comparison of some countries' populations, including China, the US, Japan and some EU members. Thanks to the healthy and rapid economic development during the past two decades, Chinese consumers' interests in fashion items have been well developed, sending an exciting signal for domestic and foreign manufacturers.

At present, the fiber consumption per capita is only 5.5 kg in China, 2 kg lower than the world average level [6] and much lower than those in the US and Japan, According to some researches, when GDP per capita reaches US$ 1000, fiber consumption experiences a quicker growth. In 1998, GDP per capita in China was US$ 774. Therefore, if the economic growth rate is kept at 7-8%, it will only take four years for China to enter into a fast-growing period of fiber consumption. What's more, the vast territory, different peoples and unbalanced regional economic development all render a unique opportunity for textile and apparel enterprises to establish market niches and promote a large variety of products.

B. Customer sophistication level

The upmarket image of Italian and French counterparts can be partly attributed to their sophisticated and demanding consumers. Their sensitivity to fashion trends and critical requirements for quality place heavy pressure upon textile and apparel manufacturers to pursue technological innovation. The strong fashion culture has cultivated an ideal atmosphere for consumer development. By contrast, Chinese consumers were almost "fashion illiterate" before its adoption of an open-door policy before 1978. For a very long time, textile and apparel items were mere living necessities. Though the situation has changed a lot, especially in some big cities, Chinese consumers as a whole lack fashion education as well as self-protection consciousness.

2.2 Comparison of Chinese population with some other countries

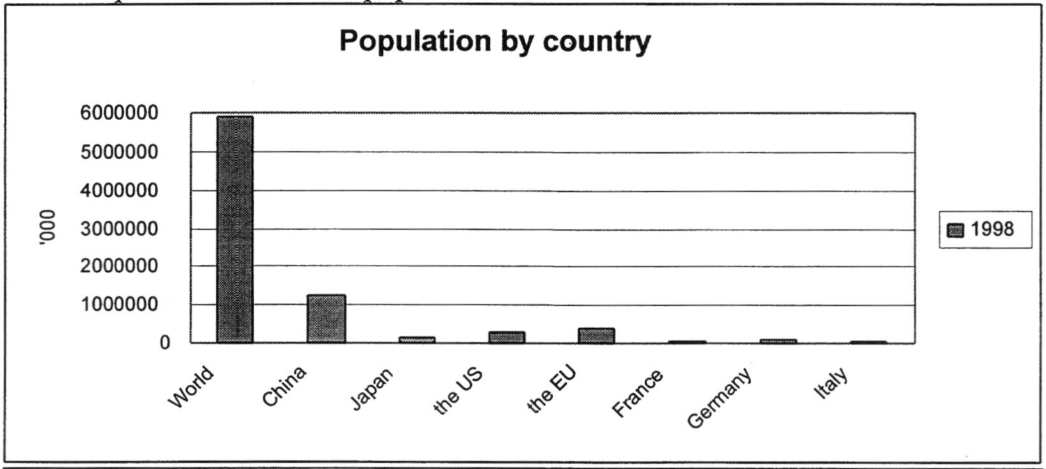

(Source: the United Nation)

China is still a developing nation with about 80% of its population living in the countryside. Low income means that the textile and apparel market is still price-sensitive. Though brand-consciousness is spreading in some coastal regions and big cities, those in the vast hinterland are still focusing upon the basic functions of textile and apparel products. According to a nation-wide survey conducted in 1997 by the State Statistical Bureau[3], price is the most important factor in apparel purchase (30.3% of the interviewees chose this factor), followed by "fitness and comfort" (24.2%), "durability" (20%), "good looks" (12.9%), "personality embodiment" (5.8%), "high grade" (1.6%), "brand" (1.6%) and "fashion" (1.3%). [16] This survey partly reflects the fact that customer sophistication is still relatively low in China. The fact that the self-sufficiency rate of domestic fabrics for local-sale apparel is as high as 90.3% also supports the above argument.

Low consumer sophistication also lies in imperfect government regulation and consumer's lack of self-protection consciousness. The Law of Consumer Interest Protection was not enforced until the year 1994. At the same time, consumers endured a seller's market for a very long time and, they haven't yet developed the necessary knowledge to protect their own interests. What's more, even if they finally decide to take steps to address the problem, they would have to find and collect proof by themselves and pay all the relevant costs at their own expense. High complaint costs and complicated processes prevent domestic consumers from being demanding and knowledgeable.

In short, China is not short of consumers, but it badly needs educated and selective consumers who can push and inspire the manufacturers to innovate and develop.

C. Segment structure of demand

China is a big country with distinct regional economic and cultural differences. It is by no means a single market. Table 2.4 shows the different GDP per capita of selected cities and provinces in 1996.

The economic development imbalance leads to varied consumer demands. On the one hand, the low-/middle-grade products remain the most popular. (See Table 2.5) On the other hand, citizens in the coastal regions and big cities with more disposable income and frequent exposure to the outside world have developed a taste for fashion products. Entry of foreign brands further cultivates brand consciousness and pursuit. Though the market is still small, it has a great potential to expand.

[3] This sample survey was conducted in 72 cities, targeting 15,600 households.

Family planning policy in China makes parents place most of their attention on their son or daughter, trying to give the best to the only child in the family. Therefore, babies' wear and children's wear have become fast growing segments.

Another point worth mentioning is the rising demand for home and technical textiles. Though home and technical textiles are still undeveloped in China (Table 2.9), they are new growth areas in the future.

D. Size and growth rate of home demand

In view of China's population, which was 1.248 billion in 1998, the domestic market demand is really huge. From 1989 to 1998, the average growth rate of apparel retailing volume in China was 1213.0%. The first five years of the 1990s experienced the strongest growth momentum, reaching an average growth rate of 20.4%.

Table 2.4 GDP per capita in selected cities and regions in 1996 (units: RMB)

Coastal Region	GDP per capita	Central Region	GDP per capita	Western Region	GDP per capita
Beijing	16735	Henan	4430	Shan'xi	3707
Shanghai	25750	Anhui	4390	Sichuan	4029
Shandong	7590	Hubei	5899	Gansu	3137
Zhejiang	10515	Hunan	4643	Qinghai	4066
Guangdong	10428	Heilongjiang	7243	Guizhou	2215

(Source: Almanac of China Statistics 1998)

Table 2.5[4] Information from some large department stores

Apparel Sales in Beijing, Shanghai, and Tianjin Department Stores with over 100 Million Sales Volume	Men's Shirt	Men's Suit	Womenswear
Low-grade (%)	54	54	75
Middle-grade (%)	32	30	19
High-grade (%)	14	16	6
Average Sales Price (RMB)	117	645	263

(Source: State Information Center, from [65],)

[4] The division standards are as follows: (units: RMB)

	Man's shirt	Man's suit	Womanwear
Low-grade	<150 (Beijing) <100 (Shanghai, Tianjin)	<1000 (Beijing) <500 (Shanghai, Tianjin)	<400 (Beijing)
Middle-grade	150-250 (Beijing) 100-200 (Shanghai, Tianjin)	1000-2000 (Beijing) 500-1000 (Shanghai, Tianjin)	400-800 (Beijing)
High-grade	>250 (Beijing) >200 (Shanghai, Tianjin)	>2000 (Beijing) >1000 (Shanghai, Tianjin)	>800 (Beijing)

2.3 Growth rate of China's apparel retailing volume between 1989 and 1998

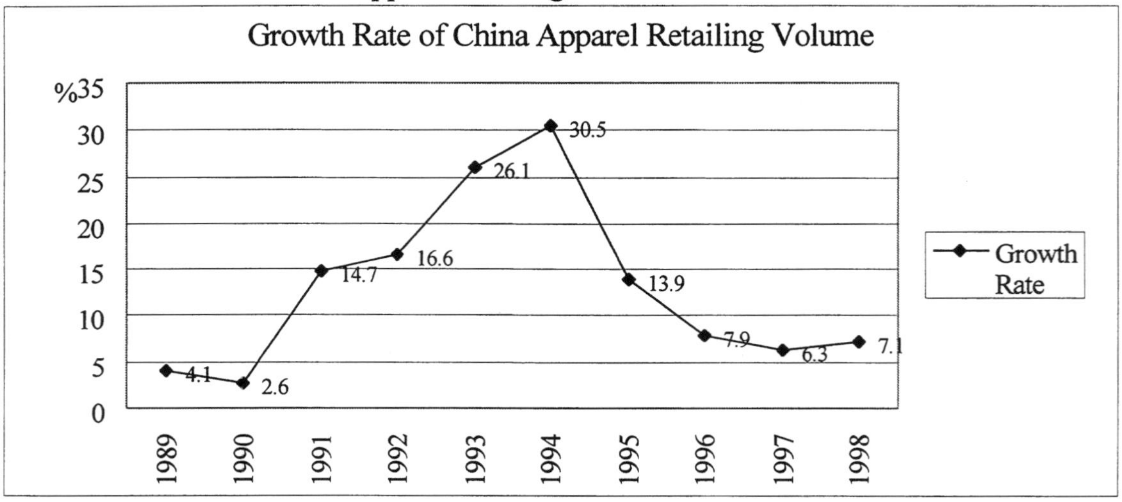

(Source: [16])

Income is still the key factor affecting textile and apparel expenditure since its demand elasticity is much higher than other categories such as food, transport and education. The actual income growth rate slowed to 3.4% in 1996 and 1997 while it was about 8% before 1995. At the same time, expectation of more expenditure on housing, medical care and education, as well as concerns about retirement life as the new social welfare system takes shape, all contributed to an anticipation of a drop in actual disposable income. Taking Shanghai as an example, apparel expenditure reduced from 9.6% in 1995 to 6.9% in 1998, while expenditure on medical services, education, entertainment, and residence increased from 16.1% to 25.7%. [66] Figure 2.3 also shows that the average growth rate of apparel retailing volume slowed to 7.1% between 1996 and 1998.

E. Distribution network

A sophisticated and well-functioning distribution system may render an additional stimulus to competitive advantages, since it puts manufacturers under intense pressure to enhance the products' quality. [7] Though efforts have been made to reform this sector in China, it still remains insufficient at present.

Prior to the 1980s, wholesale and retail businesses were monopolized by sate-run enterprises. The centrally-planned multi-tier distribution hierarchy operated as an allocation machine, rigidly separating the production sector from the distribution sector. Producers were extremely insensitive to the market changes. The textile and apparel production enterprises were neither responsible for the market demand nor concerned with the marketability of their products. The only thing they cared about was producing the planned quantity of planned goods within a scheduled time period. The wholesale and retail enterprises were also under strict central planning, merely acting as storage facilities for commodities rather than as active market players.

In 1978, China began to achieve rapid economic growth and gradually introduced a market mechanism to direct economic development. The problem of commodity shortage was alleviated in many industrial sectors. In 1983, ration coupons for fabrics and garments were abolished and, later, textile and clothing enterprises were given full autonomy to distribute their products. [17] At the same time, with the distribution door progressively opened to the outside world since 1992 (See 2.2.1.2), foreign counterparts began to make their presence felt in the domestic market. Many of the world's top brands as well as regional brands, having perceived the huge potential market, rushed to China to open their own retail outlets--which is also an effective way to circumvent the high import tariff and other import restrictions. Hot competition led to rapid market cultivation and development, particularly in the urban area. After 20-years of progressive reform, the distribution system has undergone remarkable changes. Major players in this sector include state-run department stores, foreign-funded/foreign-managed department stores, speciality stores, franchise stores, textiles/garment trade markets, and so on. Many textile and garment production enterprises are now actively engaging in retail and wholesale activities through vertical integration or other approaches in order to gain a better insight into the changeable market and consumer demand.

Despite the progress made in recent years, domestic distribution systems are still inefficient due to their malfunction for such a long time. They lack the necessary resources, such as capital, expertise and qualified personnel, to compete with the foreign enterprises. Though current liberalization in the wholesale and retail sectors is conducted on an experimental basis and subject to various restrictions, such as geography confinement and equity limitation, foreign entry has posed great challenges to their domestic counterparts.

State-run department stores occupy a key position in the distribution channels, but their problems are the most serious. Most of these stores, which enjoyed state protection and preferential treatment during past decades, failed to adjust to the new circumstances. Many still think that marketing is equivalent to selling products. In spite of the eye-catching slogans advocating that consumers are vital for a store's survival, they in fast act in the reverse direction. Therefore, the most important thing for a real change is to drop obsolete practices and cooperate closely with production enterprises based on strategic partnerships rather than the present weak and loose relationship[5]. Success of distribution reform does not necessarily depend on the capital input. It hinges upon a market-oriented attitude and an international-oriented vision of competition. In this sense, the distribution network in China is still far from maturity.

2.1.2.3 Firm strategy and rivalry

A. SOE in the textile and apparel industry

China adopted a planned economy for a long time with state-owned enterprises (SOEs) as the major players. With the establishment of a market economy, many township and village enterprises (TVEs) entered the competition arena, together with foreign-invested enterprises. Table 2.6 indicates that in 1998 SOEs account for 23.1% of the total textile enterprises, and they generate almost one-fourth of the total added value and industrial output value. The apparel sector provides a different picture, with SOEs only accounting for 9.3% in terms of enterprise number and less than 5% of total industrial added value and output value.

SOEs were once the sole players in the domestic market, facing little or no competition. They themselves weren't even concerned about their own profits and losses, both of which were under administrative control. The soft budget problem, lack of effective competition and incentives, government interference, rigid organizational structure and serious agency problems resulted in an unreasonably high debt/asset ratio, low capital utilization efficiency and poor management. Moreover, they have been over-burdened with too many responsibilities, such as pensions,

[5] Current practice of most state-run department stores is unfair for the production enterprises or sub-retailers who assume all the operational risks. Those stores just act as renters and take almost no risks.

Table 2.6 Economic indicators of China textile and apparel industry in 1998

Some Economic Indicators of the Textile and Apparel Industry									
Sector	**No of Enterprises**			**Industrial Output Value**			**Industrial Value-added**		
	Total	**SOE**	**%**	**Total**	**SOE**	**%**	**Total**	**SOE**	**%**
Textile	11276	2600	*23.1*	437.6	105.6	*24.1*	101.7	25.1	*24.7*
Fiber Processing	989	183	*18.5*	27.7	3.5	*12.6*	6.6	0.7	*10.6*
Cotton Textiles	4682	1261	*26.9*	223.0	73.6	*33.0*	52.2	17.5	*33.5*
Wool Textiles	1161	274	*23.6*	54.4	7.7	*14.2*	12.4	2.1	*16.9*
Linen Textiles	284	114	*40.1*	8.6	3.0	*34.9*	2.4	0.8	*33.3*
Silk Textiles	1732	369	*21.3*	60.1	12.0	*20.0*	12.7	2.6	*20.5*
Knitting	2018	351	*17.4*	51.4	5.0	*9.7*	12.6	1.1	*8.7*
Other Textiles	410	48	*11.7*	12.3	0.9	*7.3*	2.9	0.2	*6.9*
Apparel	5856	543	*9.27*	176.1	7.8	*4.4*	42.5	2.0	*4.7*
Chemical Fiber	803	176	*21.9*	82.7	14.5	*17.5*	18.5	2.6	*14.1*
(Unit: Billion RMB)									

(Source: http://www.ctei.gov.cn/cteinew/tongjixinxi/tjhome.asp, retrieved in November 2000)

medical care, and even housing. Though social welfare reform has been carried out, it will take a very long time to establish an improved social security system.

Corporatization has been a very popular trend for SOE reform recently, aiming to ultimately separate government functions from daily business operation, clarifying long-confused ownership/management relations, and providing a platform on which SOEs can act in a real market manner. The reform package is comprehensive, but whether it will achieve the anticipated result is still a matter of uncertainty. Since the textile and apparel industry has been chosen as an experimental sector, the final result is crucial to future SOE reform and government policy orientation.

B. Market competition

Chinese enterprises were subject to central planning for a very long time. The market itself was neglected in the decision-making process. SOEs were once the dominant players with little competition from other types of enterprises. Table 2.6 also shows that the proportion of non-SOEs in the textile and apparel industry has far exceeded that of the SOEs, especially in the apparel sector. Quick emergence of aggressive township and village enterprises (TVEs) as well as private and foreign-invested enterprises has greatly strengthened local rivalry, forcing the SOEs to be more market-oriented, especially when they lost privileges in many aspects and had to compete with other enterprises on a relatively equal footing. More and more enterprises began to realize the significance of quality and brand building. According to statistics released from the China Commercial Information Center, local brands occupied the first places in the men's suit and shirt market, taking 23.9% and 22.1% of the market share respectively in October 1998. [18][6]

However, domestic competition is still mainly limited to price wars and simple imitation, despite these positive results. Due to blind quantity expansion, the market is flooded with low-quality and low-grade mass products, many of which are unmarketable. In addition to weak competitive strength, domestic competition is not great. Commercial credit is continuously worsening among many enterprises. The inefficient distribution channel for raw textile materials makes domestic prices incompatible with international ones, placing extra cost burdens upon spinning and weaving mills and complicating the non-competition issue. For those enterprises that

[6] Man's Suit: Shanshan, Younger; Man's Shirt: Younger, Kaikai

are export-oriented, quota issues and foreign trade rights cause additional problems, such as rent-seeking. In view of the bid prices for the 21 quota categories in 1999, quota prices accounted for 10 to 30% of the export price.[7] Competition for quotas of hot-sale items is really heavy, driving their prices unreasonably high. Some enterprises even sold their quotas instead of actually using them for exportation, because this would make them a fatter profit. The other specific issue is about import and export rights. Many production enterprises still don't have the right to export their products and import some raw materials (See Section 1.2.5), artificially making them insensitive to changes in the international market and causing difficulties in adjusting their production schedule correspondingly. However, relaxation of foreign trade rights in recent years has negative consequences as well if it is not guaranteed by other effective administrative and legal measures at the same time. Take silk garment exports for example. In 1993, the average export price for a silk garment was US$6.1, which dropped to US$5.1 in 1994 because of blind expansion and a vicious price war. [19] It was only about one-tenth of the price of similar Italian products and one third of the Korean counterparts. It can be said that the reputation of Chinese silk products is destroyed by domestic enterprises themselves.

C. Firm size and strategy

Most domestic textile and apparel enterprises operate on a very small scale with limited capital and weak R&D capabilities. According to a survey on denim enterprises in 1995 sponsored by the former China Textile Council, those with one dyeing and sizing line accounted for 66.4% of the total investigated enterprises and those owning three or more lines only held 9%. Compared with some international players, domestic large enterprises find it hard to compete as regards annual output, sales volume and net profits. For example, Levi Strauss & Co. employed 1300 at its San Francisco-based headquarter and another 17000 worldwide[8]. In 1995, its net sales reached US$ 6.7 billion with a net profit of US$ 734.8 million. [20]

While foreign competitors attach great importance to strategic management, such as quick response (QR), merger & acquisition (M&A), global sourcing and relocation, many domestic companies are satisfied with gaining quick returns at the expense of long-term planning.

Unlike those in Italy which place emphasis upon product differentiation, most enterprises in China still act in an arbitrary and random manner. Lack of qualified managers with vision has not helped the situation. Moreover, upstream and downstream enterprises still stick to their own territories, lacking positive cooperation among individual links. The picture is quite different in some developed countries. For example, 65% of the local textile enterprises in the US have joined in the QR system. The Levi Strauss & Co. once declared that they could accept either an order for 100,000 pieces or 100,000 orders for one piece each. Italian firms are now looking for partnerships and strategic alliances. HdP recently purchased Valentino for around US$300m. In addition to M&A, banks now play a role in this industry to provide strong capital support, particularly instant liquid assets. The Union Bank of Switzerland (UBS) recently acquired 29.1% of the Turin sports fashion manufacturing group Basic (Kappa and Robe di Kappa). [67] There are also mergers and acquisitions in China, but most were propelled by political factors rather than by business considerations, such as realizing economies of scale or economies of scope. The most

[7] "Evolution of international textile trade system and the effects of WTO entry upon China textile exports", by Wang Shouwen (head of Quota Allocation Office in the MOFTEC), during a seminar on "WTO and China Economic Development Strategy"

[8] http://www.levistrauss.com, retrieved on August 24, 2000

common practice is that a well-managed enterprise is required by the local government to merge with a debt-ridden one in the hope of maintaining employment and stability, regardless of whether the former is willing or not. It is certain that such a forced match is unlikely to bring about economic benefits and much worse, it may involve the profitable enterprise in a very difficult situation where it begins to lose money.

As regards export-oriented enterprises, many turn to the processing trade which requires less capital and technology and is almost risk-free. In 1998, the processing trade value was US$ 18.9 billion, accounting for 46.8% of the total export value. [21] There is no denying that the processing trade contributed much to the industry's rapid development. But over-reliance is harmful to creating the industry's own competitive strength. It does save money and effort to make clothes by using foreign fabrics and brands. But, in the long run, China will fail to achieve industrial upgrading and be unable to share the biggest profits out of the whole value chain.

All in all, the strategy most domestic enterprises choose to compete with is shortsighted, unable to create long-term competitiveness and secure a healthy growth. Other political and economic factors, such as government over-interference, less familiarity with market rules and lack of specialized factors also restrain strategy formulation and development.

D. Investment climate

China's huge market is itself a magnet attracting foreign direct investment (FDI). Preferential treatment and government policies, such as tax exemption and half reduction during the previous five years, further contribute to the increased foreign capital inflow. In view of the easy entry and the comparative advantage of cheap labor, the textile and apparel industry has been a popular FDI destination. Table 2.7 shows the significant role of FIEs, especially in the apparel sector in which FIEs accounted for about a half of the sector's total added value, sales volume, profit and collected taxes.

HK is the No.1 investor, accounting for 74.3% of the total foreign investment. The traditional textile bases in the east coastal region attracted 93.9% of the foreign capital due to its comprehensive advantages in economic development, infrastructure, and investment environment.

It should be noted that future FDI will be channeled into certain priority fields under the revised "Guiding Directory on Industries Open to Foreign Investment" which divides the foreign-invested projects into four categories: "Encouraged", "Allowed", "Limited", and "Prohibited". As regards the textile and apparel industry, the following sectors are encouraged to attract foreign capital to promote development:

-Raw materials of chemical fiber which provide of economies of scale;

-Hi-tech chemical fiber products which are domestic-absent or in urgent need;

-Technical textiles with chemical fiber as the major material;

-New-type textile machinery & parts; computerized control systems;

-Dyeing & post-finishing projects aiming to develop high-grade fabrics

Table 2.7 FIE in China textile and apparel industry

Analysis of Foreign-invested Enterprises in the Textile and Apparel Industry

Unit: %

	Number	Value-added	Total Asset	Sales Volume	Profit	Tax Collection
Textile*	16.42	20.28	18.81	17.88	#1	12.76
Apparel	29.81	50.02	47.91	50.81	56.66	50.30
Leather, Fur, Down & related products	24.01	51.23	46.45	54.14	32.00	33.59
Chemical Fiber	27.23	10.02	15.06	12.65	#2	10.23

*: Not include the chemical fiber sector

#1: The whole industry suffered losses of RMB4.13 billion; while FIEs had a profit of RMB793 million

#2: The whole industry had a profit of RMB4.6 billion; while FIEs suffered losses of RMB27 million

*All the figures in the table are proportions of FIEs to their corresponding sectors in terms of enterprise number, added value, total asset, sales volume, profit and collected taxes. (Source: the Third Industrial General Survey in 1995, [68])

2.1.2.4 Related and supporting industries

An industry will not be internationally competitive without the strong backup of its related and supporting industries. The "textile complex" extends the traditional concept into new downstream industries, such as the car industry, the medical industry and the decoration industry. However, the domestic textile complex is in its preliminary stage with only 30% of the fibers consumed in these technical and home fields. (Table 2.9) Its related and supporting industries are also not in a strong position to enhance the industry's competitiveness, though the industrial system is relatively comprehensive.

A. Related industries

The textile machinery sector is the major related industry, which lays a foundation for the production of high quality and high value-added textile items. The technological content of its textile machinery production partly reflects a country's engineering and innovation level. Though the sector in China has undergone some significant changes due to relatively heavy investment in certain sectors such as spinning and weaving, its overall position in the world still lags behind the European developed nations with a technological gap of 10 to 15 years. [22] A comparison of the average import and export price in Figure 2.4 clearly illustrates the lower technological content of China-made textile machinery. It shows that while the price difference between spinning machinery imports and exports was less than two times, those for looms and knitting machines were widened to 4.5 times and 23 times.

There are about 2000 textile machinery factories with an annual output value of RMB 10 billion. [23] By the end of 1997, about 50 foreign-invested textile machinery enterprises had been established in China, with a total capital input of US$ 100 million. [23] Constrained by a lack of the necessary expertise and R&D efforts, about two-thirds of the textile machines used in China are imported. [24] The export value is much lower than the import one. In 1998, China imported US$1,166 million worth of textile machinery, almost 7 times as much as the export value.[9] As regards textile machinery exports from China, 28% were spinning machinery, 7% weaving

[9] http://textile.unet.net.cn/imp-exp98.html, retrieved in July 18,1999

2.4 Export and import unit prices in China for textile machinery in January 2000

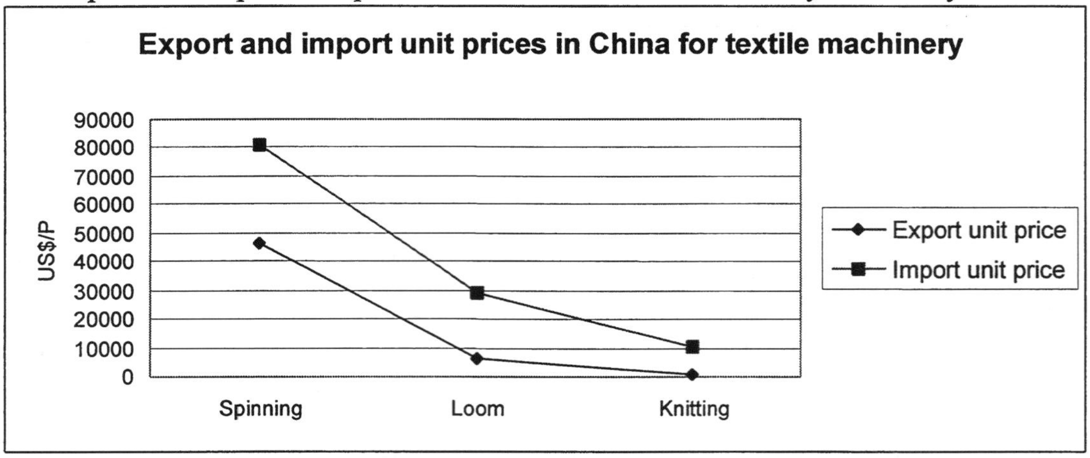

(Source: calculated based on data from Textile Asia, April 2000, p67)

machinery, 10% knitting machinery and others 55%. [22] The major export destinations are developing countries. Sewing machine production is of some strength in China. In 1999, the production output reached 10 million, of which 60% were exported to over 100 countries, accounting for about 34% of the world total exports. [25]

The technologically weak textile machinery sector can not meet the production requirements for high value-added quality textiles, forcing many textile mills to rely on imported machinery. Take shuttleless looms as an example. Domestic output was 12000 sets between 1991 and 1995, while those imported in 1994 alone amounted to 6500 sets. [22] In 1996, about a quarter of world total shuttleless loom exports were destined for China, of which Rapier/Projectile accounted for 29.2%, Air Jet 18.1%, and Water Jet 21.3%. [10] The proportion was raised to 28% in 1999. [26] Of all the imported machinery, finishing machinery accounted for about 22%, followed by spinning machinery 21%, knitting machinery 20%, weaving machinery 17%, MMF machinery 10%, and others 10%.[11] Japan, Germany, Italy and Switzerland were the major suppliers, representing 29%, 17%, 12% and 5% respectively of China total textile machinery imports in 1994. In 1999, China imported about DM 424 million worth of textile machinery from Germany, accounting for about 7% of total German exports. [27]

Since China is now dismantling obsolete spindles and enhancing control over the production of spindle frames, this will negatively affect the textile machinery industry whose focus is upon the spinning segment. In 1998, the industry suffered a loss of around RMB 100 million, making technological upgrading and innovation much more difficult. [28]

Based on the above analysis, it can be concluded that the textile machinery sector in China has failed to provide solid support for the necessary industrial upgrading in the textile and apparel industry, greatly limiting its capability to manufacture high value-added items. This is not the case in some developed fashion nations, such as Germany, Italy and France, whose textile and apparel industries have benefited a lot from their strong textile machinery sector. The German textile machinery industry holds a leading position in the world market with a share of more than 30%, followed by Japan 21%, Switzerland 11%, and Italy 11%. [29] France was the fifth largest

[10] Calculated based on data from [82]

[11] They were 1995 data from [83],

worldwide. [30] In 1997, the production of textile machines and accessories in Germany reached DM 8.5 billion, of which 93% were exported to other markets. As regards Italian and French textile machinery, about 65% and 85%[12] respectively of the turnover are destined for other countries. All of these nations attach great importance to R&D, product quality, customer service and technical innovation, which are the weakest fields in the China textile machinery industry which will take much time and effort to catch up.

Other related industries may include the press, exhibitions and models, all of which are still in the infant stage. The press industry fails to cooperate closely with the textile and apparel industry. Up to now, there are no textile and apparel-related newspapers or magazines of regional or world influence in China. As regards textile and apparel trade shows and exhibitions, China still has a very long way to go, especially compared with those held in Germany, France, Italy and the US. For example, Interstoff, held in Frankfurt, Germany, is considered one of the most important and largest international textile trade shows. It consists of 6 textile fairs each year, attracting more than 1,000 exhibitors from over 40 countries and 25,000 visitors. Premiere Vision, held twice a year in Villepinte, near Paris, draws more than 44,000 attendees from around the world. Others include Ideacomo held in Como, Italy; Texitalia held in Milan, and New York; Interstoff/Asia in HK; the International Fashion Fabric Exhibition in New York; the LA International Textile Show; and the Textile Association of Los Angeles Show, etc.

B. Supporting industries

Supporting industries refer to the transportation industry, information industry, communication industry, financial industry and so on. China has invested heavily in the infrastructure and communication facilities. The number of telephones is growing at an annual rate of 19.9%. There are 60 earth satellite stations accelerating communications with more than 200 countries and regions. The founding of the Securities Exchange in Shanghai and Shenzhen provides a good channel for domestic enterprises to raise funds. The newly emerged information industry has laid a basic foundation for the information era.

Despite the rapid growth rate of these industries, they are still not strong enough to support the textile and apparel industry due to their previous weak foundation. Take the stock market, for example; the imperfection of laws and regulations, the over-speculation of most investors, the immaturity of the market itself and the ignorance of many listed companies towards the market mechanism resulted in many problems, greatly discounting its original role.

C. Lack of a textile cluster

In the strict sense, the textile cluster has not yet been formed in China in view of its present low economic level and the weakness of the individual industries concerned. Some specific examples in Italy, France, Germany and the US may provide a deeper understanding of the weak synergies among domestic industries and the big gap with these international players.

Backed by centuries-old specialization, the Italian textile industry gains great momentum from sophisticated and advanced textile clusters. [31] Several of them can be identified in northern Italy: the cotton cluster of Brescia; the wool clusters of Biella and Prato; the knitting clusters of Carpi and Modena and the printing cluster of Como. Consider the wool cluster of Biella, for instance. About 2000 interdependent woolen companies are supported by 200 textile technology firms, a fair trade organization

---

[12] CHINA TEXTILE & APPAREL, 8/9,1999; Textile Asia, March 1999, p30

(Idea Beilla), a specialized magazine, and a research and development center (Texilia). The PV International Fabric Exhibition held in Paris, which is called "the first visual exhibition", is the world's leading fabric fair. It renders solid support for the promotion of French fashion items. The German textile industry has benefited great from the strong textile machinery industry. The US textile industry has also gained synergy from its suppliers. BASF is the global leader in nylon. Technological breakthroughs from the chemical giant "Du Pont", such as nylon in 1938, Dacron Polyester in 1950 and Lycra Spandex in 1960 which successfully replaced silk, wool and rubber respectively, have blazed new roads for domestic textile and apparel industry.

2.1.2.5 Strong government support

The rapid increase of Chinese textile and apparel exports during the past two decades can be largely attributed to the big changes in government policy since the 1980s. This shifted from a heavy industry orientation to a light industry orientation, from inward looking to outward looking, from central planning to market-based, and from SOE predominance to co-existence of diversified ownership forms.

Though the central government is now trying hard to enhance macro-economic control rather than over-interference in the micro-economic operation, the current political and administrative systems determine that active government support is still very crucial to an industry's future development. Some measures may have dual effects and even obstruct productivity improvement, but strong policy support from the government is indispensable when the textile and apparel sector is now undergoing painful structural adjustment and when China is still in the transformation process from a planned economy to a market economy.

Policy support at the present stage includes a compensation scheme for spindle elimination, asset re-arrangement, debt write-off and conversion into shares, export quota allocation reform, increase of the export rebate rate, and so on. Most of them aim at SOE reform, which is arguably the single most important issue upon which China's economic future hinges. [32]

Compensation scheme for spindle elimination

The first step is to eliminate 10 million obsolete spindles and lay off 1.2 million employees by the year 2000. [11] To overcome the immediate difficulties textile enterprises may face from the industrial restructuring, the government will offer capital guarantees by offering the firms RMB3 million in subsidy and RMB2 million in interest-free loans for every 10,000 spindles removed. [12, 33] As regards the lay-off, the government will provide guarantees for their minimum living expenses, usually ranging from RMB180 to 220 per month for three years. [34]

Asset rearrangement

Merger, acquisition and bankruptcy are employed to realize asset reshuffles. The domestic textile and apparel industry features high fragmentation, small scale and outmoded management. Asset rearrangement is designed to readjust the structural form of enterprises in the manner of capital flow, thus leading to the optimization of the asset reallocation. In view of the ninth "five-year" plan, it is now placing its focus on the establishment of 50 large-sized group enterprises and the development of collective and private entities.

Some enterprises with strength have been transformed into corporations and listed in the stock market. In 1996, 58 textile enterprises were listed in the stock market, attracting RMB8 billion. In 1998, five listed textile companies underwent asset reshuffle to replace the bad assets with good ones, successfully expanding their business into new lines such as home/technical textiles. [69] This move helped these enterprises expand into new textile-related fields with a better capital structure. The future trend for SOE reform is to transform more enterprises to corporations with private capital playing a more important role. The state will not be a controlling shareholder except in some strategically important sectors.

Debt write-off and conversion into shares

In 1997, RMB9.73 billion was earmarked to write off bad debts for state-run textile enterprises. In 1998, the government set aside another RMB12.6 billion, focusing upon the bankruptcies of debt-laden companies. [11] This step helped to decrease the loss of the textile sector by RMB2.6 billion, reducing the asset-debt ratio of these enterprises from 77% to 73%.

Another new idea is "converting debt into equities". The Shandong Ru Yi Wool Textiles Group signed an agreement with two asset management companies [13] recently. After debt conversion, these two companies will hold 43% of the equities and the debt/asset ratio will reduce from 78% to 58%, saving RMB22 million financial expenses every year.

In 2000, China plans to convert RMB 15 billion to stocks in 50 textile enterprises in the hope of reducing the debt/asset ratio by 3 percentage points. Moreover, through merger and a bankruptcy and debt write-off of RMB 15 billion, it is expected that the debt/asset ratio could be lowered by another 3 percentage points. [70]

Preferential loan and export credit arrangement

Banks provide a financial package of exporter's credit to those having import and export rights. For example, the Export and Import Bank signed an agreement with China Textile Machinery Co., Ltd. (CTM[14]) in March, 1999 for an amount of US$181 million, designed to support the company's exports. Banks also extend loans with lower and favorable interest rates to ease the financial burden of the debt-laden industry.

Export quota allocation reform

Under the quota allocation reform, enterprises will no longer own fixed "life-long" quotas. There are now three ways of quota allocation: open tenders, open application and performance-based allocation. [35] The first of these was introduced in 1998 in 21 textile product categories, which are usually hot sale items. Relevant quotas are to be sold in a bidding process through which bids are submitted simultaneously by facsimile and mail. The deposit required of the first 10% of winning enterprises has been raised from 30% to 80%. As regards quotas with a utilization rate of less than 50% in previous years, which covers 98 product categories in 2000, quotas are open to the enterprises on a first-come-first-served basis. The remaining quotas are allocated based on actual export performance of individual enterprises during the previous year, of which 30% are allocated to manufacturing plants with autonomous export rights in 2000, 10 percentage points higher than in 1999. [36] At the same time, MOFTEC has entrusted China International E-Commerce Net (CIECNET) to develop a feedback system to effectively evaluate the quota usage level. Those who fail to utilize over 30% of their purchased quota for the current year will be completely disqualified from participation in future auctions and those whose quota utilization rate is below 70% will not be allowed to bid for the same categories in the following year. [37]

---

[13] China Oriental Asset Management Company and China Hua Rong Asset Management Company

[14] CTM is the largest state-owned textile Machinery Company in the country and one of the 120 pilot enterprises chosen in a trial program to be restructured into conglomerates.

Change of the export rebate rate

The export tax rebate is among the few permitted by the WTO as long as they do not exceed the actual tax imposed upon the exported products. China has simplified its present export rebate system to four levels: 5%, 13%, 15% and 17%. Textile machinery and apparel export rebate rates were raised to 17% in 1998. The export rebate rate for Xinjiang cotton also rose to 17% in order to encourage export of domestic cotton. From January 1, 1999, the export rebate rates for other textile raw materials and products have been increased to 15% in the hope that this will stimulate the current dull textile export scene due to cost reduction.

Some of the measures discussed above may violate relevant WTO regulation, such as direct export subsidies and debt write-off, which have to be eliminated after China enters WTO. This point will be further explained in the following sections.

2.1.3 Competitive position of China's textile and apparel industry in the World context

In view of the above analysis of the current diamond model and comparison with counterparts in some developed countries, it is clear that China's textile and apparel industry are only competitive in traditional low-/middle-end markets, with apparel production as the most competitive sector. According to the international competitive coefficient[15] in apparel, China was 0.93 in 1997, much higher than Korea (0.50), Italy (0.47), HK (0.21), France (-0.34) and Germany (-0.51). [15]

2.1.3.1 Low-/middle-grade apparel production is internationally competitive.

This sector is still labor-intensive, catering to China's present factor strength. The export composition also indicates that apparel is much more competitive than textiles in the low-/middle-end markets.

In 1999, the total textile and apparel exports were US$ 43.1 billion, of which 70% was attributed to apparel exports. The average export price is US$2.54 per unit of apparel--US$1.74 per unit of knitting apparel and US$3.92 per unit of woven apparel.[16] Figure 2.5 is a breakdown by various material types or categories. According to authoritative US organizations, the average unit price of Chinese garments exported to the US in 1998 was 26% higher than that of 77 countries; and in 1999 the gap has expanded to 33%. This means the overall level of Chinese apparel exports has been progressively upgraded, though it remains in the low-/middle-end markets.

The developed countries have turned to higher value-adding activities, such as R&D on new-type fabrics and fashion design. Other low-cost suppliers, though developing very fast, find it hard to compete with China at the present stage, because China has accumulated more than 50 years experiences in this aspect. Table 2.8 shows that among the four category groups, China registered the highest in three of them, the exception being the fabric category. However, the threat of other countries will be more prominent as time goes on, because these countries are also the major players in the low-/middle-end markets. Head-on competition cannot be avoided and will become hotter when the current quota mechanism ends in 2005.

---

[15] The international competitive coefficient = (export value – import value) / (export value + import value)

[16] Calculation is based on the total quantity and total export value from January to November 1999.

2.5 Chinese apparel export prices

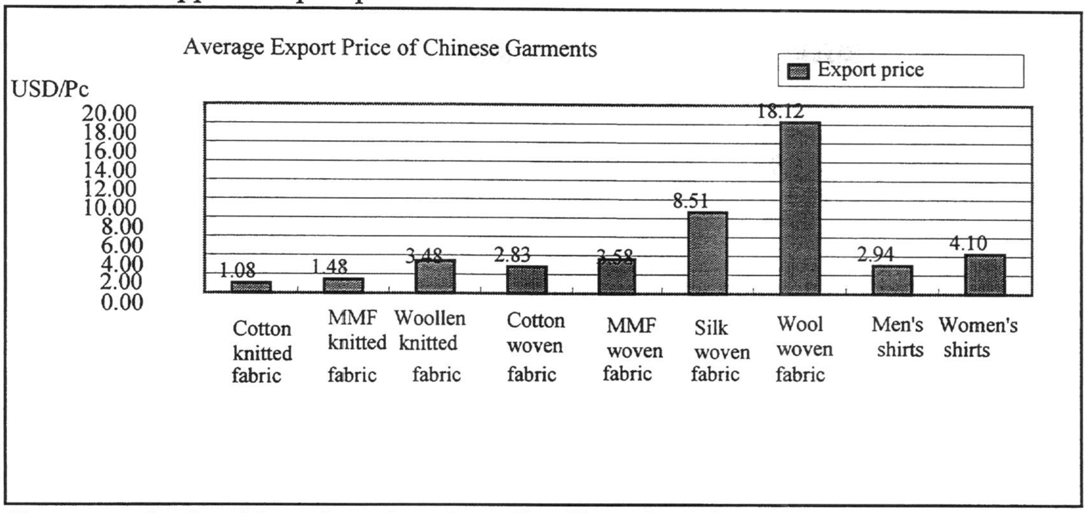

(Source: calculated based on data from Textile Asia, July 1999, p66)

2.1.3.2 Traditional low-technology-content textile products remain competitive

The cotton spinning sector is the pillar of the whole Chinese textile industry both in terms of industrial output value and export volume. During the first 11 months in 1999, cotton fiber, yarn and fabric exports accounted for 76.1%, 29.5% and 40.3% of the total fiber, yarn and fabric[17] export volume respectively. As regards the export value, the ratios were 29.1%, 24.7% and 31.3%, indicating the low-technology content.

2.1.3.3 Local brands have their voices in the low-/middle end market

Hot domestic rivalry has positively stimulated the brand building process. A number of local brands have been established in men's suits, shirts and knitted underwear, occupying a large market share. But they are still players in the low-/middle-end market. Take men's shirts, for example. Foreign brands, either directly introduced from abroad or manufactured through foreign-invested companies in China, still take the lion's share in the high-end market, with prices five to ten times higher than the local brands. [18] Such a pattern is an exemplar of the present domestic market for most product lines, indicating that local brands still have a long way to go.

2.1.3.4 Relatively comprehensive industrial system

Though many sectors in the industry are still weak and not internationally competitive, the industrial system is comprehensive, ranging from cotton to synthetic fiber; from intermediary products to branded end products. The production output of chemical fiber, yarn, silk fabrics, knitted fabrics and apparel all rank No. 1 in the world. The economic development imbalance and the status of "developing country" in nature will make low-grade products appealing for a long time.

[17] The total fabric export volume here doesn't include the chemical fiber fabrics.

Table 2.8[18] US import prices from China and some developing countries (US$/unit)

US Import prices from Some Developing Countries				
Country	Yarn	Fabrics	Clothing	Made-ups
Mexico	0.36	0.94	3.25	0.62
China	**0.46** [a]	**0.70**	**4.83**	**2.07**
Thailand	0.35	0.56	4.22	1.12
Indonesia	0.32	0.54	3.76	0.76
India	0.37	0.87	4.03	1.18
Turkey	0.39	0.99	3.74	0.92
Phillipines	0.39	0.70	3.60	1.12

a: This figure is the 1997 figure, from China Customs

(Source: calculated based on data from [38])

2.1.3.5 An "assembly line" rather than a fashion leader

The developed countries embrace almost all the advanced textile-related technologies. The geographic re-location strategy provides them with a better opportunity to concentrate upon high-value-added activities. One good example is that the output of non-woven fabrics in the US, EU and Japan totally captures 86-92% of the world total output. [39] Therefore, when putting the Chinese industry in the global context, the whole picture is not encouraging.

The industry's success is still dependent upon traditional factor advantages which are less and less decisive in today's globalization trend. The weakest links lie in chemical fiber, dyeing & finishing, fashion design, product development and innovation, as well as new-type textile machinery resulting in a concentration upon low-value-added activities. The wide gap in the value-added processes between China and the developed countries is very difficult to bridge within a short time.

Figure 2.6 draws a comparison between Italian and Chinese textile and apparel export prices. As regards yarn exports, the average price difference is about US$0.7/M^2 while widened to US$1.4/M^2 for fabric exports. The gap in wool apparel exports is even more notable, reaching US$14.8/M^2.

Figures 2.7 and 2.8 provide a more comprehensive picture of the position of Chinese textile and apparel exports. The average export price of French fabrics to the US was US$20/M^2, the highest among the six countries. If silk fabric are considered for example, the US average import price from France was US$28.3/M^2, three times higher than the average import level and four times higher than that from China, a country known as the "No. 1 silk producer" for so many years.

Table 2.9 is a comparison of fiber consumption ratios in China and other markets, showing the big difference both in domestic consumption patterns and the export product mix.

[18] Unit: US$/SME (square meter equivalent)

2.6 Textile export prices in China and Italy in 1997

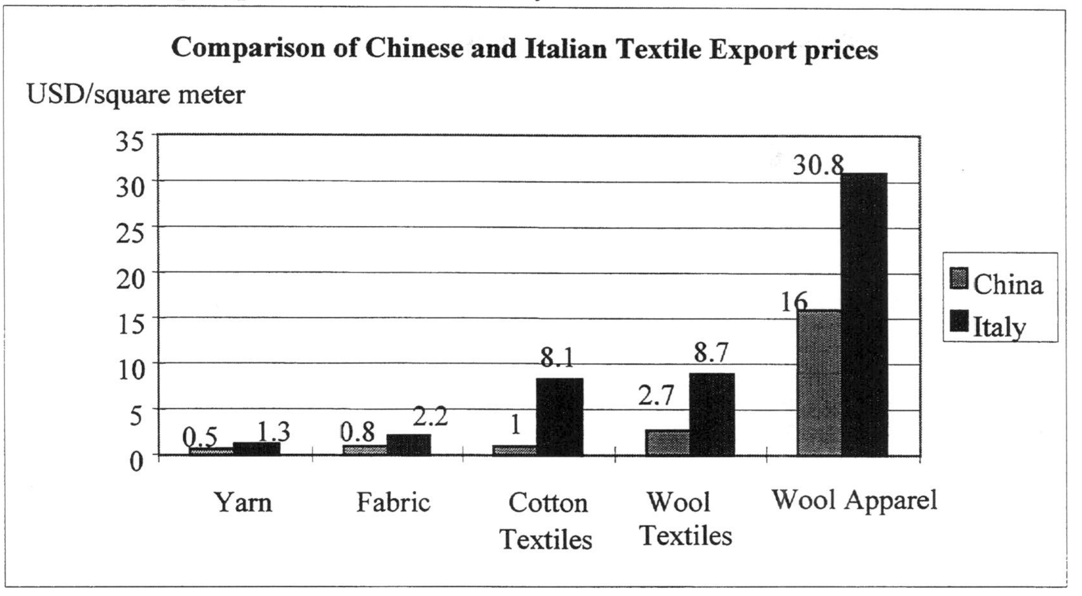

(Source: calculated based on data from [40])

2.7 US fabric import price from some selected countries

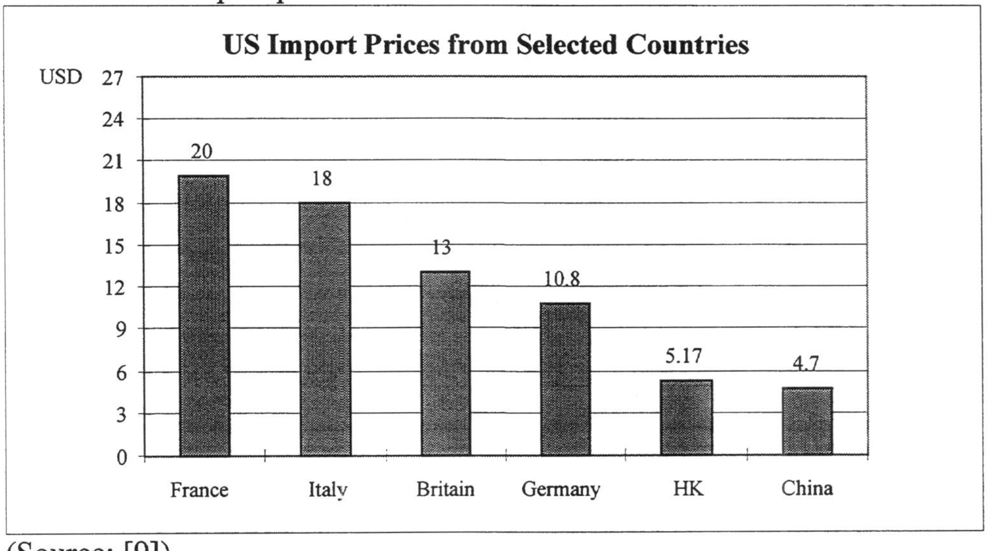

(Source: [9])

2.8 US silk fabric import prices from China, Germany, Italy and France

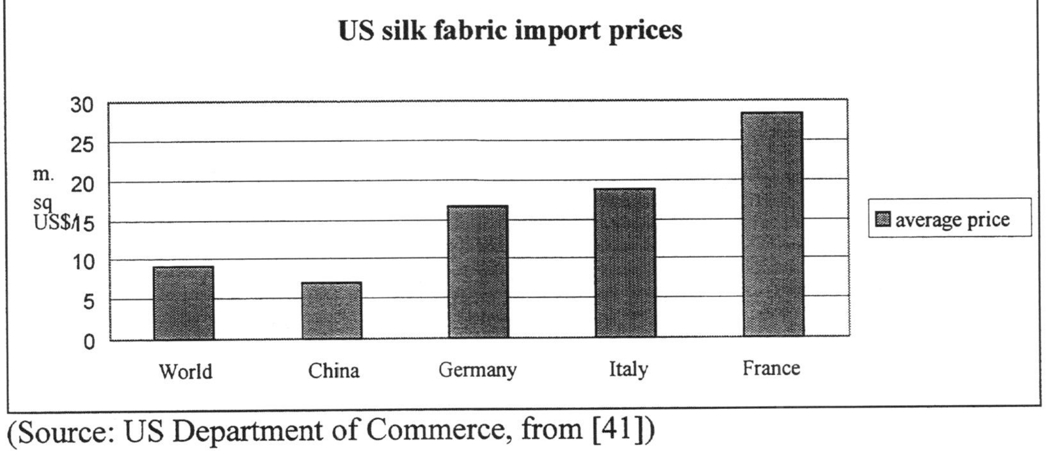

(Source: US Department of Commerce, from [41])

Table 2.9 Fiber consumption in the US, West Europe, Germany, China and Japan

Fiber Consumption in the Apparel, Home and Technical Textiles					
	US	West Europe	German	China	Japan
Apparel	39%	47%	45%	70%	30%
Home Textiles	39%	35%	30%	20%	31%
Technical Textiles	22%	18%	25%	10%	39%

(Source: [42]; [71])

The developed countries have shifted their production efforts to niche products with high added value in order to avoid head-on competition with low-cost manufacturers in the developing countries. For example, the US textile industry has moved away from staple fabrics for the apparel sector and moved toward niche products. Technical fibers that satisfy the sportswear and activewear industry such as surfing have witnessed outstanding growth. [43] Exportation of specialty fabrics that are not available in other countries offers the industry a unique competitive advantage. Many world famous brands have penetrated into the Chinese market through various routes, such as direct export, franchise and direct investment. However, the most important value-added links such as design, product development and marketing are retained firmly in foreign hands, which explains why China is just an important "assembly line" rather than an influential fashion leader, despite its large export volume.

2.2 PROSPECT OF CHINA'S WTO ENTRY

2.2.1 China's continuous efforts concerning WTO entry

2.2.1.1 Historic review of China's application for entry into WTO

China first applied to accede to the GATT in 1986 and its effort to join this multilateral mechanism has never been subdued since then. With the establishment of the WTO and its more and more important role in the world, it is essential for China to become a player itself in order to integrate better with the global economy.

The accession process involves preparation of a protocol “package”, consisting of the protocol, the working party report and the market access schedules for industrial and agricultural goods and services. While the first two are discussed multilaterally, the last one is negotiated bilaterally with Members of particular interest.

China has entered into bilateral talks with 36 member countries (EU as one member), among which Japan was the first nation within the Group of Seven to officially approve China's application into WTO during Japanese Premier Keizo Obuchi's visit to China in July 1999. Australia was the second after Japan among the industrialized nations to finish negotiations on China's bid to join the WTO. Talks with the US and EU have not been smooth because of the political sensitivity and its related issues apart from economic concerns. The successful conclusion of US-China and EU-China talks paved the way for preparing the final "package"[19].

[19] According to the accession process, only after China successfully concludes negotiations with the above-mentioned 36 Members can the market access schedules form a part of the whole “package” for final approval.

During the negotiation process, China insists on three principles with respect to the entry requirements: (1) China will join the WTO as a developing country; (2) There should be a balance between obligations and rights for new members; (3) There must be no discriminatory clauses. China has taken great pains to become a WTO member by putting serious offers on the table during the past 13-years of negotiations, including tariff reduction, non-tariff removal and other liberalization measures. Some have been progressively enforced in practice. The following is a brief explanation, with the textile and apparel sector as the focus.

A. Tariff reduction (also see relevant points in Section 1.1.2)

China has cut import tariffs several times during recent years. In April 1998, China proposed a comprehensive tariff offer to the Working Party, promising that the trade-weighted average tariff rates for industrial goods will reach 10.8% by the year 2005. From January 1, 1999, China began a new round of tariff reduction covering 1014 commodity categories, an initiative to lower the average import tariff rate by 2 percentage points by the year 2000. The textile and apparel sector has experienced the most drastic adjustments involving more than 90% of textile and apparel imports. The tariffs imposed upon textile raw materials were lowered by the largest extent. On January 1, 2000, a total of 819 textile products listed on the tariff schedule underwent reduction by another 0.6 to 2.0%.

B. Non-tariff barriers removal

Most textile products are subject to compulsory prior import licensing managed by MOFTEC. China is committed to adopting a tariff-rate quota system instead of the present quantitative restrictions in some particularly sensitive sectors, such as wool and cotton.

China in July 1995 offered to progressively liberalize the trading monopoly within eight years after accession, aiming to bring its domestic foreign trade regime into line with prevailing international practice. [72] MOFTEC began to carry out reforms, entitling more manufacturing enterprises, either state-run or privately-owned, to conduct exports by themselves. It has further lowered requirements for obtaining foreign trade licenses, including: (1) A preferential system is designed for 120 key state-owned pilot conglomerates and 1,000 selected key state enterprises and large state industrial enterprises registered with MOFTEC. Production enterprises dealing in non-electromechanical products will have the right to set up wholly-owned or majority-owned foreign trade companies to handle the export of their own products or other related products, providing their annual export value surpass US$10 million. (2) The minimum registered capital of trading companies qualifying for import and export has been reduced to RMB5 million in the coastal area and RMB3 million in the central and western areas. (3) For foreign trade companies with annual exports of over US$100 million in the previous year, their wholly-owned or majority-owned subsidiaries will be entitled to import and export provided that their export volumes are not less than US$10 million. Since 1999, MOFTEC has extended foreign trade rights to the private sector on condition that they have a minimum of RMB8.5 million in net assets, RMB 50 million in annual sales and US$1 million in annual exports. About 142 private enterprises with a total export volume of US$450 million in 1998 were granted foreign trade rights up to mid-1999. In addition, 5 joint-venture foreign trade companies were established in Shanghai and Shenzhen with local partners holding 51% shares. According to the latest EU-China bilateral talks, China has promised to completely remove the state monopoly on silk exportation by 2005. [44]

A new "Contract Law" incorporating the original three relevant laws has been in force since October 1, 1999. This move sends a positive signal that the domestic legal frame work will be clarified and improved step by step.

C. New market access in the distribution sector

Joint ventures in China have to export most of their manufactures to meet the requirements of foreign exchange balance with only an approved portion for domestic sale. On the other hand, foreign investment in the distribution sector was prohibited before 1992.

However, the central government is now considering expanding the domestic sale ratio for these foreign-invested companies if the relevant products meet the national industrial policies and are not subject to quota limitations. The State Council approved 16 foreign-invested retailing enterprises with import and export rights and 2 commercial chain stores co-invested by local and overseas companies from 1992 to 1998, not to mention those approved by local governments. According to the recent pilot program, China has decided to further open up its domestic retailing and wholesaling market. Foreign companies can now enter the Chinese retailing market in the form of joint ventures or cooperatives with local partners in the capitals of all provinces, autonomous regions and other major cities. [45] Preferential treatment will be granted if foreign-based commercial entities pool investments in the underdeveloped central and western parts of China. (For example: 40-year operation period compared with 30-year in the coastal area). Chinese partner should hold dominant shares in principle, except those particularly approved by the State Council. Though solely foreign-owned commercial enterprises engaging in retailing and wholesaling are still forbidden at the present stage, the green light is given to the four cities, including Beijing, Shanghai, Tianjin and Chongqing, to each establish one pilot joint-venture wholesale enterprise.

2.2.2 US-China WTO-entry negotiations

When Premier Zhu Rongji visited the US in April, 1999 with a package of attractive offers, the US President Clinton failed to accept it due to huge domestic pressure. Bilateral talks between the two sides were suspended after a series of political events until October, and the two struck an agreement by the end of 1999. Detailed information about the agreement is not available at present stage. Therefore, we will focus upon the major concerns between the two sides over the textile and apparel issue.

2.2.2.1 Quota phase-out issue

The US once insisted that quotas imposed upon China's textile and apparel imports should be phased out over ten years starting from the date of its accession to ensure "fairness to other countries". Though, the US side did not put this clause in the final agreement, they still insist that there be soft standards upon China's textile and apparel exports into the US market. According to the 1997 Sino-US agreement, if China accedes to the WTO, the consultation mechanism will remain in force for an additional 4 years after the quota phase-out for WTO members in 2005. This indicates that China textile and apparel exports will be still subject to temporary quota impositions until the year 2009.

2.2.2.2 Non market economy issue

China is the major target in most anti-dumping and anti-countervailing cases raised by developed countries. US always labels China as a "non-market economy" regardless of its remarkable achievements in the building of market economic system. It can thus apply punitive measures not on a case-by-case approach and make decisions apparently randomly based on data from a third economy and other incomparable information.

2.2.2.3 Effective protection of "intellectual property rights"

China is often accused by the US for its failure to effectively protect intellectual property rights, including designs and trademarks concerning textiles and apparel. The US domestic "Super 301"Act often targets China as a serious violator on a priority list which is the ground for US to apply trade sanctions. Therefore, the US pressed China to take effective steps in addressing this tough issue, including legal and administrative measures.

2.2.2.4 Permanent MFN (NTR) issue

In terms of WTO basic principles of non-discrimination, the US shall grant permanent MFN status to China after its entry into WTO instead of the current annual renewal. Nevertheless, the US always regards its domestic laws as superior to international regulations. Under the American Jackšon-Vanik Amendment to the Trade Act of 1974, MFN (now called normal trade relations, NTR) may not be extended to any "non-market economy" unless the President grants a waiver certifying that the country does not restrict free emigration. This can be overridden by the Congress through a joint resolution of disapproval. On May 25, 2000, the House of Representatives passed the PNTR bill agreeing to grant permanent NTR to China after it enters WTO and in September, the Senate approved, too. With the final sanction from President Clinton, one of the most uncertain issues was at last settled, signaling a happy message for enterprises in both countries.

2.2.3 EU-China WTO-entry negotiations

The EU is one of the major importers for Chinese textile and apparel products. In 1999, it imported 59.4 million Euro worth of textile and apparel products from China, accounting for 7.7% of the total imports from China. The bilateral textile agreements between China and the EU consist of two types: one is an MFA-type Agreement, and the other is an Agreement on Silk, Linen and Ramie. As regards textile-related issues, negotiations between the two sides are mainly focusing upon the following points.

2.2.3.1 Consolidation of tariff rates at a reasonable level

According to an EU report, the present level of Customs duties imposed upon EU textile and apparel products by China are high, ranging from 25% to 100%. [46] It has requested China to reduce and bind its applied tariffs for textiles and apparel as a basis for negotiation: 0% for raw materials, 5% for yarns and fibers, 10% for fabrics and 17.5% for apparel.

2.2.3.2 Abolition of non-tariff barriers after a transitional period

The EU insists that China should guarantee the application of the fundamental rules of the WTO under a specific safeguard mechanism. The above-mentioned non-tariff barriers practiced in China should be addressed and eliminated within reasonable transitional periods, such as the right of enterprises to participate in foreign trade, export subsidies in textiles and apparel products, etc.

2.2.3.3 Legal transparency and increased market access

Many EU manufacturers have often complained about the low import penetration level the EU textile and apparel products in China's domestic market. The trade balance with China in this sector is always negative both in value and in quantity. Therefore, in agreement with the US strategy, the EU demands more market access from China as well as enhanced legal frameworks. The principle of "reciprocity" was first raised by the European industry, and is now a powerful weapon by linking MFA phase-out with EU access to foreign textile and apparel markets.

2.3 GENERAL RIGHTS AND OBLIGATIONS AFTER CHINA ENTERS WTO

2.3.1 Rights

2.3.1.1 MFN (or NTR) status

When China enters WTO, it will immediately and automatically enjoy MFN status with every Member country. The permanent MFN status granted by the US is of particular importance to China since the US is one of China's most important trading partners. However, it should be noted that entry into WTO doesn't mean China will enjoy equal treatment from all foreign countries. There are exemptions from MFN under the WTO regime, including frontier trade, free trade areas and customs union.

2.3.1.2 Active participation in rules set and multilateral talks

China has been outside the multilateral trading mechanism for such a long time that it cannot voice its opinion upon many international issues, which adversely affect China's international profile as well as its economic and political interests. Entry into WTO enables China to actively participate in the multilateral talks and articulate its views in the global arena. China will no longer be an onlooker when setting international rules and regulations, which is of great importance to China's future economic performance and development path.

2.3.1.3 Dispute settlement in the multilateral mechanism rather than unilateral retaliation

WTO set up a special division—the Dispute Settlement Body (DSB)-- to handle worldwide trade-related disputes. Because of the particular role of the textile and apparel sector in almost every Member nation, the Textile Monitoring Body (TMB) was established to address controversial issues in this sector. After China joins this global organization, trade disputes will be settled within a fixed frame rather than by arbitrary and unilateral decisions. China's textile and apparel exports often suffer unexpected losses because of unilateral actions taken by the importing nations. For example, the US frequently resorted to a quota-cut by a third for Chinese textile and apparel products since 1994, totaling US$94 million against Chinese quotas. [73] Such action is obviously incompatible with the WTO rules and has to be discontinued after China becomes a Member. Dispute settlement procedures will be more predictable and transparent to domestic enterprises.

2.3.2 Obligations

Rights and obligations are the two sides of a coin—no obligations, no rights. China has to assume its own obligations in order to enjoy the benefits. Some measures may be detrimental to domestic industries in the short run.

2.3.2.1 Reduction of its own tariff rates and non-tariff barriers

Though China has continuously lowered its import tariff rates, current duties are still high, particularly in some highly-protected sectors such as automobiles and apparel. Non-tariff obstacles, as mentioned above, are the major concerns to many trading partners. Therefore, it is necessary and important for China to make improvements in this aspect despite the possible pressure from fragile domestic industries which are well protected partly by current non-tariff barriers.

2.3.2.2 More open market access

China has committed to opening up its service and distribution markets progressively. The pace not only depends upon the economic situation, but also hinges upon current bilateral talks with other countries. With more foreign companies entering the domestic market, national ones, which lack information and technological support, may be forced out of the market when competing with their foreign counterparts. This is the price China has to pay in order to find its way out of the present tight circumstances and integrate with global economy.

2.3.2.3 Non-discriminatory and national treatment

Some current industrial policies and practices favor local companies, an approach not consistent with the WTO rules which require treating foreigners equally with the nationals. On the other hand, preferential treatment granted to one country should also extend to other countries on the basis of non-discrimination. Therefore, China has to review and adjust many of its current practices in order to comply with its new obligations.

2.3.2.4 Stricter adherence to WTO regulations

Entry to WTO requires China to establish a more transparent legal and operational framework. The MOFTEC gazette has issued to carry official texts of national-level trade-related laws and regulations. Several official web sites have also been established to make domestic information available to all interested parties, such as http://www.moftec.com.cn and http://www.cei.gov.cn. However, the process is far from being complete and thorough. Transparency does not simply refer to information availability. It involves significant changes in the present legal framework and daily practices as well as the corresponding ways of thinking of every person. The transformation is by no means an easy task. There is a set of WTO rules to manage the world trade in a comprehensive way. For example, some trade-related investment policies currently enforced in China may violate the WTO spirit, such as local content schemes, import balancing requirements, and various incentives designed to direct investment to priority development sectors. All of these measures have to be reconsidered within the relevant WTO frames to create a more predictable and stable investment environment.

Therefore, successful entry into WTO is just a starting point for China. It has a lot of work to do after the entry. The most difficult part is to ensure that WTO-based practices are adhered to and to work out appropriate laws and regulations.

2.4 SPECIFIC IMPACT UPON THE CURRENT 'DIAMOND' STRUCTURE

China's entry into WTO itself is a normal issue, just as China became a member of the IMF or the World Bank. However, the 14-year pursuit of entry brought many non-economic factors into the unusually long application process. As a result, it is now one of the biggest issues not only to China but to the world as well. Even the declaration of the Sino-US agreement upon WTO entry by the end of 1999 triggered strong reactions from the world stock markets. Those for and against, though totally different in views, are all agreed that WTO entry is a great influential force upon the national economy. It is hard to predict its actual impact at the present stage, but an analysis of its possible influence upon the diamond structure of the textile and apparel industry may give some insight into the future picture.

2.4.1 WTO entry itself as a chance event

When President Clinton did not accept the accession package offered by Premier Zhu Rongji in April 1999, the prospect for China's WTO entry was clouded. The following political events between the two sides cast another shadow on the bilateral talks, together with huge domestic pressures, especially from some WTO-sensitive industries such as telecommunications and car industry. At that time, few people could predict whether and when China was able to become a WTO member considering its past 13-year pursuit during which politics played an overwhelming role. In this sense, China's WTO entry can be viewed as a chance event, which is conducive to shaping a nation's competitive advantages. [2]

Entry into WTO means China will truly compete in the interwoven global environment with WTO-based rules and regulations, triggering a series of both internal and external reactions. Generally speaking, acting in an international manner can better integrate the nation with the rest of the world, enhancing its international competitiveness in the long run. Significant changes are expected in domestic legal and trade regimes as a result of China's WTO entry. As regards the textile and apparel industry, the ATC alone indicates drastic changes in the competition platform, accompanied by new uncertainties. For example, urged by strong lobby forces, what new protectionist measures would foreign governments take to safeguard their domestic textile and apparel industry? To what degree would imports increase because of improved access and what would be the possible consequences? Would domestic enterprises be competitive enough to with stand increased competition both in the export and the domestic markets? Would domestic currency undergo

Table 2.10 US textile and apparel exports to China (in US$ million)

US textile and apparel exports to China					
Product	1997	1998	January-August		
			1998	**1999**	Percent increase
Yarn	4	6	4	6	50.0%
Fabric	40	40	28	36	28.6%
Made-up	16	19	9	9	0.0%
Apparel	8	6	4	6	50.0%
Total	67	72	45	57	26.7%

(Source: http://otexa.ita.doc.gov/tbrbal.htm, retrieved on November 5, 1999)

depreciation or appreciation with the progressive opening of the capital market? All of these questions cannot be answered at present, but they become more and more important when China's WTO entry is drawing near. It can be concluded that this event would change the game rules upon which domestic enterprises used to play, forcing individual firms to adopt corresponding strategies in the new competition picture.

2.4.2 Impacts upon the four aspects of the Diamond

2.4.2.1 Upon the factor conditions

A. Increased imports mainly due to raised market access

China enjoys a very favorable trade surplus as regards textile and apparel items with the export volume far exceeding the import one. From January to May 2000, the surplus was US$ 14.0 billion. WTO entry indicates a balance of rights and obligations for China. While Chinese textile and apparel exports may gain greater penetration in foreign markets, its domestic market has to be more open to foreign products as well. The continuous fall in import tariff rates and the removal of non-tariff barriers are leading to increased market access, and it is easier for foreign products to find their way into the domestic market.

Table 2.10 shows US textile and apparel exports to China from 1997 to August 1999. Among the four categories, only made-up products maintained the same export volume in the Jan-Aug period in 1998 and 1999. The other three categories enjoyed a rapid growth rate--this is one direct result of the increased access to China's domestic market.

MFA quota liberalization may also boost importation of intermediary inputs and textile machinery as a result of increased exports. The USITC report suggests that the elimination of textile and apparel quotas would result in US$ 2 billion of additional machinery and equipment being imported to China. [47]

Since most Chinese textile and apparel products are low value-added items, they are more sensitive to changes of raw material cost, which is relatively important in the cost structure, compared with fashion ones. Generally speaking, downstream enterprises gain most benefits from increased imports of raw material and intermediary products. In 1998, woven and knitting apparel exports reached 3.31 billion and 5.85 billion pieces respectively, among which 59.8% and 47.7% used imported fabrics. [6] With the reduction of tariff rates and the removal of non-tariff barriers, apparel manufacturers will be able to use less expensive fabric imports for their final products, further enhancing their cost advantage. As mentioned earlier, the chemical fiber sector is one of the most difficult links in the textile industry. High domestic prices for the raw materials are the major problem. For example, in 1999, the domestic price for PTA was about RMB 800/ton higher than the CIF price. [15] Therefore, when China lowers tariff rates as one of the WTO commitments, the chemical fiber manufacturers will gain benefits from such a move. While some sectors benefit from increased imports, some will suffer, such as the fabric manufacturers and apparel producers engaging in middle- or high-range products. As a result, the resources can be allocated to the more efficient and competitive sectors and the market mechanism will play a larger role instead of the current trade-distorting measures.

B. Rising labor cost

Economic development will no doubt raise labor costs. China's WTO entry is expected to bring more foreign capital into the country, especially to the east coastal region in which most textiles and apparel bases are located. Consequently, the labor cost will be pushed up to a higher level, making production of labor-intensive products more costly in this region. Some production facilities may have to be moved to less developed areas in the middle or western regions. (This point will be explained in detail later.)

C. Capital and human resources diversified to more profitable industries such as information industry

The past decade has witnessed the magical growth of computer-based industries. Its quick returns on investment and promising future blueprint have attracted much more capital inflow and active participation of qualified personnel. In sharp contrast to the disappointing enrolment of textile-related majors, courses with "computer", "information" or "communications" titles are the most popular with university applicants nowadays.

It is well recognized that WTO entry will have a strong impact upon fragile domestic hi-tech industries. But its positive affects cannot be ignored, either. More foreign investors will rush to tap China's huge domestic market, especially those newly emerged but still undeveloped sectors. Their strength and higher-than-average salary payments will significantly tilt the balance and capture a larger proportion of capital and human resources.

Management levels will be enhanced as a whole, since managers will be exposed to more open competition in the post-WTO era. Market-oriented ways of thinking will be facilitated and strengthened.

2.4.2.2 Demand conditions

A. More sophisticated customers due to economic development and increased access to world fashion

WTO entry will provide domestic consumers with a good opportunity to access world famous fashion brands and quality products because of their easier entry. The open-door policy has cultivated fundamental brand consciousness while the complete economic integration between China and the outside world will further boost this trend. Take Shanghai for example. Consumers in Shanghai are more sophisticated and demanding than those in other parts of China. This can partly be attributed to its traditional fashion center status. Many multinational corporations opened speciality stores in Shanghai due to its economic strength in China and attractive market potential. The easier access to world-class products has helped to develop Shanghai consumers' fashion tastes and boost demand for high-range products. It can be expected that domestic consumers will learn more as a result of the economic development. Though WTO entry is not the sole cause, it does provide a unique opportunity for more foreign counterparts to tap the Chinese market, with domestic consumers being the largest beneficiary.

B. Regional differences will be further widened

From Table 2.4, it is clear that regional differences are very distinctive. Rapid economic development in the east coastal region is due to its traditional economic strength and preferential government policies. The report entitled "China Sustainable Development Strategy" issued by the Chinese Academy of Sciences shows that Shanghai still occupies the first place as regards sustainable development capabilities, followed by Beijing, Tianjin, Guangdong and Jiangsu.

Figure 2.9 shows that the regional economic differences have been further widened. The average GDP per capita in the coastal region was 3.78 times that in the west in 1994 and increased to 4.14 times in 1997; The same is true when comparing the central and the west. In 1994, the difference was 1.37 times while it rose to 1.55 times in 1997. The contrast is greater if individual cities are taken into consideration. GDP per capita in Shanghai was 9.79 times than that in Guizhou in 1994, 10.64 times in 1996 and 11.63 times in 1997.

According to statistics from the former China Textile Council, 93.9% of FDI was concentrated in the east coastal region. [68] Though China has declared a series of policies, aiming to encourage investment in the west and central regions, the weak economic foundation and less favorable investment environment will hinder investors from taking further steps. Therefore, the east coastal region, backed by its relatively strong economic development momentum and well-established infrastructure, will continue to be the hot investment spot after China enters WTO. Most of the foreign capital inflow will be channeled into this region and will help boost economic development, further widening the regional differences.

C. Improved efficiency of distribution network

China's WTO entry will greatly change the current distribution network. The presence of foreign entities will not only sharpen local competition, forcing mal-managed enterprises out of the market, but will also provide useful expertise and examples for domestic enterprises to learn. Competition brings about progress. With the improved distribution network, manufacturers will feel the need to change as well. Closer cooperation between the two sides can be expected.

2.4.2.3 Firm strategy and rivalry

A. Hotter local rivalry

First, there will be more foreign presence in the domestic market. Some people point out that after China enters WTO, foreign enterprises may prefer direct export instead of direct investment. Their argument is mainly built upon three factors—the lifting of non-tariff barriers, reduction of import tariff rates, and withdrawal of non-WTO-based preferential treatment such as the preferential tax policy. They think that easier market access makes direct export more attractive than direct investment. However, more people view WTO entry as a new chance for foreign capital inflow, a more predictable and stable investment environment and improvement of the legal system. For example, in view of the trade-related investment agreement under the WTO regime, FIEs in China shall not be subject to the domestic-sale ratio or local-content requirement after China enters WTO. Therefore, to most potential investors, China's entry into WTO is a very positive signal.

2.9 GDP per capita in the coastal, central and western regions[20]

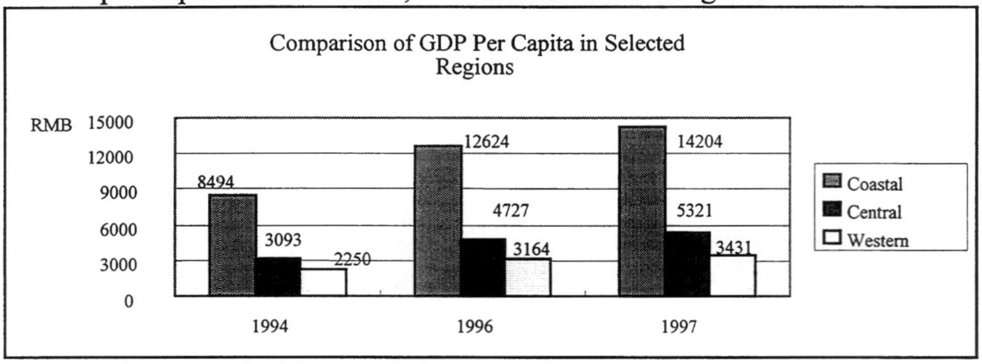

(Source: Almanac of China Statistics, 1996-1998)

Second, the competitive environment will be more open and fair to the non-state-owned enterprises. The state will gradually withdraw its control except in some strategically important sectors. As regards the textile and apparel sector, it can be predicted that the ratio of traditional SOEs will continuously decrease. More SOEs will be transformed to corporations and governed by a modern enterprise system. Loss of the state protection umbrella will expose SOEs to open market competition. In addition, government subsidies to the state sector is a practice not accepted by the Agreement on Subsidies and Countervailing Measures under the WTO regime.[21] Therefore, with the withdrawal of preferential treatment of SOEs after China's WTO entry, all the market players will enjoy a relatively free and fair competitive environment. The "Private solely-owned Enterprises Law" issued recently will also encourage entry of more private enterprises as well.

Third, China's commitment to liberalizing foreign trading regimes after its accession into WTO and the ten-year phase out schedule in the ATC will greatly alleviate the rent-seeking problem and other non-competitive factors in China. State trading will be relaxed progressively with more enterprises engaging in foreign trade. This point alone will bring about positive changes to the current market structure and competitive patterns.

B. More focus upon strategic competition

Strategic competition was totally ignored in the past since it was not important in winning the competition, which was subject to many non-competitive factors. With more market players and an improved competitive environment, competition will be upgraded to a higher level. "The jungle survival rule" will be justified with more real facts. Enterprises are bound to shift their efforts to more technical aspects such as product quality and innovation to survive the competition.

2.4.2.4 Related and supporting industries

WTO entry will no doubt be a competition-catalyst, because the door is scheduled to open wider to the outside world. All domestic industries will more or less feel its

[20] The average GDP per capita is calculated based on the 15 cities mentioned in Table 2.3.

[21] According to the Agreement of Subsidies and Countervailing Measures (SCM Agreement)--an integral part of WTO, "subsidy" refers to a financial contribution by a government or any public body within the territory of a Member which confers a benefit. There are three basic types of subsidies: prohibited subsidies (such as export subsidies and local content subsidies); actionable subsidies (such as production subsidies; and non-actionable subsidies. (including basic research and pre-competitive development subsidies, assistance to disadvantaged region, and assistance to adapt existing facilities to new environmental requirements)

impacts. The textile-related and supporting industries are no exception. The previous analysis shows that these industries are either newly burgeoning or fragile. Therefore, in the short run, it will be a painful experience for these enterprises, accompanied by increased bankruptcy and a rising unemployment rate.

Consider the textile machinery industry, for example. At present, the average import duty for textile machinery is around 8% to 10%, lower than in many other countries. [48] Therefore, the impact due to tariff reduction upon the domestic sector will not be very large. What may pose a bigger threat to local enterprises will be the increased foreign direct investment. As mentioned in Section 2.1.2.3 (D), new-type textile machinery is in the "encouraged" category under the "Guiding Directory on Industries Open to Foreign Investment". With the improved investment environment and anticipated good performance in China's textile and apparel exports, more foreign investors will be lured to play in this huge market. The big technological gap between China and foreign investors will make local machinery production less appealing, especially when the foreigners can enjoy the same labor-cost advantages, more easily access the domestic market and play on a more equal footing with domestic ones. Based on these arguments, textile machinery producers in China will face fierce challenges after it enters WTO, particularly in the short run.

On the other hand, competition stands for progress. The pains are the necessary price for future vigorous growth, otherwise, most domestic enterprises would continue to be uncompetitive and narrow-minded. From this perspective, WTO entry will, in the long run, facilitate the current industrial restructuring process with active foreign participation to secure the necessary capital and technology inputs as well as invaluable management experience.

2.4.2.5 Summary

WTO entry mainly changes the game rules, the macro environment and corresponding government policies. Its impact upon an individual industry is progressive and takes a long time, especially for the positive influences. In the short run, the industry will feel stronger negative impacts upon its fragile sectors and enterprises. So long as they can survive the first round attack, they can move toward a more stable and favorable development environment.

Table 2.11 is a summary of all the points mentioned above. It indicates that WTO entry has some direct and some indirect influences upon the current four constituents of the diamond. However, it is impossible to draw a definite conclusion that it is a 100 percent positive force in shaping the new competitive advantages of the textile and apparel industry.

2.4.3 The most possible scenario in the next five years

The above analysis depicts a general picture of the Chinese textile and apparel industry before and after China's entry into WTO. Evolution of the diamond model is a very long and slow progress. The four constituents interact, and each plays a role in this integrated and complicated system. Any drastic change in one factor will trigger a series of chain reactions within the model, leading to a new evolution direction. Because WTO impacts are progressive and slow, the four constituents will not experience drastic changes in the next few years, and there should be a relatively stable export performance. Some quantitative changes, such as quality enhancement and hotter market competition, will help strengthen the industry's present competitive

Table 2.11 Diamond structure summary

Diamond Structure	Present Condition		WTO-Effects	
	Positive	Negative	Positive	Negative
Factor Condition	1. Low labor Cost	1. Fast growing rate		1. Further push up the labor cost
	2. Available unskilled and semi-skilled labor force	2. Lack of technicians and advanced talents	2.Improved management level	2. Further diversion of advanced talents
	3. Abundant Natural Resources	3. Lack of high-grade and quality intermediary input	3/4. Easier access to foreign market and cheaper imports / benefits to downstream enterprises	3/4. Impact upon the fragile sectors due to increased imports, especially on upstream enterprises
		4. Obsolete machinery and low technology		
	5. Continuously improved physical infrastructure	5. Weak information & communication infrastructure.	5. Increased FDI inflow in the information sector	5.Impact upon domestic enterprises
Demand Condition	1. Huge domestic market	1. Less sophisticated and demanding consumers	1. More sophisticated and demanding	
	2. Rapid growth of apparel retailing volume	2. Current deflation situation--anticipated drop of disposable income on apparel		2. Possible intensification of this psychological anticipation
	3. Regional difference catering to varying demands	3. Too small demand for high-grade/quality products	3. Widened regional differences and economic imbalance (both positive and negative)	
		4. Insufficient distribution network	4. Improved distribution network	4. Impact upon uncompetitive enterprises / increased bankruptcy
Firm Strategy and Rivarly	1. Establishment of market mechanism	1. Imperfect market mechanism	1.Market mechanism improvement	
	2. Formation of brand and quality consciousness; trading right liberalization	2. Low-level competition, featuring malicious price war and deteriorating commercial credit	2. Hotter but more orderly market competition	2. Increased bankruptcy of domestic enterprises and temporary rise of unemployment rate
	3. Large FDI investment		3. More predictable and transparent investment environment	
Related Supporti-ng Industry	1. Comprehensive industrial system	1.Weak foundation, especially in technology	1. Service/hi-tech industry will be the hot investment fields	1. Strong impact upon the fragile industries, especially in the short run
		2. Lack of cluster	2. Dependent upon too many factors	
Govern-ment Policy	Strong, but some are not WTO-based		Remain strong and more in line with WTO rules	
	Focus on SOE reform, some involving direct protection		Strategic control rather than direct interference	
Oppor-tunites	Rising demand for home/technical textiles		More predictable dispute settlement mechanism	
	Increased fiber consumption level		Participation in rules-set and multilateral talk	
			Export growth based on ATC regulations before 2005	
Threats	Sharper international competition with entry of more low-cost exporters		Loss of secured market access in major export markets after 2005	
	Regional trading blocs		More serious	
	Non-quota barriers		More serious	

position in the middle-/low-range products. Therefore, the most possible scenario in the next five years is a relatively solid diamond model with international competitiveness in the middle-/low-end market.

2.4.3.1 Export volume will be relatively stable

The domestic textile and apparel industry is export-oriented. Therefore, export performance is very crucial to the whole industry. Many predict that domestic textile and apparel exports will enjoy a very quick increase after entry into WTO because of the immediate benefits of ATC and the future quota-free environment. The argument has a ring of truth, but over-optimism may prove to be wrong.

The ATC is, in some respects, a word game concerning its ten-year quota elimination plan. Big suppliers such as China benefit less than expected. Take the current Sino-US textile agreement for instance. The agreement reduces quotas in 14 apparel and fabric product categories in which the US believes that China has repeatedly violated quotas through illegal transshipment via third countries. What's more, the average annual quota growth rate (base rate) for China under this agreement is 1.3% for apparel and 2% for made-up textiles[22], ranging from 0.1% for wool sweaters to 4.1% for gloves and mittens of silk blends and non-cotton vegetable fibers.

Table 2.12 illustrates the expected annual quota growth rate if China enters WTO, which is not desirable at all. It should also be noted that quota-imposing countries leave most of the sensitive products, such as apparel, until the last stage. For example, the US quota integration schedule delays 89% of the apparel quotas until January 1, 2005. In addition, domestic exports will be subject still to temporary quota impositions for an additional 4 years after 2005. This clause casts uncertainty on future domestic export growth. Therefore, it is clear that China will not gain much from the liberalization process. Even when the quota is completely freed in 2005, China may fail to witness a sudden rise in export volume, because the MFA-type exports only account for one-fourth of total textile and apparel exports.[23]

Also, world demand will not change drastically as well. In 1998, world textile and apparel trade volumes were US$151 billion and US$180 billion respectively, down 5% and 1% over 1997. Trade diversion out of regionalism has posed a great challenge for China textile and apparel industry. (See relevant points in section 1) Mexico has replaced China as the leading exporter to the US and has continued to increase its share in the US textile and apparel import market.

There is another thing meriting special attention. In 2000, there are 32 commodities requiring quota bidding, 11 of which are truly voluntary export restraint. These 11 commodities are usually non-regenerated resource products whose international

Table 2.12 Expected annual quota growth rate in China after its WTO entry*

Annual quota growth rate		
Base rate(%)	1998-2002(%)	2002-2005(%)
0.10	0.13	0.16
1.30	*1.63*	*2.06*
2.00	*2.50*	*3.18*
4.10	5.13	6.51

*The figures in italics in the table means that most of Chinese MFA textile and apparel exports are subject to the base rates between 1.3% and 2%

[22] From the report "Assessment of the Economic Effects on the US of China's Accession to the WTO" issued by the National Retail Federation.

[23] The total textile and apparel export volume is around US$42-45 billion, among which the MFA-type exports are about US$10 billion.

market capacity is limited and whose supply exceeds demand. The major purpose is to effectively control under-price dumping activities. Domestic textile and apparel exports always face anti-dumping investigations. Though some of the export prices are really based on the low production costs, a number of others are the result of malicious price wars or simple pursuit of foreign exchange earning rather than profit considerations. Therefore, execution of VER is an appropriate approach to regulate the export market and shortsighted actions of some domestic enterprises. With quota liberalization in 2005, adoption of VER will be the natural choice for some textile and apparel exports to realize efficient resource allocation and orderly market competition.

According to the above explanations, domestic textile and apparel exports will not experience a rapid increase, neither will they slump after China enters WTO. The quota mechanism will continue to maintain current export market orders until 2005.

2.4.3.2 Basic factor conditions will remain relatively unchanged

Though WTO entry may push up labor costs, especially in the east coastal region, it will not increase by a big margin during the next five years. China will still enjoy comparative advantages in labor cost and available natural resources. As regards advanced factors, it takes time and efforts to achieve some qualitative breakthroughs.

2.4.3.3 Domestic demand will be further diversified, with middle-/low-end products remaining as the most popular choice

China will enforce its market access commitment step by step with its entry into WTO. In addition to planned tariff reduction, non-tariff barrier removal and foreign trading rights extension, its commitment in the service sector, including distribution, telecommunication, insurance, banking, securities and professional services, is significant. Foreign investors have long directed their attention to this well-protected sector, which generates high profits and has a very promising prospect. In view of China's current economic situation, the east coastal region will continue to be an attractive investment place for foreign investors because of its stronger economic foundation and favorable investment climate. The increased inflow of foreign capital and technologies will help develop the tertiary industry and boost the performance of the regional economy. As a result, regional economic differences will be further widened in terms of per capita GDP and disposable income. Though the government has set a series of industrial and investment policies in the hope of channeling more funds to the west and central regions, it will take quite a long time for them to catch up with the east coast region. Therefore, although demand in some big cities in the east coastal region will move upmarket and become more diversified, its market proportion will remain small and subject to regional limitation. Low-grade products will continue to find a ready market and cater to potential consumers in the vast hinterland and the middle-/low-income groups in the more developed areas. However, low grade doesn't necessarily mean low quality. The gradual establishment of market mechanisms will secure fair competition among enterprises of various ownership forms with a focus upon profit maximization—an approach centering upon quality improvement and product innovation. Consumers in tomorrow's buyer's market will be more selective, partly stimulating this development trend. For the next several years, domestic demand for low-grade products with improved quality will still account for the largest proportion of sales while demand for middle-/high-end products will be on the increase, especially in the bigger cities.

2.4.3.4 Firm-level competition will focus more on quality and low-/middle-range brands

In the textile sector, foreign investments are now focusing upon the chemical fiber projects which are capital- and technology-intensive. In the apparel sector, foreign entry usually brings along high-grade brands whose market segment is small at the present stage and for at least the next five years. Because of the lack of strong capital support, as well as a low technology level, domestic enterprises are mainly focusing upon the mass-product market. Thanks to vigorous domestic competition, some domestic brands in men's wear and underwear have attained a firm and large share in the low-/middle-end markets. It is difficult and unrealistic to predict now that local brands will become internationally competitive, considering the current technological level and weak support of the various processing links. But it is true that more and more enterprises have realized that brand is a powerful weapon in the fight to survive the hotter competition.

With this consciousness in mind and backed by the huge domestic market, domestic enterprises will make efforts to boost competitiveness through quality improvement and brand establishment. Therefore, if taking every aspect into consideration in a five-year framework, such as labor cost, technological level, capital support, demand conditions and firm strength, domestic enterprises will have a better performance in the mass product market and be competitive in the middle-/low-range products.

2.4.3.5 Some positive changes in cluster forming

After China enters WTO, the new game rules will provide a favorable external environment for domestic enterprises; at the same time, entry may accelerate industrial restructuring through more vigorous competition in a more open and rule-based environment.

However, cluster upgrading is no easy task. It requires close cooperation in every aspect. As mentioned earlier, lack of an effective textile and apparel cluster is a main reason for China's current competitive position in the international market. Though some measures have been taken to strengthen cooperation between the downstream and upstream enterprises through vertical integration or other policy support, it is still very hard to achieve close cooperation within the industry after actual separation for more than 50 years. In addition, even if the textile and apparel industry is successfully upgraded to a higher technological and productivity level, lack of strong support from its related and supporting industries would make such development unsustainable.

China is a capital-scarce country. Though foreign capital inflow has greatly alleviated this problem, it is not the solution to the problem per se because foreign investors have their own interests which may contradict or be incompatible with the national development plan. How to allocate the limited capital in the most efficient way is a question of key importance. Time is another important factor. Cluster forming and upgrading is a slow progress. It cannot be completed in one or two years, especially when the current foundation is weak and not solid. In this view, the textile and apparel cluster will remain in its current situation, with some positive changes, for several years to come.

2.4.4 Strategic implications for China's textile and apparel industry

In view of the above analysis, the primary task for the textile and apparel industry in China is to develop new competitive advantages, such as design and marketing strengths as well as product upgrading.

2.4.4.1 Transfer pattern of world textile and apparel industry

The transfer pattern of the World's textile and apparel industry may sketch a meaningful picture of the domestic industry. It follows the way in which the labor cost plays a dominant role.

The labor cost of the apparel sector in the US was 63 times higher than that in Indonesia in 1998. (Figure 2.11) Such sharp differences of labor cost among countries is the major driving force underlying the gradual re-location of the textile and apparel industry from the developed to the developing countries during the past decades as illustrated in Fig. 2.10. The textile and apparel industry in the developed countries, in order to circumvent the disadvantage of much higher labor costs, had to pursue high value-added processes through technical upgrading and innovation. They attach great importance to R&D and capital investment. Thus, though they may lose the race in the middle- low-end markets, they successfully maintain their monopoly presence in the high-end market and own most of the world-famous brands and trademarks. In some respects, the globalization trend enables MNCs to work out their strategy in a worldwide context, effectively combining the low labor cost of the developing countries with their research, design and marketing strength. The current prevalent practice of OPT arrangement is a good illustration of this cooperation pattern. The ultimate winners are those MNCs headquartered in the developed countries because they control the key links of the whole production chain. Most developing countries, including China, simply function as "assembly lines".

2.4.4.2 Geographic relocation within China as an effective way to boost competitiveness

Figure 2.11 shows that the labor costs in the Chinese apparel sector have already been higher than in some developing countries such as India, Bangladesh and Indonesia. The production cost is and will continue to be the key consideration for middle- or low-range products. Exports from China have been facing serious challenges from these lower-cost suppliers. Take the EU textile and apparel imports for example. In 1998, India and Indonesia ranked the third and ninth respectively with their share reaching 9.6%, compared with China of 12.2% of the market share. [49] It is estimated that 40% of Chinese textile and apparel exports have encountered head-on competition with those from Southeast Asian countries. [50] Therefore, if the industry fails to acquire new competitive advantages and continues to depend upon the basic factor strength, it will ultimately lose the competition.

The relocation strategy adopted by some developed and newly industrialized nations provides a useful example for the Chinese textile and apparel industry. China is a very big country, with one-fifth of the world's population and different economic development levels among various regions. This offers a unique opportunity to realize "relocation" within a single nation. Geographic proximity, the similar language and cultural background and a uniform legal framework make this cooperation pattern workable. However, geographic relocation is far more than simple spindle relocation.

It contains profound implications, including corresponding policy support, market development and efficient firm-level strategies.

A. Factor specialization vs. generalization

The traditional textile and apparel bases, such as Shanghai, are located in the east coastal region where the labor cost has been increasing continuously during recent years. It was natural for these bases to transfer some production facilities to the smaller cities in the east region or to more remote areas in the central and west areas in recent years. However, this move failed to achieve expected results. Though the central and west areas are rich in natural resources and low-cost labor, their weakness in other important aspects such as management and marketing are fatal to their development. Therefore, in order to make geographic relocation effective, the industry should, at the same time, equip these regions with qualified talents and machines based on local conditions. Transfer of production facilities is just the start of the relocation process. Cultivation of qualified managers and other marketing talents is most important, together with infrastructure improvement.

With the shift of production facilities to other regions, a traditional textile base such as Shanghai will place emphasis upon high-value added activities. Today's Shanghai is somewhat like HK in the 1980s, when HK textile and apparel manufacturers began to shift simple and low-value added processes to the Chinese mainland. After two decades' development, the HK textile and apparel industry has emerged as one of the famous fashion centers enjoying a high reputation for its design and high-quality products. The undergoing industrial upgrading is an opportunity for Shanghai and other traditional textile bases to transform their·textile and apparel

2.10 Transfer path of the World textile and apparel industry

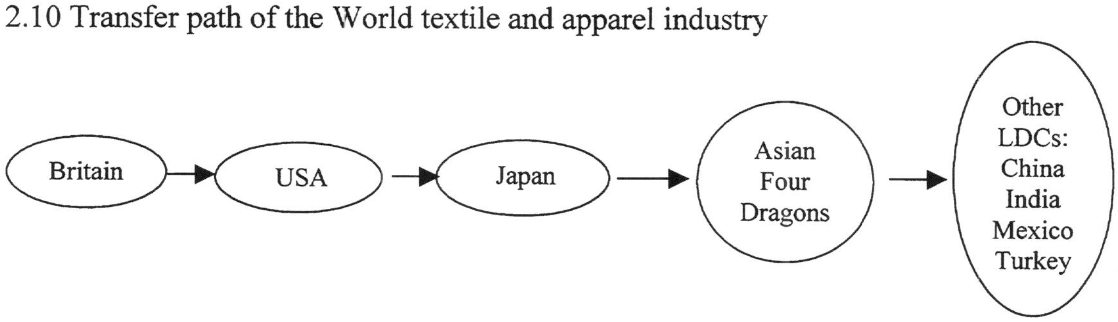

2.11 Comparison of labor costs in selected countries

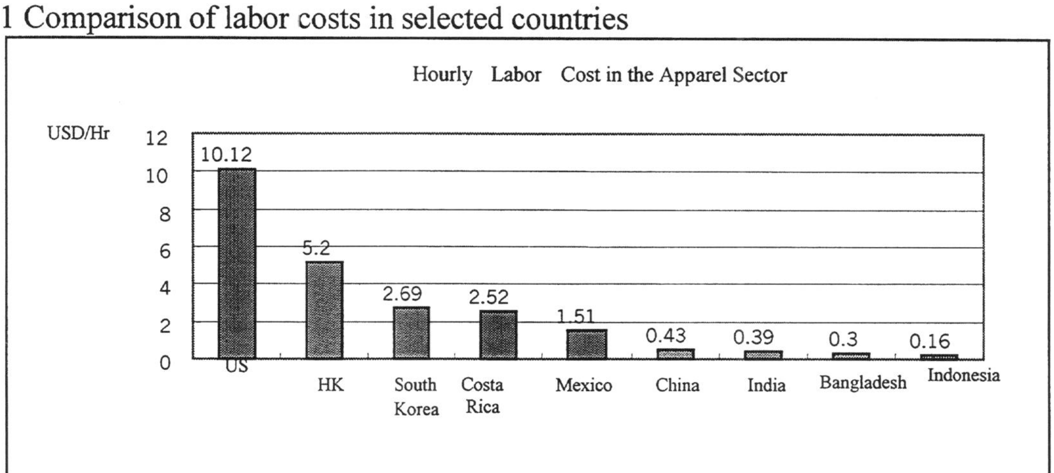

(Source: Werner International Inc.)

industries. From January to November 1999, the textile and apparel exports and imports handled through Canton, Shanghai and Jiangsu accounted for 56.0% and 70.3% respectively, indicating their extensive foreign trade channels and network. Owing to the quicker and sounder economic development, talented people, especially new graduates, are very willing to work in these regions, which has further reinforced their human resource bases. Combined with relatively stronger research capabilities, improved infrastructure, and more sophisticated local demand, these cities/provinces have the potential and conditions to develop to an advanced level, focusing upon past weak links such as dyeing and finishing. The textile sector should focus upon home textiles and technical textiles, together with innovation in high-quality and new-type fashion fabrics. The apparel sector should attach greater importance to brand images, market promotion and fashion design.

More importantly, these enterprises should provide the necessary human and capital resources to help less developed regions to accumulate market experiences and gain market access. The merger of the Shanghai-based Hailuo Group and the Sichuan-based Chendu Apparel Group is the first case concerning transregional group integration. The Chendu Apparel Group, which is entitled to import and export, has 14 subordinate companies with total assets of 240 million RMB. However, its imperfect operational mechanism and weak management level make it run at a loss. The manager with the Hailuo Group, which has strength in marketing and brand management, is an attempt to achieve synergy in the long run. Effective end-market division and management cooperation are the prerequisites for its smooth operation.

B. WTO-based government policy

It is extremely important for the Chinese government to progressively shift its focus from direct supervision to strategic orientation. Direct government intervention has rarely generated a competitive advantage in textiles (Singleton 1997). Government's major functions are to create and maintain a favorable macro-environment within which the enterprises are responsible for their own actions and decisions.

- Preferential tax policy to all the investors

From January 1, 2000, a new preferential tax policy has been in force to encourage foreign investment in the west and central regions. These FIEs will continue to enjoy 15% income tax for another three years after the former 5-year preferential period is over. This policy is designed to attract more foreign capital inflow to boost these regions' economy. However, FIEs may be less willing to channel capital into these regions in view of their present investment environment. The government should extend this preferential policy to investors from the east coastal region as well. Domestic textile and apparel enterprises in the eastern region have been hampered by rising labor costs and limited expansion space during recent years. This tax policy will certainly help develop their interest in the vast hinterland and draw more local investors of strength to promote regional economic development. Such practice is also in line with the WTO rules which advocate national treatment.

- Equal emphasis upon SMEs

Current government policy places more emphasis upon large enterprises or company groups. They either encourage horizontal and vertical integration or give special attention (such as loose bank credit) to the present large enterprises, most of which are SOEs. By contrast, small and medium enterprises, though they account for the largest ratio in the industry, receive little or no support despite word commitment.

The healthy development of SMEs is of key importance to the diamond foundation. Government support does not mean direct interference in their daily operation. What SMEs need are a relaxed expansion space and an equal competition environment guaranteed by corresponding laws and regulations. What's more, government should help them to gain access to advanced technology, and to exchange with their counterparts in other countries to acquire more experience. To put it simply, government should not prejudice SMEs in policy setting and should play an active role in their technical upgrading.

- Secured market access for enterprises in the central and west

One of the biggest obstacles for enterprises located in the central and west regions is that they fail to have a sufficient market. Without market support, they lack the foundation to develop to a higher level, even if they are equipped with more advanced machinery. The government should play a role here. It should help the enterprises in less developed areas to improve their marketing capabilities for their manufactured resource-intensive products while encouraging those in the advanced areas to give up the low-end market and move upmarket. Though the ultimate aim is to let the market mechanism have the final say, it is still necessary for the government to take some steps at the very beginning in the hope of getting a good start, with sustainable development momentum. Otherwise, the central and west regions will still be reduced to simple assembly lines, which will greatly thwart the geographic relocation process and the macro aim of "developing the central/west region".

- Encouragement of overseas investment

With the progressive quota liberalization, major importing nations figure out other ways to protect their domestic textile and apparel industry. The NAFTA arrangement among Mexico, Canada and US paved the way for OPT activities between Mexico and US. The "yarn-forward" rule of origin makes non-member market access extremely difficult. Therefore, for domestic enterprises, it will be better to carry out foreign direct investment in some developing countries, such as Mexico, Turkey and the Caribbean Basin Area. Government should play an active role in this aspect, including information gathering, analysis of target market profiles, advice on efficient market access and training about important laws and regulations in the target market.

- WTO-based government protection

--WTO-allowed safeguard mechanism

With import tariff reduction and the removal of non-tariff barriers as well as direct state protection, it is now imperative for the government to set up a WTO-permitted safeguard mechanism. For example, the WTO permits a member country to take some unilateral remedies, such as anti-dumping and countervailing duty measures to ensure fair play. China did not publish relevant laws until the year 1997. Domestic chemical fiber suppliers have always claimed that some foreign producers dumped their products in China's market, causing material injury to the domestic industry. However, no further steps have been taken except oral warnings of possible anti-dumping investigations. At the same time, when facing foreign charges of dumping activities, most of the involved parties either didn't know how to handle this issue or adopted a non-cooperative attitude. As a result, they lost the case and had to endure all the consequences. All of this reflects the fact that domestic enterprises still don't know how to effectively protect their interests with rule-based weapons. After China enters into WTO, the government should make efforts to advocate WTO-based rules and regulations and help domestic enterprises to get familiar with these through various seminars and training programs. What's more, it should support and help develop trade associations to enhance exchange and cooperation among enterprises and organize enterprises to settle trade disputes with foreign partners.

--Important points in the trading regime

WTO entry means that domestic enterprises will compete in a more open environment. Reform of the foreign trading regime is more than simple relaxation of foreign trading rights. Stricter regulation is necessary to ensure the market order and to avoid blind and ill-considered price wars. The current performance evaluation standard, based on foreign exchange earning level rather than actual profit realization, should be dropped immediately. Furthermore, the government should help gather international market information, hold world-class exhibitions and fashions shows, and issue information about the latest development trends at regular intervals.

"Green consumption" is now another hot issue in the world, together with the formation of so called "green barriers". Germany's prohibition of azo dye usage in textiles has affected 15% of China's relevant exports. [51] Compared with the international "green" trend, China lags far behind. ISO14000 may be new to domestic enterprises. The government should take prompt steps to deal with the new situation, otherwise, more exports will be denied entry into the export market. The primary task is to advocate and enforce "cleaner production" [24] at each manufacturing step. Environment-friendly regulations should also be in force as early as possible.

--Follow-up of individual policies

The government should also establish a complete system to follow up the enforcement of individual policies. Take current SOE reform as an example. The government has carried out a reform package to help SOE out of the loss mire. Statistics show that there were some improvements in the SOEs performance. But there was a lack of detailed and in-depth analysis of each policy's contribution and the real implication of various statistics, such as loss reduction percentage and its underlying reasons. Close follow-up and continuous correction can ensure the overall coherence of government policies, which is crucial to its final effects.

C. Firm strategy

International success of a certain industry, in the final analysis, depends on firm-level strategy. To individual enterprises, WTO entry means that they should work out firm strategies based on new game rules and a new international environment. However, member status will not change the basic competition strategies.

- Genetic strategy

Many domestic enterprises don't have an insight into their own strengths and weaknesses and know little about the market operation. [2] As a result, current competition often ends in a malicious price war. Therefore, with or without WTO entry, domestic enterprises should always keep the market in mind and be strategic-minded.

There are two generic strategies with different emphases—one is upon cost, the other is upon differential products. [52] Since products in most domestic enterprises are still low and middle-range, production cost is the most important factor in future market competition, especially for standardized products. With the geographic relocation of the textile and apparel industry, those located in a raw-material supplying base should attach priority to cost control while downstream enterprises should focus more on product differentiation to gain their place in a more specific market niche.

[24] This idea was first put forward by the United Nations in 1989.

At the same time, no matter what strategy a firm may choose, product quality is always an issue of prime importance, together with quick and prompt reaction to the market demand. Large buying houses in Europe have reduced the number of their suppliers to concentrate their purchases on more efficient ones based on their products and their logistical performance as well. Enterprises should pay close attention to this buying trend, trying to shorten their usual delivery time from 5-6 months to 8-12 weeks after order placing to meet the QR requirements.

- Focus upon "two-end" activities

Domestic enterprises often ignore "two-end" activities, including product development and marketing. This phenomenon is closely related to the old distribution system in China. Prior to the 1980s, wholesale and retail businesses were monopolized by sate-run enterprises. The central-planned multi-tier distribution hierarchy operated as an allocation machine, rigidly separating the production sector from the distribution sector. Producers were extremely insensitive to market changes. The textile and apparel production enterprises were neither responsible for the market demand nor concerned with the marketability of their products. The only thing they cared about was producing the goods based upon a planned quantity within a scheduled time period. With foreign direct investment in China, a common thing is that the "two-end" activities are retained in foreign hands. As a result, though China has enefited a lot from this foreign capital inflow, many enterprises are trapped in the lowest value-added links in the whole value chain. Thus, relocation in China should pay much attention to this aspect. With production facilities moved to the hinterland, the associated "market" should be given to these enterprises as well. This can help enterprises in the central and western regions to be really market-oriented.

- Long-term vision

There are few competitive advantages that cannot be imitated. Though a detailed and good firm strategy may be simple in nature, it should be drafted with a long-term vision.

The processing trade is very important to the quick emergence of China's textile and apparel industry. However, it is also an obstacle to building local brands in the long run. Many enterprises are satisfied with their processing activities, using buyer's samples and imported materials, as it is virtually risk-free. It is unreasonable to give up this processing trade immediately and it will stay for quite a long time. The important thing is that the enterprises, especially those of strength and ambition, should embody the idea of brand building in their long-term development plan. Engaging in higher value-added activities is the ultimate approach to survive the competition and acquire hard-to-copy competitive advantages.

The shortsightedness of some enterprises also casts a shadow upon the effectiveness of current reforms. For example, spindle elimination aims to reduce redundant production capacity by dismantling obsolete spindles. However, owing to some enterprises' myopia, some of the obsolete spindles were not eliminated but sold to the TVEs and private enterprises to continue production. These enterprises gained a little profit from the sale, but they brought a very negative impact upon the reform process.

The long-term vision of an enterprise ultimately means to its relentless pursuit of product innovation and technical upgrading. Enterprises of strength should establish their own R&D department, while small ones should learn to absorb from others. The development history of some Japanese industries after WWII provides a very good example to domestic enterprises. Limited capital should be used to create maximum profits. The past practices such as repeated introduction and blind expansion must be discarded right away. Size is not important, but economic efficiency is most important. Downstream and upstream enterprises should cooperate closely to share common information and technology. This is an inevitable trend and a necessary way to boost the performance of the whole industry.

D. Consumer market development

The latest government report emphasizes that China will continue to boost domestic demand through various measures, such as policies concerning housing, pensions, and income growth, especially for peasants. In the diamond model, consumer sophistication and market conditions play a role in forming an industry's competitive advantages. Consumer market development largely depends upon two aspects: government policy orientation and the enterprises' role.

- Government policy orientation

Local demand is sometimes influenced by government policy. The concept of "green consumption" first appeared in countries where government places great emphasis upon environment protection and promulgates many relevant laws. China for a long time was in a seller's market. It was not until the 1990s that consumers were able to be selective and demanding. Domestic law on consumer interest protection was only published in the early 1990s. Therefore, there is much to do for the government to cultivate consumers through policy orientation, such as nationwide environment-consciousness through promotion of environmental protection.

- Enterprises' role

Enterprises play a more direct role in forming market demand and upgrading consumer sophistication. Domestic enterprises usually focus on simple product advertising and neglect cultivation of their consumers. Better communication between enterprises and consumers not only helps enterprises to know better about the market conditions but also to enrich the consumers' knowledge about products and enterprises. Regular exhibitions and fashion shows also enable consumers to have access to the new products and new technology, thus broadening their horizons.

2.5 IMPACT UPON THE HK TEXTILE AND APPAREL INDUSTRY

2.5.1 Mutual importance of China and HK to each other's economy

2.5.1.1 Role of the mainland in HK economic transformation

HK is a place lacking the necessary natural resources for agricultural and industrial development. However, it is endowed with an ideal port and a strategic location acting as the doorway to South China. It is also in the international time zone that bridges the time gap between Asia and Europe. The HK economy has undergone some big and fundamental transformations during the past half-century. The Chinese Mainland, either passive or active, plays a crucial role in its economic development.

A. Before 1978

Before the 1950s, the economic activities of HK focused upon a transit trade. It was a bridge linking the Chinese Mainland and the outside world. There was almost no established industrial sector in HK at that time.

However, after the breakout of the Korean war and the subsequent trade boycott of China by the United Nations, HK lost its role as an entrepôt port for China's major importation from and exportation to the West. Because of the hostile external environment towards the newly founded Chinese regime during this period, the mainland turned to independence and a self-sufficiency policy for its economic construction. Foreign trade was mainly conducted with the East European bloc and the former Soviet Union. Therefore, the economic and trade link between the Chinese Mainland and HK stagnated. HK domestic exports and re-exports to the Mainland were HK$31 million and HK$175 million respectively in 1977, representing only 0.1% and 1.8% of HK's total domestic and re-export volume. Because of the limited availability of natural resources, there was almost no agriculture sector in HK. It had to import the indispensable living substances and other raw materials from the Mainland. That was the major reason for the relatively larger share of its import volume with the Mainland, which was 16.6% in 1977. (HK External Trade 1977)

It was during this period that HK began to develop its own manufacturing sector in order to fill the blank of the loss of its entrepôt port role between China and the outside world.

B. After 1978

HK started exporting its own products with the quick emergence of the manufacturing sector after the 1950s, which were labor-intensive and export-oriented. Such a prompt move gave a new growth momentum for the economy when the entrepôt trade was declining sharply. However, with the continuous rise of labor and land costs, as well as more competitive new entrants, this sector faced a dilemma by the end of 1970s. The increased operational cost forced the manufacturers to find alternative approaches to survive the hotter international competition.

It was right after 1978 that the "open-up" reform in China ushered in the economic integration between HK and the Chinese Mainland, especially between HK and the Pearl River Delta. The imbalance of economic strength and different needs between the two sides at that time laid a foundation for close economic cooperation. HK was troubled with its soaring labor and land costs while the Mainland was abundant in these factors; the Mainland was in urgent need of foreign capital and advanced technology to boost its fragile economy, which HK was in a position to provide based on a decades' accumulation and development. What's more, the geographic proximity and cultural similarity were also a plus. HK manufacturers began to shift most of their production facilities to the Mainland to exploit its low labor and land costs as well as its huge potential market and various preferential policies designed to attract overseas investment. At the same time, HK retained the higher value-added activities such as quality control, product design, market promotion and distribution. The so-called "Front Store and Back Factory" pattern well illustrates the cooperation between the two sides. Backed by the production support from the Chinese Mainland, HK again resumed its entrepôt port role (Figure 2.12) and upgraded to a higher and more comprehensive level. According to the HKSAR government statistics, out of HK's total re-export volume, which was HK$1159.2 billion in 1998, 95% either originated from or were destined for the Chinese Mainland. [53]

2.12 Proportion of HK re-exports to its total exports between 1960 and 1998

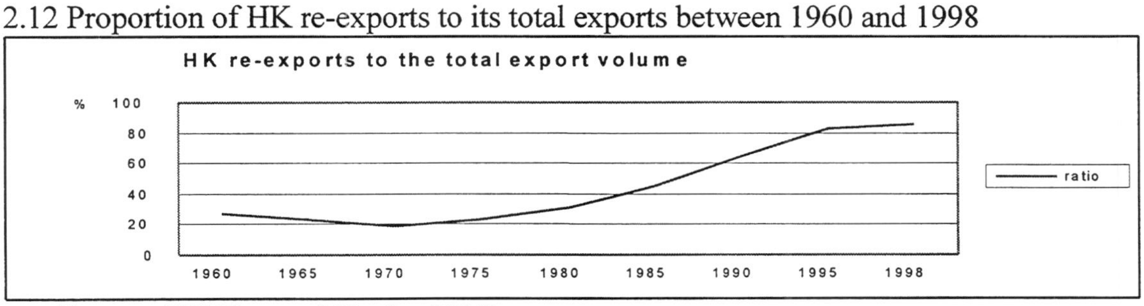

(Source: HK External Trade, 1998)

After twenty years' development, the mainland is now the most important offshore production base for HK companies, while HK has developed into a very important international financial and trade center with a strong and well-established service sector. This new positioning of HK can be mainly attributed to the "open-up" policy adopted as the Chinese Mainland, which provided an excellent opportunity for HK to grow into an international service center.

2.5.1.2 Role of HK in China economic restructuring

A. The largest investor and major trading partner

HK has been the largest investor in the Chinese Mainland since the latter began to carry out its open door policy. In 1998, HK signed more than 7,805 investment projects in the Chinese Mainland, with contracted and utilized capital of US$ 18.8 billion and US$ 19.4 billion, representing 39.4%, 34.3% and 40.8% of the national total respectively. By the end of April 1999, there were 329,637 foreign-funded projects registered in the Chinese Mainland, among which 54.8% were tied to HK interests. Contracted and utilized capital inflow from HK reached US$ 300.7 billion and US$ 142.7 billion, accounting for 51.6% and 50.7% of the national total. More than 120,000 enterprises involving HK interests were registered in the Mainland.

At the same time, HK was the Mainland's fourth largest trading partner after Japan, the US and the EU in 1998. According to China Customs' statistics, bilateral trade between the Chinese Mainland and HKSAR was US$ 45.4 billion in 1998, accounting for 14% of the Mainland's total trade volume.[25]

B. Rapid emergence of the Pearl River Delta

With the shift of HK production capacities to the Pearl River Delta, which is close to HK and enjoys various preferential policies designed to attract foreign investment, this region experienced a very rapid and dynamic economic development. The Pearl River Delta is the hottest spot of HK-based investment. According to the HK Economy Yearbook 1996, about 80-90% of the plastics industry, 85% of the electronics industry and 90% of the toy industry have moved to this region. It now engages more than 5 million local people working for these HK-backed companies or factories, which is over 19 times the size of the local manufacturing workforce in HK. The capital, technology, management expertise and extensive international distribution channels introduced from HK are the major driving force for the quick emergence of the Pearl River Delta and help promote the economic integration between the Mainland and the rest of the world.

C. Intermediary between the Chinese Mainland and the outside world

In addition to its important role in the Mainland's economic development and restructuring, HK is a unique gateway of the Chinese Mainland to the outside world. Thanks to its strategic location, well-established banking and financial facilities, government non-interventionist policies and rich knowledge of the Mainland market, it has become a very important venue for multinational corporations to access and tap the Mainland market. Many foreign enterprises have moved their regional headquarters or head offices to HK as a way to further penetrate into the Mainland market, which is expected to be more open and vigorous in the future. Furthermore, as an important sourcing center, more than 75% of overseas buyers in a survey conducted by HKTDC replied that over half of their Chinese imports were sourced via HK. [55]

[25] All the data in this part are retrieved from this web site: http://www.tdc.org.hk/main/china.htm

HK is also an efficient channel for foreign capital entering the Mainland. More than 70% of foreign capital was introduced into the Mainland of China via HK. On the other hand, because of HK's extensive external relations with the outside world, it is an ideal place for Mainland enterprises to raise capital. Almost every province, autonomous and municipality in the Mainland now has resident representative offices, agents or distributors in HK. According to the Chinese official estimate, there are over 1,856 Mainland-backed enterprises registered in HK, among which over 72 were listed on the HK Stock Exchange.

Another indication of the significance of HK's intermediary role is the average annual growth rate of 16.3% in HK's ordinary re-exports to China (not related to outward processing) between 1992 and 1997. [74] In 1998, re-exports of Mainland origin products through HK rose from HK$545.8 billion in 1994 to HK$691.2 billion. Goods re-exported to the Mainland via HK were HK$407.4 billion in the same year, 26.2% higher than that in 1994.[26] About 35% of HK's re-exports are destined for the mainland and 60% of the re-exports are of Chinese origin. [56]

2.5.1.3 Implications for HK trade and investment

A. Trade relationship with the Chinese Mainland

The Mainland has been HK's largest trading partner since 1985. The Mainland's share of HK's global trade jumped from 9.3% in 1978 to 37.7% in 1998. It is currently HK's largest import source and export market, accounting for 40.6% of HK's total imports and 34.4% of HK's total exports. (Table 2.13)

The shift of HK production facilities to the Mainland contributed to the drastic rise of the re-export volume between the two sides, which at the same time, boosted the prosperity of all re-export activities in HK. In 1998, HK re-exports to China accounted for 35.1% of the total re-export value. (Figure 2.13)

HK's trade with the Chinese Mainland mainly focuses upon outward processing activities. In 1998, 77.4% of HK's domestic exports and 44.1% of re-exports to the Mainland were concerned with outward processing activities; 82.7% of HK's imports from the mainland and 87.6% of HK's re-exports of Mainland origin to other countries and regions other than China were associated with outward processing.[27] Therefore, the manufacturing cooperation pattern between the two sides not only helps maintain

Table 2.13 HK imports and exports with China between 1975 and 1998

HK imports and exports with China, in HD$ bn

	1975	1980	1985	1990	1995	1996	1997	1998
Total Trade	63.30	209.89	466.57	1282.41	2835.25	2933.50	3071.04	2766.74
With China	6.97	28.20	120.18	394.51	987.08	1049.81	1116.17	1044.11
share %	*11.01*	*13.44*	*25.76*	*30.76*	*34.81*	*35.79*	*36.35*	*37.74*
Total imports	33.47	111.65	231.42	642.53	1491.12	1535.58	1615.09	1429.09
From China	6.81	21.95	58.96	236.13	539.48	570.44	608.37	580.61
share %	*20.3*	*19.7*	*25.5*	*36.8*	*36.2*	*37.1*	*37.7*	*40.6*
Total exports	29.83	98.24	235.15	639.88	1344.13	1397.92	1455.95	1,347.65
To China	0.17	6.25	61.21	158.38	447.60	479.37	507.75	463.44
Share %	*0.6*	*6.4*	*26.0*	*24.8*	*33.3*	*34.3*	*34.9*	*34.4*

(Source: HK External Trade 1998)

[26] HK External Yearbook, 1998

[27] HK External Trade, March 1999

2.13 HK re-exports to China between 1975 and 1998

(Source: HK External Trade, various issues)

2.14 FDI in HK by major investors in 1997

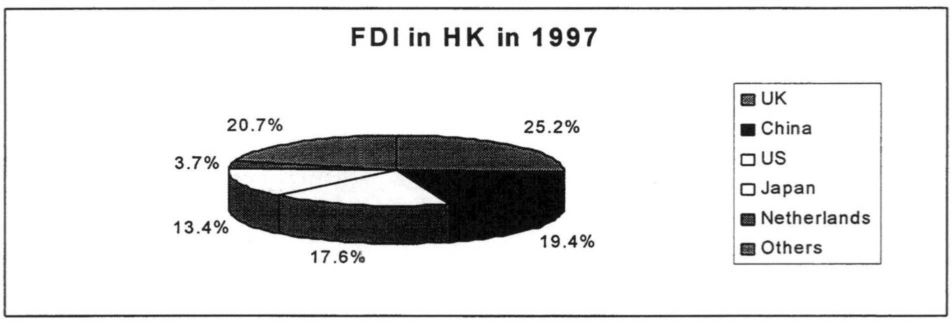

(Source: HK Economy Yearbook, 1999)

2.15 FDI in HK by China between 1994 and 1997

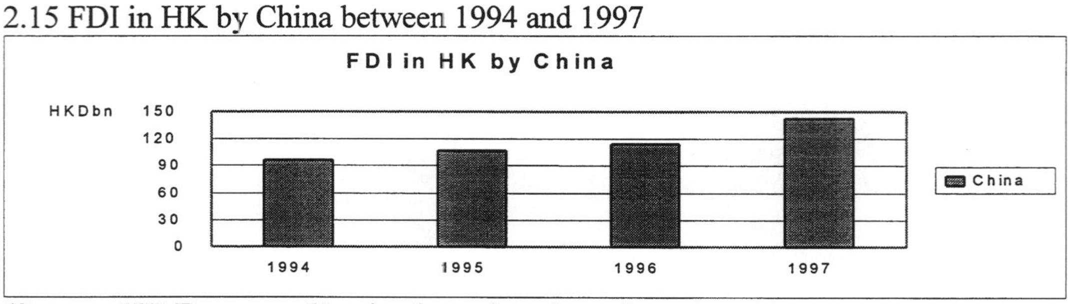

(Source: HK Economy Yearbook, various issues)

the competitiveness of HK-labeled products, but also boosts the development of the service sector, especially trade-related services. The entrepot trade in HK nowadays is endowed with new characteristics, far beyond the traditional sense.

B. Close investment linkage

Trade and investment always grow hand in hand. As mentioned above, HK is by far the largest investor in the Mainland. However, such investment is not one-way traffic. China is also a major investor in HK, second only to the UK in 1997. (Figure 2.14) From 1994 to 1997, the Mainland investment in HK increased from HK$ 96.6 billion to HK$ 142.4 billion. (Figure 2.15) In a word, HK and the Mainland are each other's very important trading partner and investor. The close and complementary economic relationship is crucial to both sides.

2.5.2 Profile of the HK textile and apparel industry

2.5.2.1 Importance in HK economic development

A. Key importance to HK industrialization

The textile and apparel industry plays a very important role for HK economic development. In 1955, its employment accounted for 37.2% of the local manufacturing workforce and in 1975, the proportion rose to 51.8%. [75] Though it has faced many troubles and difficulties since the late 1970s, it remains as the backbone of the HK manufacturing sector. Today, the apparel industry is still the largest manufacturing sector in HK in view of its gross output, its employment or its domestic exports. Together with the textile industry, they totally employ 31.4% of the local manufacturing workforce. [75] In 1998, the textile and apparel domestic exports were HK$ 85.6 billion, amounting to 45.4% of the total domestic export volume; their re-exports topped HK$ 187.0 billion, or 16.1% of the total re-exports. If the two figures are combined, the sector's total export value reached HK$ 272.67 billion, accounting for 20.2% of HK's total exports. [76]

B. Present situation

Though the textile and apparel sector still plays a key role in the HK economy, it is facing a lot of difficulties and challenges. In addition to the reasons mentioned above such as rising labor and land costs, the limited domestic market and fierce competition in the international market, the overwhelming challenge is the inevitable declining trend of the manufacturing sector in HK. The territory itself is now engaging in fewer manufacturing activities and shifting to a service-oriented economy. Table 2.14 gives some insight into the present situation of the textile and apparel sector. Both the export and import share of the textile and apparel sector shrank by various degrees, accompanied with descending employment from 16.5% in 1980 to 3.6% in 1998 of the total workforce.

The HK economy is now more interwoven with the outside world, especially with the Asian region. It is prone to suffer from world or regional crisis. The 1997 Southeast Asia Financial Turmoil dealt a heavy blow to HK, causing a setback in the economy until the second quarter of 1999 when GDP experienced a small positive growth of 0.5% in real terms.[28] The increasing dependence upon other economies makes HK more sensitive to the external environment and changes. There are various

Table 2.14 Textile and apparel sector in HK's economy

Textile and Apparel Industry in HK's Economy

	1970	1980	1990	1998
T & A exports to total exports (%)	39.8	34.3	28.8	20.2
T & A imports to total imports (%)	18.7	16.4	20.8	15.1
T & A employment to total employment (%)	NA	16.5	11.7	3.6

(Source: Calculated based on data from HK Economy Yearbook and HK External Trade, various issues)

[28] Half-Yearly Economic Report 1999

Table 2.15 Breakdown of HK textile industry in 1998

Textile Industry Composition By Some Major Sectors

Sector	Spinning				Weaving		
Sub-sector	*Cotton spinning*	*Wool Spinning*	*Synthetic Fiber Spinning*	*Other Fiber Spinning*	*Cotton Woven Fabrics*	*Wool Woven Fabrics*	*Silk Woven Fabrics*
Establishment	7	1	3	20	96	1	3
Employment	2128	100	94	169	3329	9	8
Average	304	100	31	8	35	9	3
Sector	**Dyeing & Printing**				**Knitting**		
Sub-sector	*Yarn*	*Weaving Fabrics*	*Knitting Fabrics*	*Clothing*	*Cotton*	*Wool*	*Others*
Establishment	38	70	31	26		298	
Employment	671	2063	376	296		4461	
Average	18	29	12	11		15	

(Source: HK Economy Yearbook 1999)

arguments with respect to the future road the manufacturing sector should take, among which are the "Shell" versus "Balance" argument. The former stresses that HK should focus on service-related activities and further re-locate the manufacturing sector to other regions; while the latter points out that HK should maintain a manufacturing sector featuring hi-tech and industrial upgrading in order to keep the economy healthy and sound. It is beyond this paper's scope to discuss these arguments in detail. However, they indicate, from certain aspects, that the textile and apparel industry is now at the crossroads. To shrink or expand depends not only on the local conditions, but also on the stance the whole manufacturing sector will hold in the future.

2.5.2.2 Hong Kong's Textile industry

A. General conditions

The textile industry in HK has a history of more than 50 years. The first cotton mill was established in 1947. [75] Table 2.15 is a recent breakdown of this industry by some major categories. The 1950s and 1960s saw the rapid development of the textile industry. In 1973, it employed 17.7% of the total manufacturing workforce. [75 (1975)] At the beginning of the 1970s, the cotton spinning industry turned to using the open-end spinning approach instead of ring spinning. Automation and computerization also play a role in the knitting, dyeing and printing industry. Absolute employment in the industry experienced a downturn from the late 1970s in the face of soaring land, labor and energy costs. Figure 2.16 illustrates the employment changes between 1995 and 1998. It should be noted that though the employment reduced to 18533 in 1998 from 28811 in 1995, its proportion to the total manufacturing sector was relatively unchanged. The average number hired in each establishment was around 13 in 1998.

B. Major characteristics

- Products

The textile industry is export-oriented. Most of the products are for export, among which woven fabrics accounted for approximately 48% in 1998. (Table 2.16) It is also a major supplier to the local apparel industry, in addition to its export orientation. It specializes in cotton yarns and natural fabrics. With the shift of low-end products to the Chinese Mainland and other regions such as Southeast Asian countries, the

2.16[29] Employment changes in HK textile industry

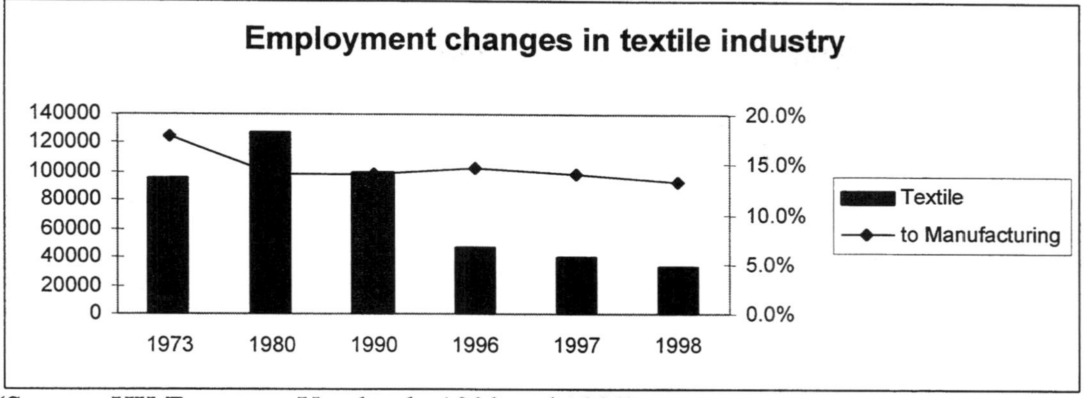

(Source: HK Economy Yearbook, 1998 and 1999)

manufacturing operations in HK mainly concentrate upon high value-added items.

- Foreign investment

As mentioned earlier, most of the foreign investments coming into HK have been poured into the more advanced sectors such as electronics or the auto industries. The textile industry attracted foreign investment of US$ 182 million (equivalent to HK$ 1.42 billion), with 21 establishments with a total of 2,063 workers by the end of 1997, accounting for 2.8% of the total foreign investment in the manufacturing sector. Liberia was the major source of investment, followed by Japan and Taiwan. [57]

- Strengths

--Computer-aided design and computer-aided manufacturing are widely used to offer quality fabrics to the apparel industry.

--The local textile industry, based on close cooperation and a complete information network, is able to meet promptly orders from the apparel industry, at a very short notice. At the same time, an intranet communication infrastructure funded by the SAR government has been launched to ensure efficient electronic communication and trade between the two sectors.

--The textile enterprises pay much attention to product innovation and upgrading in order to compete with low-cost suppliers such as Pakistan, Thailand and India. New features are developed and incorporated into products, including anti-bacteria and anti-UV agents, crease-resistance, water-repellence, as well as easy-care characteristics.

- Weaknesses

--Because of its export-orientation feature, this sector is prone to be negatively affected by an unfavorable outside economic environment. In 1998, textile exports registered a drastic year-on-year drop of 10.7% mainly due to the Southeast Asian Financial Crisis which resulted in a sluggish demand in HK's major export markets.

--Though the local enterprises make great efforts to introduce state-of-the-art technology to the textile sector, which is relatively capital-intensive, the research & development capabilities in HK are not strong enough to provide a firm foundation for sustainable future development. Most of the textile enterprises are of small size, except the cotton and wool spinning sectors, hampering long-term capital as well as R&D investment.

[29] In Figures 2.16 and 2.25, original data are from HK Economy Yearbook, 1998 and 1999. Then, they are calculated with some adjustments. Here, knitwear is deducted from the textile sector and added to the apparel sector. The same adjustment is made in Tables 2.16 and 2.18.

Table 2.16 HK textile exports by categories

Textile exports by categories		
Category	**1997%**	**1998%**
Textile Yarn	21.8	22.1
Woven Fabrics	49.0	47.9
Cotton	*22.7*	*24.0*
Man-made	*21*	*18.6*
Others	*5.2*	*5.3*
Knitted or Crocheted Fabrics	13.9	15.1
Special Yarns and Fabrics	7.4	7.0
Finishing Accessories	3.3	3.5
Textile Made-up	4.1	4.0
Floor Coverings	0.5	0.4

(Source: [58])

C. Textile trading

- Import and export volume

In 1970, the total textile export value was HK$ 1.67 billion, and it increased to HK$ 101 billion in 1998—60 times more than that in 1970. The total textile imports increased from HK$ 3.0 billion to HK$ 104.4 billion during the same period. (Figures 2.17 & 2.18) Textile domestic exports topped HK$ 17.2 billion in 1992 and then declined on a year-on-year basis for the next six years. By sharp contrast, the re-export value jumped from HK$ 0.4 billion in 1970 to HK$ 90.2 billion in 1998 accounting for 89.3% of the total textile exports.

The textile trade in HK was clouded in 1998 mainly due to the aftermath brought about by the Southeast Asian Financial Turmoil. The sluggish demand from Asian countries, especially from the Chinese Mainland, contributed to the 13.9% drop of the total textile trade volume in 1998, of which the textile imports and exports experienced a 16% and 10.7% decrease respectively.

- Major textile trading partners[30]

According to Figure 2.19, approximately half of the textile domestic exports were absorbed by the Chinese Mainland,[31] followed by the US and Indonesia. The Chinese Mainland was also HK's largest textile re-export market, holding 71.4% of the total re-exports, 33 times more than that re-exported to Bangladesh, which is HK's second re-export destination. (Figure 2.20) As regards textile imports, the Chinese mainland was again HK's biggest trading partner, contributing 39.1% of the total import volume. (Figure 2.21) If exports and imports were combined together, the Chinese Mainland accounted for 53.8% of HK total textile trade.

- Textile trade pattern

--The Chinese Mainland is HK's No. 1 textile trading partner, both in terms of export and import value. From Figures 2.22 to 2.24, it will be seen that trade volume registered a rapid increase from 1975 to 1985 and then developed at a relatively moderate speed. Section 2.5.3 will give a more detailed illustration discussion of this aspect.

[30] Data in Figure 2.19, 2.20 and 2.21 are from HK industrialist, November 1999

[31] In the figures, China refers to the Chinese Mainland.

2.17 HK textile domestic exports and re-exports

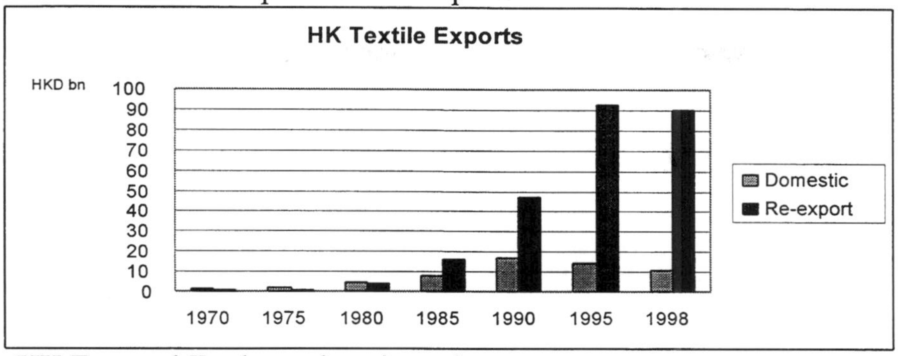

(Source: HK External Trade, various issues)

2.18 HK textile and apparel imports

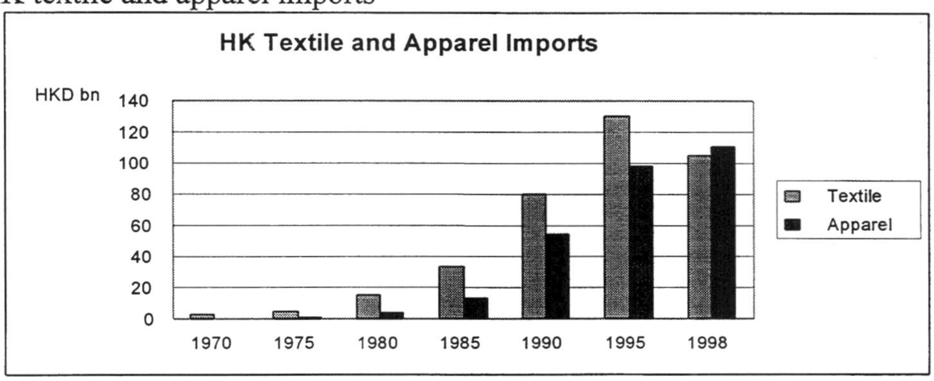

(Source: HK External Trade, various issues)

2.19 Textile domestic exports from HK to major markets in 1998

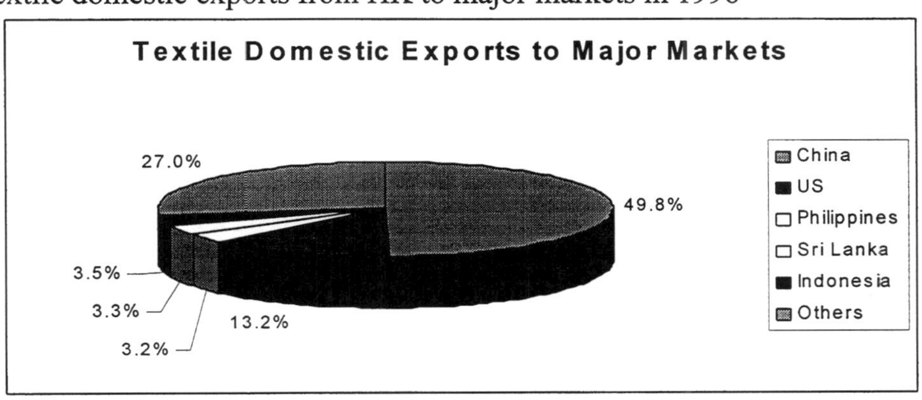

2.20 Textile re-exports from HK to major markets in 1998

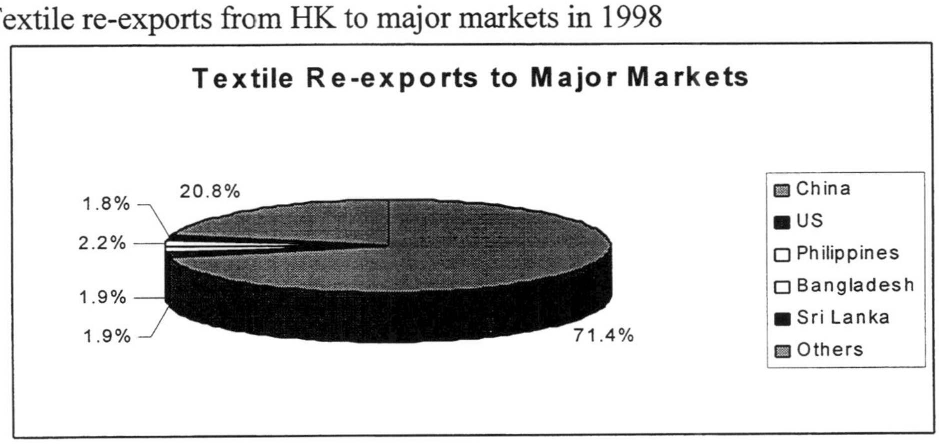

2.21 Textile imports from major markets to HK in 1998

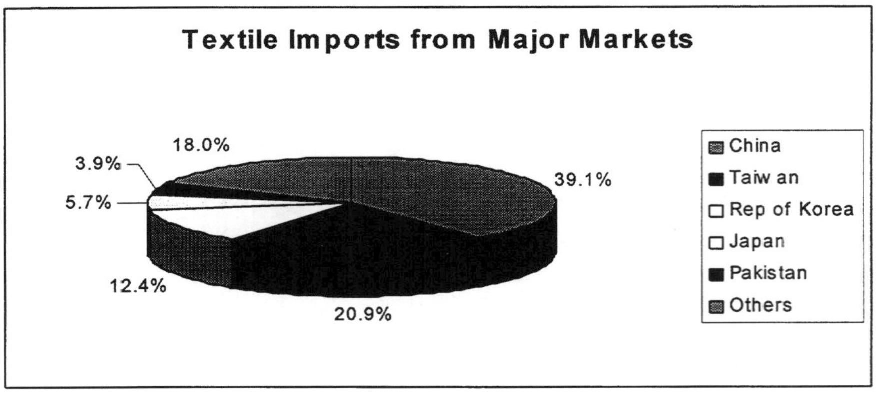

2.22 Share changes of HK Textile Domestic Exports to the Major Markets

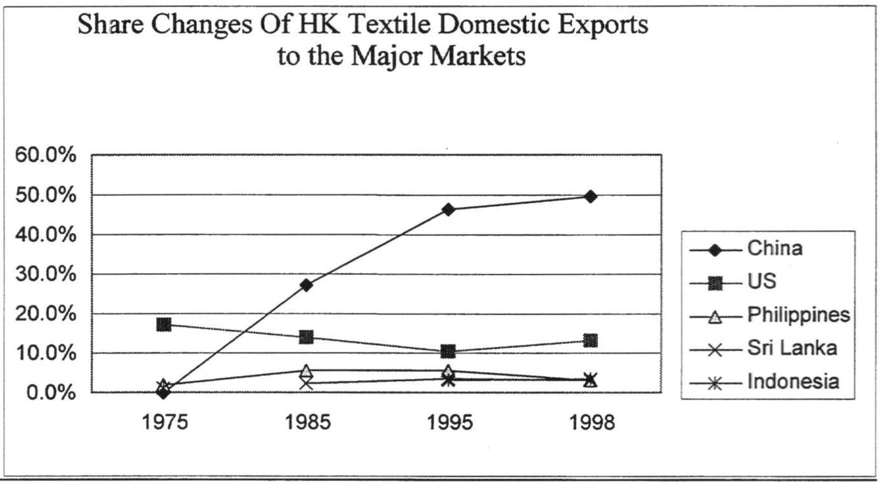

(Source: Calculated based on data from HK Industrialist, various issues; HK External Trade, various issues)

--Except for the US, several Asian developing countries, including the Mainland, the Philippines, Sri Lanka, Bangladesh and Indonesia, were HK's major textile export destinations; at the same time, Taiwan, Korea and Japan were the major textile suppliers to HK apart from the Chinese Mainland. They supplied 39% of the textiles to imported into HK while the Mainland provided another 39%. Such a pattern shows that the HK textile trade is concentrated upon the Asian region.

--Though HK is still the world's leading textile exporter, its domestic exports play a less significant role. Textile domestic exports only accounted for 10.7% of total textile exports in 1998, compared with 76.6% in 1970. This indicates trend of the local textile industry during the past three decades. Based on these figures, it can be said that HK is now a world textile sourcing center rather than a manufacturing base.

2.5.2.3 Hong Kong's apparel industry

A. General conditions

Hong Kong's fame as an famous apparel-manufacturing center has developed in less than five decades. It has been a very important pillar of the HK manufacturing sector since the early 1960s with the largest employment, the highest gross output value and biggest domestic export volume. Over the years, the apparel industry has

2.23 Share Changes of HK Textile Re-exports to the Major Markets

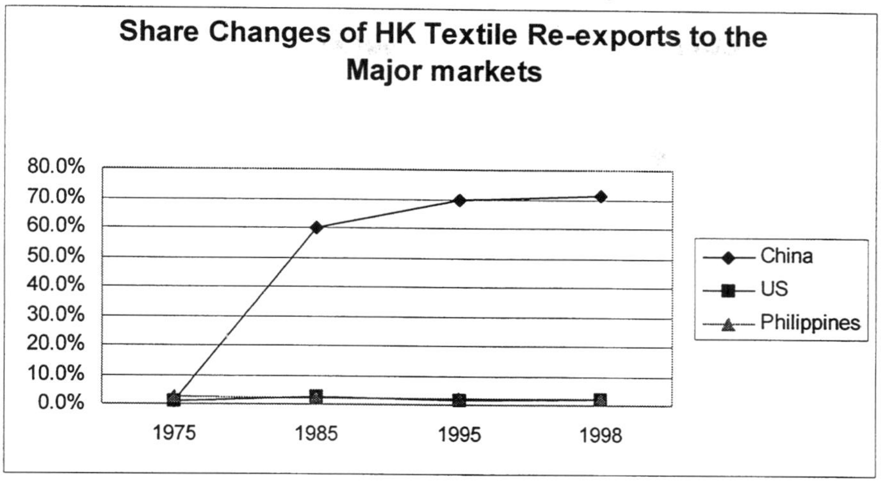

(Source: Calculated based on data from HK Industrialist, various issues; HK External Trade, various issues)

Table 2.17 HK apparel industry structure

Apparel Industry Composition						
Sub-sector	Wovenwear	Knitwear	Underwear & Pajamas	Fur	Other	Total
Establishment	1,938	742	120	15	239	3,054
Employment	36,855	14,142	1,297	123	6,095	58,512
Average	19	19	11	8	26	20

(Source: HK Economy Yearbook 1999; [16])

2.24 Share changes of HK Textile Imports from the Major Markets

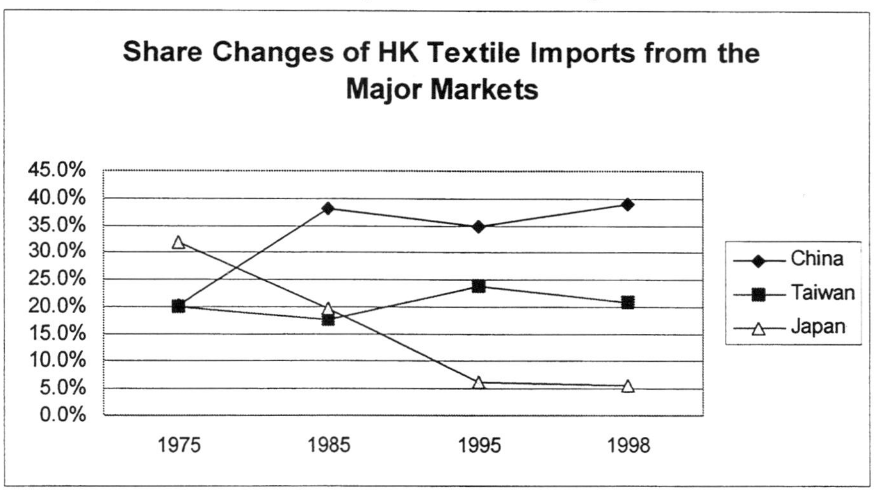

(Source: Calculated based on data from HK Industrialist, various issues; HK External Trade, various issues)

successfully evolved from a low-cost supplier to a regional fashion center featuring good quality, prompt delivery, quick response and excellent fashion sense.

The HK apparel industry mainly comprises the following sub-sectors: woven wear, knitwear, underwear, and other small divisions including leather & fur apparel and apparel accessories. (Table 2.17) The wovenwear sector accounts for about 63%

Though many production facilities have been shifted to other regions since the 1980s, the sector is still very active in providing one-stop services ranging from design and production to marketing and retailing. It focuses upon high value-added activities, functioning as a strategic control and sourcing center for its extensive production network. Though troubled by rising costs and protectionist measures adopted by other countries, the apparel industry is still the largest employer in the manufacturing sector, despite the sharp drop of 47% in its absolute employment level between 1995 and 1998. (Figure 2.25)

B. Major characteristics

- Products

HK is the world's second largest apparel exporter after the Chinese Mainland. It has successfully built up its upmarket image based on design, workmanship and quality. Since most enterprises have set up offshore production facilities outside HK, particularly in the Chinese Mainland, the local enterprises are mainly responsible for quality control, logistic arrangement, fashion design, sales and marketing.

The apparel sector is also export-oriented with a variety of products ranging from woven and knitted apparel to clothing accessories. Table 2.18 shows that woven wear is the largest export item, holding 39.4% of the total apparel exports in 1998.

- Foreign investment

FDI in the apparel sector amounted to US$ 110 million, or 1.7% of total FDI in all the manufacturing sectors by the end of 1997, with 34 establishments of with a total 5039 employees. Switzerland was the largest investor with a 55.3% share, followed

2.25 Employment changes in HK apparel sector to the manufacturing industry

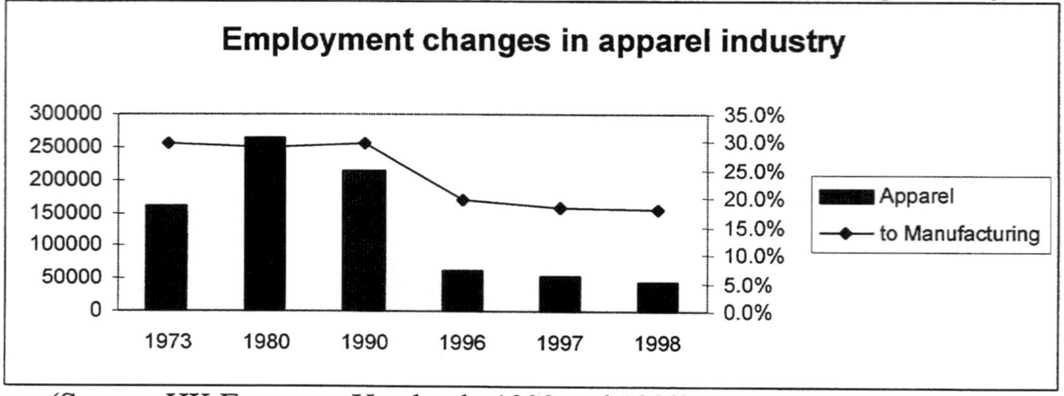

(Source: HK Economy Yearbook, 1998 and 1999)

Table 2.18 HK apparel exports by category

Apparel exports by category				
Category	1997%		1998%	
Woven Wear	40.8		39.4	
For Men or Boys		*16.8*		*15.9*
For Women or Girls		*24.0*		*23.5*
Knitted Wear	16.6		17.2	
For Men or Boys		*5.6*		*5.6*
For Women or Girls		*11.0*		*11.6*
Clothing Accessories	9.6		9.6	
Of Textile Fabrics		*3.5*		*3.9*
Of Non-textile fabrics		*6.1*		*5.7*
Other Apparel Articles	33.0		33.8	

(Source: http://www.tdc.org.hk/main/industries/ipclot.htm)

2.26 FDI in HK apparel sector by 1997

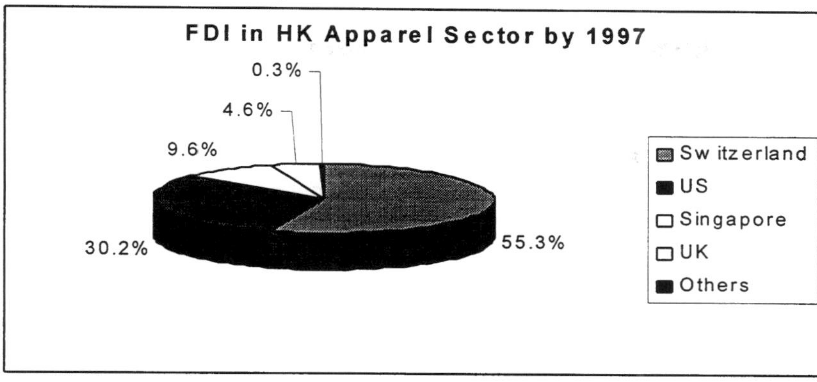

(Source: http://www.igsd.gov.hk/investment/html/m_sector_industry_11.html)

by the US, Singapore and the UK. (Figure 2.26)

- Strengths

--Given the importance of "Quick Response" in the worldwide textile and apparel trade, a Quick Response Center for Textiles and Clothing Industries was set up in 1995 to help deal with customer needs and strengthen awareness towards quick response. Many manufacturers have applied "supply chain management" to their daily operation, especially to the offshore production network. Electronic Data Interchange (EDI) is employed among various parties to facilitate the coordination process and working efficiency. A full EDI service for production notification has been implemented since February 28, 2000.

--Like the textile sector, computer-aided design and computer-aided manufacturing approaches are adopted to boost productivity and value-added volume. A three-dimensional pattern design technique has been developed to reinforce the apparel quality, fitness and silhouette in the manufacturing process. The apparel samples can also be designed on three-dimensional mannequin images with wearing effect visualization and then sent electronically to buyers abroad.

--Fashion changes with time. Therefore, "time is money" is the golden rule of the apparel sector. Almost all the enterprises in the apparel sector are small, with the average employment of 20 per establishment. Such a small size enables the enterprises to adapt swiftly to market demands and make corresponding adjustments easily and flexibly. It can produce a broad range of products both in bulk and in customized items within a very short lead time, backed by an advanced transportation and communication system.

--The international competition nowadays covers many more aspects in addition to production cost. Over the years, the apparel sector has been making great efforts to build up an international profile for quality and fashion items with identifiable images and brand names. There are 400 to 500 fashion designers in HK with an average annual billing per firm of HK$ 2.3 million. Four local design schools turn out 70 to 80 graduates each year. The fashion designing strength is one of the key elements in the current status of the HK apparel sector.

- Weaknesses

The apparel sector has some common weaknesses with the textile sector, such as relatively inadequate R&D foundations and a lack of long-term vision for capital and technological investment. On the other hand, it has its own difficulties and dilemmas.

--Clothing exports to the United States are continually affected by trade disputes over illegal transshipments, and US Customs has published a list of HK companies that were prosecuted for illegal transshipment. Due to the unclear and sometimes confused information of the list, imports from an innocent company might also have suffered unnecessary losses. At the same time, US Customs officials have made several factory visits to ensure legal exports since 1996. Such action will no doubt disrupt normal production activities.

--The apparel industry is more labor-intensive than the textile industry. Therefore, the production cost is crucial to the product competitiveness. The HK apparel sector has found an efficient way out of the cost dilemma during the past years by OPA arrangements with low-cost regions; however, the changing trade policies of major importing nations poses a great uncertainty towards overall manufacturing plans. For instance, the change of the origin rule of the United States with respect to apparel origin forced local manufacturers to make corresponding adjustments in order to conform to the new rule. Under the old rule, a manufacturer could cut the fabric in HK and send it to the Chinese Mainland for assembly, then bring it back to HK for exportation, with HK as the country of origin; but now it will be categorized as of Chinese Mainland origin under the new rule.

--The major export destinations for HK apparel exports are the US and EU. Though the apparel manufacturers work very hard to shorten the delivery time with these major markets, the emergence of Mexico and Turkey as important apparel suppliers to the US and EU respectively poses a great threat to the HK apparel sector. The comprehensive information network and short lead time cannot completely make up the relatively longer geographic distances to these markets, while the countries mentioned above enjoy this natural advantage together with preferential arrangements between them. Therefore, it can be imagined that the international competition faced by the HK apparel sector will be hotter and heavier in the coming years.

--With the rising demand for protectionist trading policies in the major importing nations, various new trading barriers were designed to protect domestic industries. The Green Barrier is one of them, carrying a lofty slogan of environmental protection. To meet these extra and new requirements, such as "green labels" and the ban on azo dyes in certain European countries, the manufacturers have to make additional efforts, which constitute an extra cost burden upon them.

2.27 HK apparel domestic exports and re-exports between 1970 and 1998

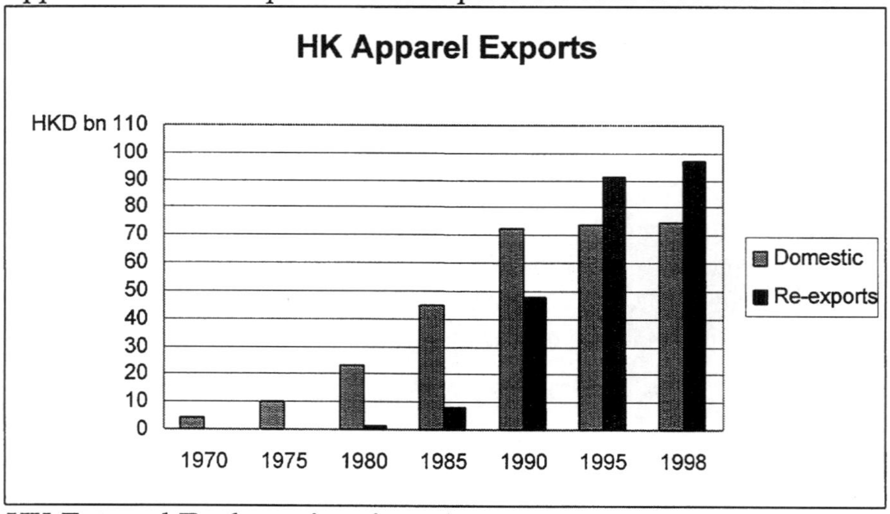

(Source: HK External Trade, various issues)

C. Apparel trading

- Import and export value

Total apparel exports increased 39 times between 1970 and 1998, rising from HK$ 4.39 billion to HK$ 171.67 billion; the import performance is even more remarkable, increasing from HK$ 0.28 billion to HK$ 110.74 billion during the same period. (Figures 2.27 & 2.18)

The apparel domestic exports reached record high of HK$ 77.2 billion in 1992 and fluctuated between HK$ 72 and 74 billion for the next six years. The re-exports value topped HK$ 106.7 billion in 1997 but experienced a 9.3% drop in 1998 for the reasons mentioned previously. However, the decreasing margin for apparel imports and exports were much smaller than the for textile trade--4.8% and 4% respectively.

- Major apparel trading partners

Figure 2.28 shows that 47.1% of the apparel domestic exports went to the US market in 1998, followed by the Chinese Mainland and the UK. The US was also HK's largest re-export market, accounting for 26.6% of the total re-exports. Compared with the domestic exports, the export markets for HK apparel re-exports are more diversified. The top five markets only accounted for 58.5%. (Figure 2.29) As regards apparel imports, the Chinese Mainland was the single most important partner, holding 92.5% of the total apparel imports. (Figure 2.30)

2.28 HK apparel domestic exports in 1998

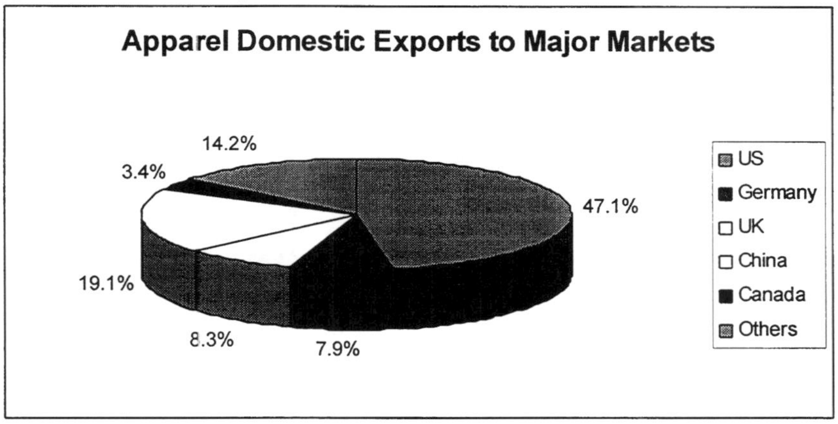

(Source: HK Industrialist, June 1999, p52)

2.29 HK apparel re-exports in 1998

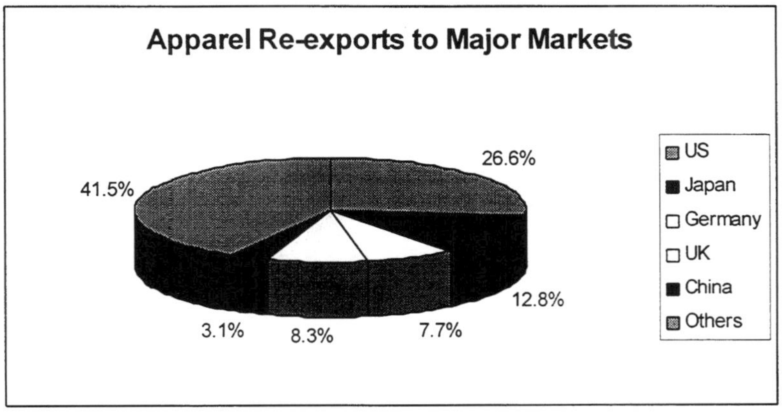

(Source: HK Industrialist, June 1999, p52)

2.30 HK apparel imports in 1998

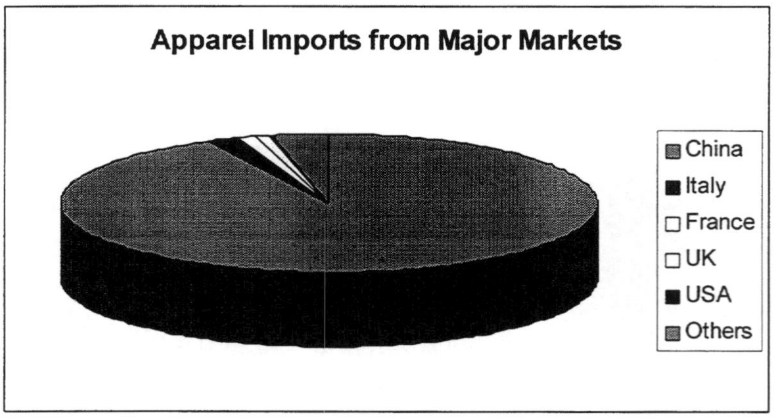

(Source: HK Industrialist, June 1999, p52)

2.31 Share changes of HK Apparel Domestic Exports to the Major Markets

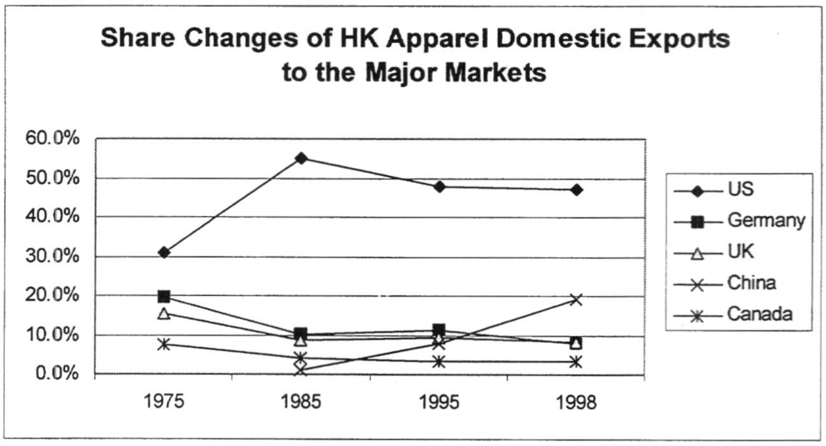

(Source: Calculated based on data from HK Industrialist, various issues; HK External Trade, various issues)

2.32 Share Changes of HK Apparel Re-exports to the Major Markets

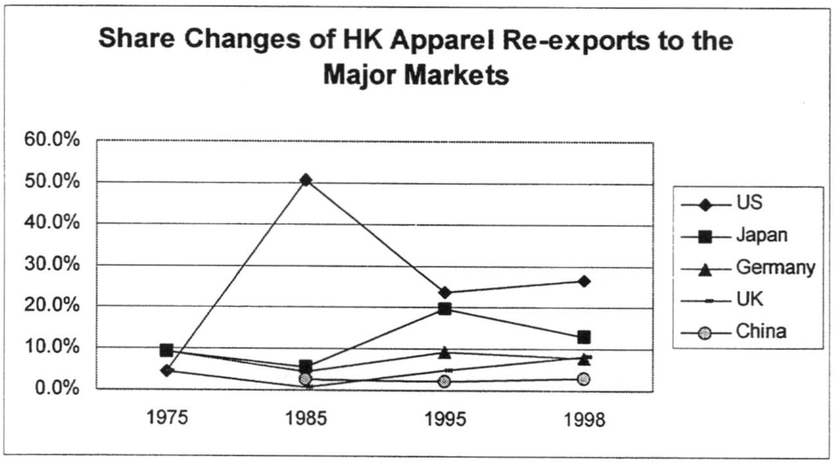

(Source: Based on data from HK Industrialist, various issues; HK External Trade, various issues)

- Apparel trading pattern

--In contrast to the textile trade pattern, the US and EU are the major export markets for HK apparel. This reflects the labor-intensive feature of the apparel industry. China only accounted for 1.1% of the domestic exports in 1985, but its share jumped to 19.1% in 1998. This partly demonstrated the market potential of the Mainland for fashion items because of its rapid economic development.

--The Chinese Mainland is the single most important apparel supplier for HK. Considering its unparalleled role in HK's textile and apparel trade, one can come to the conclusion that the textile and apparel trade links between the two sides are unique and vital to their textile and apparel industries.

--The re-export value first surpassed that of the domestic exports in 1992. The gap between the apparel domestic exports and re-exports volume is not so striking as in the textile trade. The former accounted for 43.6% of the total export volume while the latter accounted for another 56.4%. This indicates that the local apparel industry is still an important player in the world apparel trade regardless of its re-location trend to the Chinese Mainland in recent years.

--The HK textile and apparel trade is subject to quota restrictions set out in individual bilateral agreements with the US, EU, Canada and Norway. Table 2.19 shows the quota utilization rates for HK textile and apparel domestic exports to the four markets. The US is the largest importer for HK domestic apparel exports as well as the second largest importer for textiles, so its quota utilization rate is the highest among the four, over 81.9% in 1998. Except for some hot-sale items, such as woven blouses and trousers, the quotas for many categories are somewhat meaningless, as the utilization rates are low.

The year 1998 marked the beginning of the second stage of quota liberalization. The US removed quantitative restrictions upon 23 HK textiles and clothing categories, the EU liberalized quotas on 4 categories and Canada on two.[32] However, since it is the importers that decide the liberalization process at each stage, the most sensitive categories are deliberately left to the last minute. In this sense, the phase out effect is really of limited significance to the HK textile and apparel industry until the year 2005.

Table 2.19[33] Quota utilization level for HK textile and apparel exports to four markets

Quota Utilization Rates		
Markets	1997	1998
US	75.4%	81.9%
EU	53.8%	53.0%
Canada	49.4%	48.7%
Norway	68.9%	60.5%

(Source: http://www.tdc.org.hk/main/industries/ipclot.htm, retrieved on November 17, 1999)

[32] http://www.tdc.org.hk/main/industries/t2_2_38.htm

[33] As regards Norway, the ratio is calculated based on data from HK Trade Department, Notice to Exporters, Series 3A (Countries other than USA & EU) No.31/98, "Export of Restrained Textiles to Norway: performance against quota limits as at 31/12,98", Miss Wong Ying for Director-General of Trade

2.5.3 Impact of China's WTO entry upon the HK textile and apparel industry

2.5.3.1 Impact upon HK textile and apparel exports to the Chinese mainland

A. Textile domestic exports

Figure 2.19 shows that about half of HK textile domestic exports went to China in 1998 and HK is the fourth largest textile supplier to the mainland as well (Figure 2.33). In view of the mainland's great need for textile items, which accounted for about 91.6% of its total textile and apparel imports in the first five months of 2000 (Table 2.20), HK suppliers will have much room for future development. Various studies indicate that after China enters the WTO, the apparel industry will be one of the biggest beneficiaries due to quota liberalization and tariff reduction in intermediaries. (Section 2.6)

According to Table 2.21, the first 11 items are among the top 50 HK exports to the mainland and the remaining three are of interest to HK. It shows that tariff rates upon most of these items will undergo a more than 50% cut by 2004 or 2005. With the expected increase in apparel export volume and the big tariff rate cut for many HK export items to the mainland, demand for HK quality textiles such as fabrics will rise as well. Moreover, HK textile manufacturers have the edge to meet the need based on

2.33 China textile imports from major markets in 1998

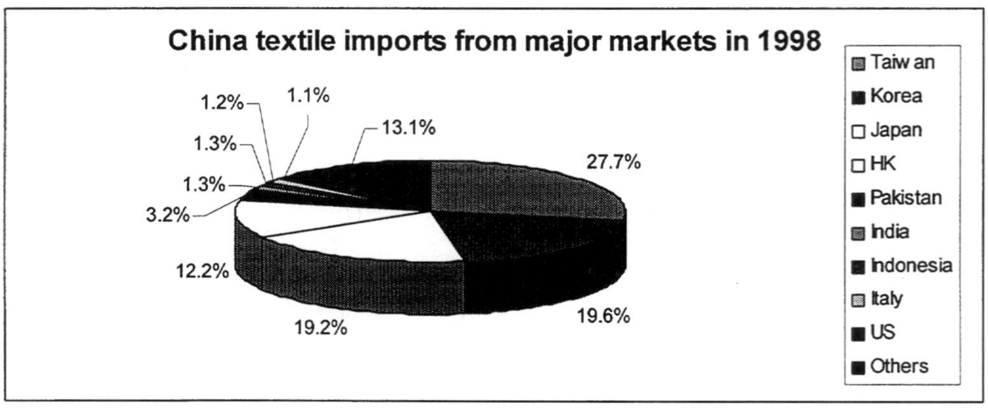

(Source: based on data from "Almanac of China's Foreign Economic Relations and Trade, 1999/2000; the same source was also the basis for Figures 2.34-2.36)

Table 2.20 China and HK textile and apparel import and export mix during January to May 2000

China and HK textile and apparel exports and imports				
Item	**Domestic exports**		**Imports**	
	China	HK	China	HK
Textiles	33%	12.7%	91%	49%
Textile Yarn	*6%*	*1.0%*	*20%*	*13%*
Textile Fabric	*27%*	*11.7%*	*71%*	*36%*
Apparel and Accessories	67%	87.3%	9%	51%

(Source: based on data from "HK external trade, May 2000" and "http://textile.unet.net.cn/imex05.htm, "Import and Export of Textile and Garments in January-May Period 2000, Table 3 & 4)

Table 2.21 Tariff cut schedule between 2000 and 2005 in the Sino-US agreement

Tariff reduction schedule of some selected textile and apparel items

Item	Base rate %	Bound rate %	End date
Cotton Yarn	10--11	5	2002
Woven cotton fabrics weighting not more than 200g/sq.m	19	10	2004
Woven cotton fabrics weighting more than 200g/sq.m	19	10--12	2003--4
Synthetic filament yarn	17--23	5	2004
Woven synthetic fabrics	36	10	2005
Synthetic yarn	20--23	5	2004--5
Plastic impregnated textile fabrics	15--26	10--12	2002--4
Other knitted or crocheted fabrics	24--32	10--12	2004--5
Knitted or crocheted jerseys, pullover, cardigans, waistcoats and similar articles	28--33	14--16	2004--5
Knitted and crocheted clothing accessories	33	14	2005
WG suits, ensembles, dresses, trousers, etc. (woven)	31--36	16--25	2004--5
MB suits, ensembles, trousers, etc. (knitted)	31--36	16--25	2004--5
WG suits, ensembles, dresses, trousers, etc. (knitted)	31--36	14--25	2004--5
MB suits, ensembles, trousers, etc. (woven)	31--36	16--17.5	2004--5

(Source: [59])

2.34 China apparel imports from major markets in 1998

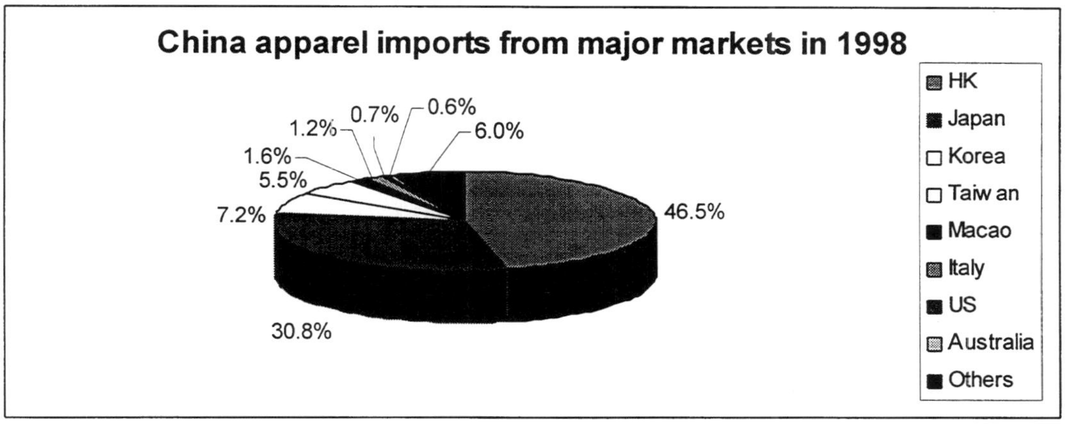

geographic proximity, available quality products and rich experience in trading with the mainland.

B. Apparel exports

Though the mainland is not the largest export market for HK apparel exports, it has become more and more important during recent years, rising from 1.1% of HK apparel domestic exports in 1985 to 22.0% in 1999. At the same time, HK is the largest export supplier to the mainland market, accounting for 46.5% of the mainland's total imports in 1998. (Figure 2.34) The huge population and rapid economic development in the mainland provide a golden opportunity for HK apparel manufacturers. The improved market access, including the significant cut in tariff rates (Table 2.21), will make exports to the mainland easier and more attractive. Consumers in the mainland market, especially in some big cities such as Beijing, Shanghai and Shenzhen, have developed a sense of fashion and brand awareness. HK, as a fashion center, has some regionally-famous brands which cater to the middle and up markets. Based on this argument, it can be expected that domestic apparel exports will enjoy an increase after China's WTO entry.

However, there is one thing worth attention: With full quota liberalization in 2005, the country of origin will definitely lose its current significance. As a result, HK textile and apparel manufacturers will have more incentives, based on the sharp comparison of the cost structure between HK and the mainland, to re-locate their low-value added production facilities such as apparel assembly lines to the mainland. Because of this, HK apparel domestic exports may experience a decline in the long run while textile exports may take the opposite direction due to the well-anticipated increased outward processing activities.

C. Re-exports

As an entrepot port, textile and apparel re-export volume via HK each year is very large. In 1998, about 71.4% of HK textile re-exports and 3.1% of HK apparel re-exports were destined for the mainland market. Based on TDC's survey of offshore trade, about 37.5% of textiles and 53.8% of apparel were sourced from the mainland. [60] In 1997, there were US$ 72.4 billion worth of textile and apparel exports from the mainland, of which 37% were re-exported via HK. Considering those destined for Japan, the US and EU markets, about 18.7%, 45.3% and 51.7% respectively were re-exported via HK. [47]

As regards China's WTO entry, there are two different opinions concerning HK's future role as an intermediary between the mainland and the outside world. One is optimistic, predicting that HK will acquire more opportunities with China's further integration into the world economy; the other is quite pessimistic, arguing that trade liberalization and improved market access on the mainland will overshadow HK's re-export performance. Both views have a certain ring of truth. China's WTO entry means more trade opportunities, particularly for labor-intensive products. Based on extensive trade channels, qualified personnel and a well-developed service sector, HK has competitive advantages in providing quality and prompt service for the mainland's small and medium enterprises, which lack the necessary expertise and skills in imports and exports. These are invaluable assets which can be maintained for a long time. Of course, with more enterprises on the mainland obtaining foreign trade rights, it is inevitable that some shadows will be cast upon HK re-exports. The key is whether HK manufacturers and traders can adjust quickly to the new competition picture and take positive steps to bring their competitive strength into full play.

2.5.3.2 Impact upon HK textile and apparel exports to other countries/regions

A. Comparison of HK and the mainland's textile and apparel exports between the two sides

- Major export markets

Both HK and the mainland's textile and apparel industriesy are export-oriented. Figures 2.19 and 2.35 show that the two sides are each other's most important textile export market and the US market is of importance for both sides. However, compared with their apparel export markets, the export markets for textiles are quite different and diversified for HK and the mainland.

As regards the two sides' apparel exports in 1998, the mainland was HK's third largest export market (if EU is taken as a whole) and HK was the mainland's leading export destination.[34] (Figures 2.28 and 2.36) Both the US and the EU are important markets for the two sides' apparel exports, though to various extents. More than 70% of HK domestic apparel exports went to the two markets in 1998, while the mainland

[34] But in 1999, Japan replaced HK as No.1 importer for the mainland's apparel exports.

exported less than 13% to the two[35]. Another difference is that Japan held a large share of the mainland's apparel exports, accounting for about one-fourth in 1998; but of HK apparel exports, Japan only imported 1.0%.

- Export products[36]

The proportion of textile exports to apparel ones is 1:2 in the mainland and 1:7 in HK, indicating that HK relies more on apparel exports. (See Table 2.20)

2.35 China textile exports to major markets in 1998

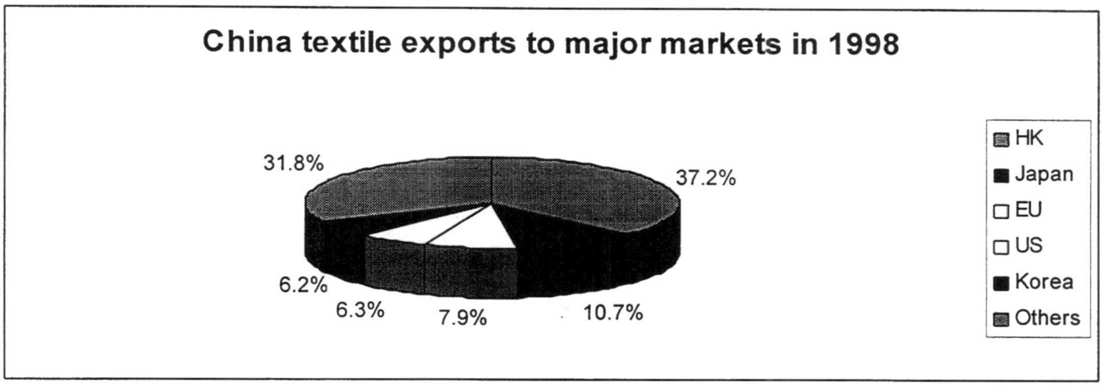

2.36 China apparel exports to the major markets in 1998

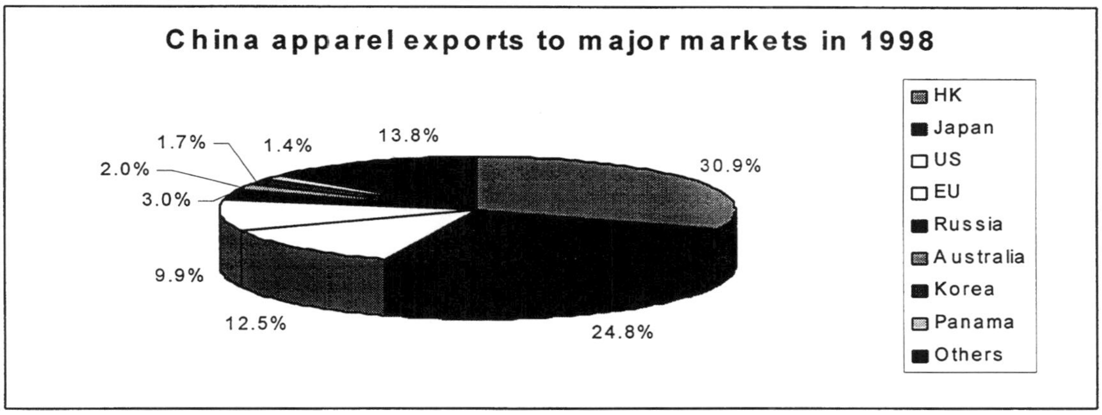

Table 2.22 Apparel exports from the mainland and HK during January to May 2000

Apparel exports from the mainland and HK by category

	China	HK
Total apparel and accessory exports	100.0%	100.0%
Knitted apparel & accessories	37.6%	43.4%
Woven apparel & accessories	54.2%	55.8%
Of which, accessories	7.9%	5.0%
Other apparel articles and accessories, other than textiles	8.3%	0.8%

(Source: calculated based on data from "HK External Trade, May 2000" and http://textile.unet.net.cn/imex05.htm, retrieved on August 1, 2000)

[35] One thing should be noted is that, of apparel exports from the mainland to HK, some are further re-exported to other markets including the US and EU, though the exact proportion is very difficult to determine due to lack of data support. Therefore, the actual apparel exports to the two markets are larger than the statistics' indicate.

[36] Proportions in this part are calculated based on data from "HK External Trade, May 2000" and "http://textile.unet.net.cn/imex05.htm". retrieved on August 1, 2000

As regards the mainland's textile exports, textile fabrics accounted for 82.8%, of which cotton and MMF woven fabrics held a relatively large share. In the first half of 2000, these two categories accounted for 25.1% and 35.6%[37] of the mainland's total textile exports. Textile fabrics from HK constituted 92.1% of HK total textile exports. Cotton woven fabric was the most important export category with 65.3% of HK total textile fabric exports. [76]

Woven apparel exports are larger than knitted ones on both sides. If apparel exports are divided by material, about 59.1% and 26.7% of HK total apparel exports during January to May, 2000 were apparel of cotton and man-made fiber; while the corresponding proportions in the mainland were 38.2% and 37.7% respectively in the first half of 2000.[38] Compared with HK apparel and accessories of non-textiles exports, China had a much larger share of 8.3%.

Knitted and woven apparel made from textile fabrics constituted the major export categories. (Table 2.22) Woven apparel exports are larger than knitted ones on both sides. If apparel exports are divided by material, about 59.1% and 26.7% of HK total apparel exports during January to May, 2000 were apparel of cotton and man-made fiber; while the corresponding proportions in the mainland were 38.2% and 37.7% respectively in the first half of 2000.[39] Compared with HK apparel and accessories of non-textiles exports, China had the much larger share at 8.3%.

- Export price

Since the US and the EU are key markets both for the mainland and for HK, it is necessary to compare the export prices of some hot export items from both sides. As regards US MFA textile and apparel imports in 1997, China was the largest supplier in value, with 11.2% of the market. HK ranked third with 7.6% of the market. The average import prices from the two were US$ 2.9 and US$ 4.8 per square meters equivalent, both higher than the world average level of US$ 2.4. [77] China was the fourth largest supplier of MFA textiles to the EU, accounting for 8.2% of total EU imports. Its far as EU MFA apparel imports in 1997 and 1998 are concerned, China was the No. 1 supplier both in terms of value and volume. Its export value reached 4.8 billion ECU in 1998. HK came third, with a value of 2.5 billion ECU in 1998. The average import prices were 17.2 ECU/kg for Chinese apparel exports, a little lower than the average import price (17.3 ECU/kg) from non-EU suppliers. Those from HK were 20.5 ECU/kg. [78]

Figures 2.37 to 2.39 show a comparison of some apparel categories, for each of which the two sides were suppliers of consequence. As regards US imports of the catogories mentioned in Figures 2.37 and 2.38, those from the mainland and HK were imported at prices higher than the world average level. Except for Categories 338 and 339 where the price differences between the two sides were relatively large, the two were among the same price-range suppliers for the other apparel categories. As to EU imports from the mainland and HK, import prices of the two sides were above the world average level except for Categories 6, 15 and 27[40]. The price differences between the two sides were small except for Category 26. (Figure 2.39)

[37] The definition for MMF-type fabrics is obscure in the mainland. According to the data retrieved from http://www.ctei.gov.cn/cteinew/haiguan/hghome.asp, MMF-type fabrics mentioned here actually refer to chemical (17.8%), synthetic (15.1%) and man-made (2.7%) ones.

[38] These two proportions are roughly calculated based on the data from "http://www.ctei.gov.cn/cteinew/haiguan/hghome.asp, retrieved on August 4, 2000

[39] These two proportions are roughly calculated based on the data from "http://www.ctei.gov.cn/cteinew/haiguan/hghome.asp, retrieved on August 4, 2000

[40] As regards Category 27 imports from the two sides, only those from HK were imported at a price lower than the world average level.

B. Possible impacts

Based on the above analysis, it is clear that competition between the mainland and HK will be much hotter after the quota liberalization process, especially in the apparel sector.

Chinese textile and apparel exports have been subject to very restrictive quota limits in the US and EU markets during the past two decades. Take Category 335 for example. In 2000, the quota imposed on Chinese exports to the US was 390,241 dozen with a mere 0.5% growth rate; while that on Turkey was 421,166 dozen with a growth rate of 8.7%. The tax equivalents for Chinese textile and apparel exports to North America were evaluated at 18.9% and 40.4% respectively; while those from Latin American countries were much lower, 9.6% for textiles and 20.2% for apparel. (Table 2.22) However, exports from the mainland have still managed to enlarge their market share in the restrictive importing countries. In 1983, MFA textile & apparel exports to the USA from China were 8.6% of total US imports, and those from HK were 19.2%. In 1997 the proportions had changed to 11.2% and 7.6% respectively [77] For MFA apparel exports from the mainland and HK to the EU, the comparison is more obvious. In 1989, those from the mainland held 8.3% of EU total MFA apparel exports from non-EU suppliers, ranking No. 3 and those from HK had 13.8%, ranking No. 1. But in 1998, the mainland replaced HK as the leading supplier to the EU, accounting for 13.1% of the market share and those from HK slumped to 6.9%. [78]

The different development track of the mainland's and HK's textile and apparel sectors indicates that the traditional comparative advantages have shifted to the mainland. In view of the export price comparison, though exports from HK as a whole have a higher export price than those from the mainland, the latter has made progress in product upgrading and has successfully raised prices for some items to a level higher than those from HK (Figures 2.37 to 2.39). The product differentiation between

2.37[41] US import prices for the mainland and HK in 1997 (1)

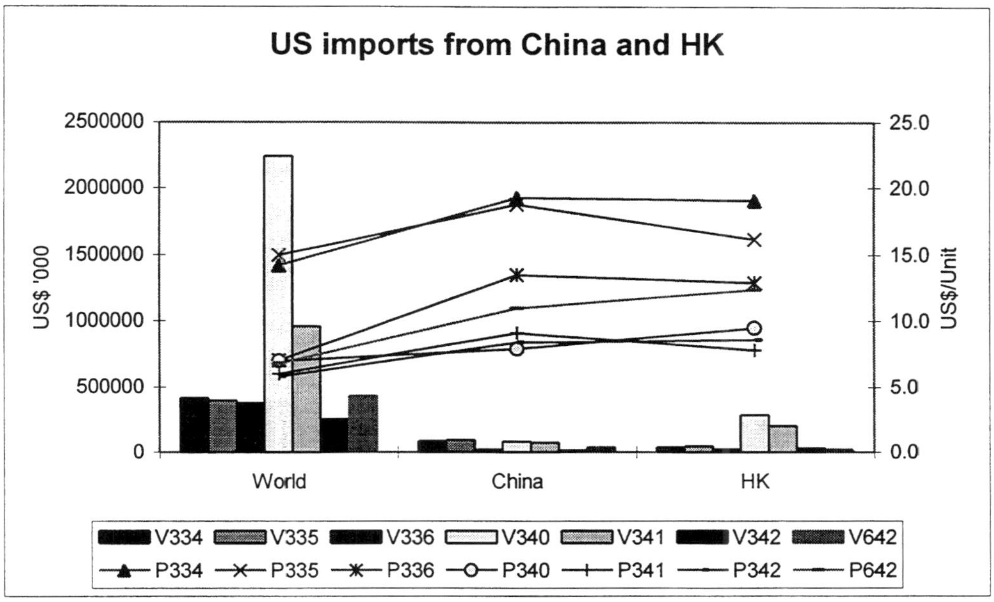

[41] Category 334: MB cotton coats; 335: WG cotton coats; 336: cotton dresses; 338: MB cotton knit shirts; 339: WG cotton knit shirts; 340: MB cotton non-knit shirts; 341: WG cotton non-knit shirts; 342: cotton skirts; 347: MB cotton trousers; 348: WG cotton trousers; 359: other cotton apparel; 435: WG wool coats; 642: MMF skirts

2.38[42] US import prices from the mainland and HK apparel exports (2)

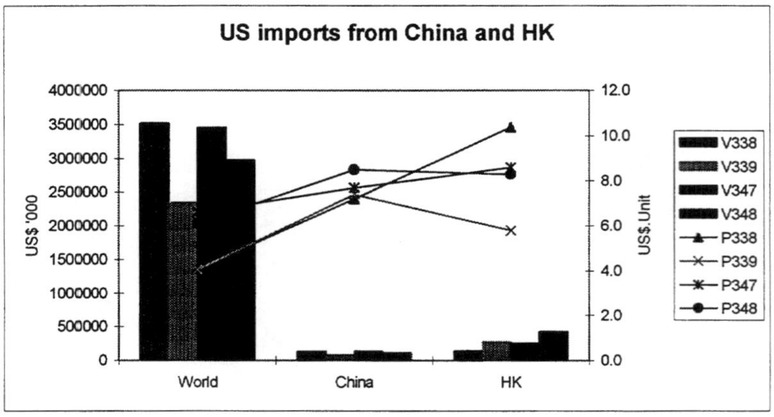

(Source: [77])

2.39 EU Import prices from the mainland and HK in 1998

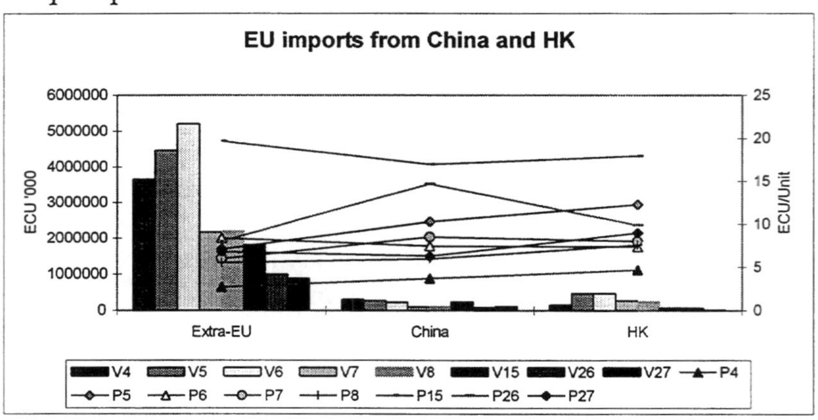

(Source: [78])

the two sides is not large enough for HK to avoid head-on competition with the mainland. What is more, as mentioned earlier, the quota elimination in 2005 may boost another round of re-location activities from HK to the mainland, a move which would contribute to HK textile and apparel industry's further decline. According to the USITC report, China filled most of its quota group limits by 95 to 100 percent from 1994 to 1998. Therefore, when the quota is fully phased out in 2005, textile and apparel exports from China, particularly the commodity products, will enjoy a relatively rapid growth in the US and EU markets. Since the HK domestic textile and apparel industry has relied more on the US and EU markets, sharper competition is inevitable. If HK fails to move upmarket and achieve more obvious product differentiation in the next few years, it will lose the edge in the competition. Generally speaking, HK apparel manufacturers will face more challenges than the textile manufacturers.

2.5.3.3 Impact upon HK's FDI in the mainland

HK has been the leading investor in the mainland for many years. It is estimated that about 80% of HK manufacturers have set up their own plants or processing trade arrangements with the mainland's counterparts [61]. HK's geographic proximity to the

[42] Category 4: T-shirts; 5: pullovers; 6: trousers; 7: Women's blouses; 8: Men's shirts; 15: Women's overcoats; 26: Women's dresses; 27: Women's skirts;

2.40 FDI in China textile and apparel industry

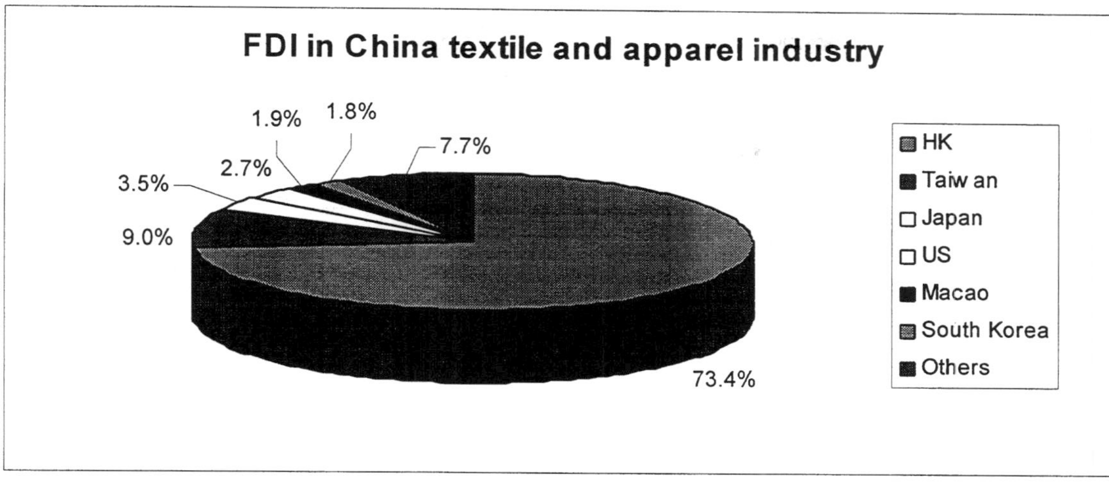

(Source: [68],)

Pearl River Delta plays an important role in its heavy investment in this region.

Figure 2.40 shows that HK investors hold a dominant share of 73.4% in the Chinese textile and apparel industry. Many HK apparel brands, particularly the casual wear ones, have established various retailing outlets in the mainland, most of which are specialty stores. These brands, such as Apple, Gold Lion, Crocodile, Bossini and Jeanswest, have won wide recognition and acceptance among the mainland's growing middle-class consumers and fashion-seeking youngsters.

Previous sections have mentioned the efforts made by China to liberalize foreign trading rights and the distribution system. According to the US-China bilateral WTO agreement, China is committed to gradually liberalizing the wholesaling and retailing of most commodities[43] within five years of accession. Quantitative, geographical, equity and incorporation restrictions on the establishment of joint ventures by foreign companies will be phased out. Moreover, requirements of foreign exchange balancing, local content and export performance will be eliminated as well. Trading rights will be progressively granted to foreign companies within three years.

Since the above commitment guarantees a more open and liberal market access, securing more investment opportunities for HK investors, China's entry into WTO has sent a positive signal to those who are interested in the mainland's huge domestic market. As Dr. Victor K. Fung, chairman of the HK Trade Development Council, pointed out, for the first time HK would have a "domestic market" of continental scale.[44] Generally speaking, the HK textile and apparel industry will grasp these opportunities to invest more in the mainland, especially when decision-making is no longer subject to quota and "rules of origin" considerations and when the huge mainland market is accessible in a real sense.

There may be negative effects upon HK re-exports. With the improved investment environment and trade liberalization drive in the mainland, HK companies may tend to establish foreign trading companies there, directly conducting exports from and imports to the mainland rather than re-exports via HK. Thus, increased capital flow to

---

[43] For example, wholesale distribution of salt and tobacco as well as retail of tobacco are not allowed even after China enters WTO

[44] This speech, with the title of "Hong Kong in China's WTO Era—Building a New Partnership", was delivered by Dr. Victor K. Fung at a luncheon following the 40th annual general meeting of the Hong Kong Federation of Industries. It was retrieved from http://www.tdctrade.com/tdcnews/0007/00070501.htm on July 25, 2000.

the mainland may be realized at the expense of shrinking re-export volume associated with the mainland. Another thing worth mentioning is that benefits enjoyed by HK investors after China's WTO entry are also open to other foreign investors. Therefore, the "big pie" will induce more fierce foreign competition, within which HK investors will feel great pressure.

2.5.3.4 Impact upon FDI in HK

In view of HK's special relationship with the mainland and its strength as a world sourcing center and "bridgehead" between the mainland and the outside world, HK is chosen by most MNCs as their regional headquarters, particularly as a base to tap the mainland market. A survey conducted by HKTDC shows that 83% of the surveyed MNCs have used HK as the springboard to boost and run their businesses in the mainland. [79]

Some people have argued that, In the future, more MNCs may prefer Shanghai as their regional headquarters instead of HK in the future, as the former is closer to the mainland market and enjoys lower cost of operation than the latter. However, according to a HKTDC report in January 2000, HK is still superior to Shanghai in terms of management levels, financial services, infrastructure, communication, and supporting services, all of which are key to investment considerations [62].

Therefore, China's WTO entry may bring about two different results concerning FDI in HK's textile and apparel sector: one an increasing trend and the other a decreasing trend. Both are dependent upon certain pre-conditions, such as the economic restructuring reform in the mainland and the sustainable edge of HK over the mainland in services and infrastructure.

In the short run, HK's role as a guide and springboard for small and middle MNCs will be strengthened because of its well-developed infrastructure, good investment environment, free market access, rich knowledge of the mainland market and easy communication with foreign partners. All of these advantages can be maintained for some time and overshadow other benefits enjoyed by direct investment in the mainland, such as low costs, direct access to the target market and better control.

2.5.3.5 Impact upon the current cooperation pattern between the two sides

As explained earlier, the cooperation pattern between the mainland and HK textile and apparel manufacturers can be characterized as "Front-store-back-factory", which is just a preliminary step and natural choice determined by the circumstances at that time. Economies in the mainland and HK are complementary in many aspects. The mainland has the labor, land, local distribution network, and solid scientific research bases while HK owns the capital, skills, overseas distribution network and necessary professionals in the service sector. However, this complementarity between the two sides has failed to realize synergy due to various limitations in past decades, such as less open market access in the mainland, quota restrictions in the export market and insufficient market information. Moreover, many of the mainland's manufacturers, which do not have foreign trade rights, are forced to go through extra and usually unnecessary links when promoting their products in foreign markets.

With China's accession into the WTO, there will be fundamental changes in the whole picture and, as a result, the cooperation pattern will undergo adjustment as well. The mainland's enterprises will no longer feel satisfactory as "back factories" for HK companies, while the latter have long been attracted by the lucrative mainland market. Consequently, a new kind of strategic partnership, on the basis of the present labor division, will be formed between the two sides' textile and apparel sectors.

First, closer R & D cooperation between HK and the Chinese mainland. Research institutes in the mainland often meet problems of how to commercialize their scientific and technological findings. Knowledge in the textile and apparel sector is of a commercial nature. For example, a fashion design, though excellent from a purely artistic view, may prove to be a market failure because it doesn't suit consumer tastes. The market success of a certain product range is not just subject to technological innovation. It also requires elaborate marketing and promotional strategies. HK enterprises are more experienced in these aspects. If the two sides can work more closely in this aspect, synergy could be generated to benefit enterprises in both sides

Second, HK textile and apparel enterprises should be more specialized. For basic items requiring low capital input, little product innovation and mass production, such as low-/middle-end men's wear and low-end women's wear, mainland enterprises have an obvious cost advantage over their HK competitors, either in the mainland market or in the export market. Therefore, HK players should completely give up these sectors and shift to manufacturing and marketing fashion wear and middle-/up-market women's wear, together with high quality fabrics. Efficient merger with the mainland's enterprises can not only realize the economies of scale, but also bring each other's strength to full play.

Third, diversification of foreign investment to other parts of China in addition to the Pearl River Delta. It is estimated that about 95% of HK's outward processing activities are carried out in Guangdong Province [61], a region of higher labor and land cost compared with the vast hinterland in the mainland. China's WTO entry means that the old special economic zones will lose many privileges and the government will attach more priority to less developed areas in the central and western parts of the country.[45] Moreover, these regions are usually abundant in raw material supplies, such as cotton, wool, and leather. In 1998, cotton produced in Xinjiang accounted for 31% of the national output, and wool and cashmere output in the Northwest region represented 37% and 25% of the national output respectively. [63] However, these regions are relatively poor in further processing activities. With preferential policy support and abundant resource bases, HK textile and apparel enterprises can find good investment opportunities on this underdeveloped area. HK has benefited much from the mainland's open-up reform since 1978. This new government policy in the mainland provide HK textile and apparel enterprises with another development opportunity. HK should become an early mover in major inland cities such as Wuhan and Chengdu, where second-tier markets accounted for 60% of the mainland's imports in 1999.

2.5.3.6 Summary

The relationship between the textile and apparel industries of the two sides, which is different from those formed by artificial preferential arrangements, such as between the US and Mexico, and the EU and Turkey, is built on the basis of close political, cultural, economic and geographic links. To the HK textile and apparel industry, China's WTO entry means far more than quota elimination. Its impact is

[45] For example, some investment projects restricted in the coastal region are allowed to carry out in central and west regions, with an open domestic market or a relatively high domestic sale ratio; Projects with foreign investment over 25% can be regarded as FIE, enjoying corresponding preferences.

comprehensive, embracing both opportunities and challenges concerning structural adjustment, foreign trade and investment. HK textile manufacturers will gain more than apparel ones.

The removal of NTBs and reduction of import tariff rates provide good prospects for HK textile and apparel exports to the mainland. Moreover, the expected growth of apparel exports to the mainland with quota phase-out indicates growing demand for imported textile items. Apparel production facilities in HK will be further re-located to the mainland, more good news to HK textile manufacturers. Consumers in the mainland have somewhat similar fashion tastes to those in HK due to cultural and geographic proximity as well as imitation psychology. Compared with some world top brands, HK brands are priced at a level more acceptable to the mainland consumers. These factors will lead to the increasing apparel exports to the mainland.

However, HK textile and apparel exports to other markets, such as the US and EU, will meet great challenges from mainland competitors, especially after 2005 when all the exporters have to compete in a quotaless environment. The comparison of export prices of some popular apparel items shows that the gap between the two sides is not large enough to avoid head-on competition. With trade liberalization and increasing quality and brand awareness in the mainland, apparel exports from the mainland will definitely pose a great threat to HK apparel manufacturers. HK apparel domestic exports to the outside world except to the mainland, will continue to shrink. As regards HK textile exports, the prospect is relatively not so serious since HK still enjoys obvious advantages in quality fabric production, which is the largest textile export category on both sides. The dyeing and finishing sector in China remains the bottleneck. As a result, a large proportion of the grey fabric produced on the mainland has to be exported, mainly to HK, for dyeing and finishing and then re-imported by China. Therefore, if China fails to solve the problem, its textile exports will still tend to be low value-added basic items.

HK's economic success in the past 20 years has been partly built upon the economic opening of the mainland in 1978. The economic gap between HK and the mainland promoted HK both as the largest investor and as the key intermediary agent between the mainland and the outside world. As many view China's WTO entry as the second economic opening, HK will no doubt grasp this opportunity to reap more fruits on a much bigger scale and scope. The mainland will be more than a manufacturing partner of HK. With the new game rules in a new competition picture, HK textile and apparel manufacturers will move beyond the former "front-store-back-factory" pattern and form strategic alliances with mainland partners in various aspects, including R & D and domestic/overseas distribution. Investment from HK in the textile and apparel sectors will not only increase, but also be diversified to more fields and geographic locations.

Past experiences have proved that HK companies have the strength to quickly adjust themselves to a new environment. China's WTO entry renders more opportunities than threats for the HK textile and apparel industry. The most significant thing is that this event facilitates structural adjustment of this sector, resulting in more efficient resource allocation. Job losses due to facilities re-location will be compensated by those created in the service sector. HK is most likely to be a typical sourcing center and logistics base, with a small but competitive textile sector and an apparel sector featuring high fashion and highly concentrated upon "two-ends"--design and distribution.

2.6 IMPACT UPON WORLD TEXTILE AND APPAREL TRADE AFTER THE QUOTA PHASE-OUT

The textile and apparel sector is one of the sensitive issues in China's WTO application. The major importing countries, such as the US and EU, are afraid of a possible import surge of textile and apparel items from China after it becomes a WTO member who can enjoy full ATC benefits. Various researches have been conducted to study the ultimate effects upon individual countries and the world as well. Though the results are not identical because of different analytical methods and focus, one view is common--China's textile and apparel sector will be one of the largest beneficiaries of China's WTO entry.

In China, opinions on China's entry into WTO are divergent. Some economists argue that WTO membership of China will boost GDP growth while opponents warn that a premature participation in globalization would not be good for China's national security until it has developed the same level of economic sophistication. However, as regards individual sectors, almost all public opinions favor the textile and apparel industry, considering it as one of the biggest winners after China enters WTO. According to an article in the official CTEI web site [15], Chinese apparel exports to the US and EU will grow by US$ 11 billion per annum after the quota liberalization, of which about US$ 5 billion is the gain from competition with other foreign suppliers and about US$ 6 billion is squeezed from domestic industries in these countries.

The US domestic market is one of the major export destinations for China's textile and apparel products. The US textile and apparel industry, which has been protected by various bilateral and multilateral agreements for more than 4 decades, views China's WTO entry as a big threat given its abundant raw materials, low-cost labor force and increased efficiency and quality. Various government agencies in the US have conducted studies to investigate the possible consequences of China's entry, particularly the impact upon its domestic textile and apparel industry. In the view of the report released by the US International Trade Commission in September 1999 [47], if China eliminates all quotas in 2005, it would gain a slightly expanded share of about 11% of the US textile import market and a whopping 30% of the apparel market. At the same time, China's world market share for apparel would rise by more than 6 percentage points in 2006. Another study conducted by Nathan Associates Inc., for the American Textile Manufacturers Institute (ATMI) drew a similar conclusion, predicting that apparel imports from China would capture 31% of US total apparel imports and that 154,500 jobs would disappear.[46] An analysis by the Ford Fund points out that Chinese textile exports will increase by 60% while and apparel exports by 200%, with the textile and apparel output growing by 25% and 74% respectively and the creation of 5 million new job opportunities.

The quota phase-out issue is not only causing upset in the developed nations, but also concern in some developing nations whose textile and apparel exports to the major importing countries have been guaranteed by the MFA regime and other preferential arrangements. Young, Martin and Yanagishima analyze the effects of quota liberalization upon the major textile and apparel exporting nations/regions, predicting that China will gain substantially at the expense not only of developed nations, but also developing ones subject to fewer MFA restrictions, such as Latin America, Sub-Saharan Africa and North Africa, particularly regarding apparel exports. [80]

[46] http://www.atmi.org/NewsRoom/releases/pr199932.html, retrieved on November 5, 1999

Based upon the above study results and analyses of China's WTO entry upon the diamond model in the previous section, several conclusions can be drawn concerning China's textile and apparel exports.

First, the export growth will not be strong before 2005, for reasons mentioned earlier. (Section 2.4.3.1) Therefore, the impact upon the world textile and apparel trade will be small and the present order will be maintained without significant changes.

Second, China's apparel exports will gain momentum after 2005 based upon the following arguments:

--Apparel exports from China have been subject to very restrictive discriminations under the MFA regime, including relatively low quota allocations compared with its large production and export potential, and much higher tax equivalents of MFA quotas compared with other regions. (Table 2.22) The year 2005 promises a quota-free competition environment in which apparel exports from China will be more competitive when competing with other low-cost suppliers.

--Though the US and EU are the major importers for China's apparel exports, they are not dominant in the whole picture. Compared with those suppliers in Latin America or Africa whose exports are highly concentrated upon these two markets, China's apparel exports are in an advantageous position, which is another plus after 2005.

--The reform of the foreign trade regime as well as the alleviation of rent-seeking problems caused by the quota allocation scheme will lead to increased economic efficiency and better resource allocations. As a result, apparel production and exports will be more market-oriented and efficiency-based, greatly enhancing the competitiveness of domestic apparel exports.

--When China becomes a WTO member, it will enjoy other countries' improved market access, especially those of developing nations. Since China's apparel exports are competitive in the low- and middle-end markets, they will find more outlets other than traditional export markets.

--China's WTO entry also provides its exports with a more stable external environment in which dispute settlement mechanisms will at least control the usage of unilateral measures by the major importing countries, such as the US.

Third, the future for China's textile exports is not so optimistic as for its apparel exports. Problems in this sector are more serious and hard to resolve in a short time frame. Compared with apparel production, price elasticity for textile goods is much lower and the technological gap between textile exports from developing and

Table 2.23 Comparison of tax equivalents of MFA quotas among major exporters[47]

Exporter	Tax equivalents of MFA quotas, % of FOB: Textiles		Apparel	
	North America	EU	North America	EU
HK/Korea/Taiwan	8.8	10.3	18.9	16.3
ASEAN	12.0	16.0	36.5	38.2
China	18.9	27.4	40.4	36.1
South Asia	19.0	27.4	40.4	36.1
Latin America	9.6	14.2	20.2	17.6
ROW	4.6	6.2	8.6	8.7

(Source: [80])

[47] ASEAN here include Indonesia, Malaysia, Philippines, Singapore and Thailand; ROW refers to Sub-Saharan Africa, Middle East, North Africa, Economies of Transitions and the rest of the world.

developed nations is too large to be narrowed in the near future. Some key production links, such as MMF and dyeing and finishing, are still weak at the present stage. Moreover, domestic apparel producers may turn to using more imported fabrics, which may become cheaper than domestic ones after significant tariff reduction. Obsolete textile machinery, lack of talented people and debt-ridden textile enterprises are another three bottlenecks for the sector's future development. Therefore, though quotas will be freed in 2005, it is not good enough to help solve these deep-rooted problems, which involve a more comprehensive reform package, such as the SOE issue, the educational system and the capital market.

Fourth, though China's textile and apparel exports are expected to gain from China's WTO entry, several conditions may have negative effects upon this scenario:

--The diamond upgrading process in China itself is an issue of overwhelming importance. If the result is disappointing, then the export prospects will be lusterless even in a more favorable external environment. Based on a World Bank report [81], between 1986 and 1987, the comparative advantage indexes of the apparel sectors in India and the Philippines were 79 and 76 respectively, given that the index in China was 100. However, between 1992 and 1994, the indexes in the former two nations rose to 94 and 84 respectively, considerably narrowing their gap with China. If China fails to move up the value-added ladder, pressure from these competitors will be sharper.

--China's textile and apparel exports will continue to be subject to possible quota imposition for additional 4 years until the year 2009 and product-specific safeguards for an additional 12 years after its WTO entry. (US-China WTO agreement) These will the potential weapons against China's textile and apparel exports in the future.

--Enhanced regionalism. Despite the political factors of today's regionalism trend, economic considerations play a key role, too. As regards textile and apparel production and trade, quick response is one of the underlying reasons for the current closer cooperation between border countries. Competition has extended to many non-price factors, of which geographic proximity, demand-oriented production and prompt delivery have become more significant. In this sense, China's relatively long distance from the US and EU markets will be a disadvantage in foreign partners' sourcing decisions.

--Entry of more low-cost suppliers. Without quota limitation, any textile and apparel supplier can export in theory. Moreover, the US and EU, for various political and economic reasons, offer preferential policies support to help some least developed countries to establish a textile and apparel sector as a preliminary step for further economic development, something like the 807A arrangement. Though these late movers don't pose a material threat to China's textile and apparel exports at present, their growth, backed by preferential Customs entry and close cooperation with manufacturers in the US and EU markets, may successfully secure a certain market share from traditional exporters, including China.

All in all, the quota phase-out arrangement completed in 2005 will be likely to trigger drastic changes in the world textile and apparel trade pattern. China, as the world No. 1 textile and apparel exporter, will be one of the beneficiaries after it becomes a formal WTO member. China's WTO entry will have the most significant impact upon the apparel sector. Apparel manufacturers in the developed nations as well as some developing nations may be negatively affected due to the increase of apparel exports from China. Corresponding adjustment processes and new policy measures will interact together to establish a new order after 2005.

REFERENCES

1. Kilduff, P., Strategic development prospect for Taiwan man-made fiber textile and apparel industry (in Chinese). Taiwan Textile Monthly, 1998, p.6-15.
2. Porter, M.E., On competition.1998, Boston, Harvard Business School Publishing.1-271.
3. Xun Chen, Y.B., China after its WTO entry: from nation to nationals, the challenges and opportunities for China in the 21st century.2000, Gan Su, China, Gan Su Ren Ming Press.144-166.
4. Corporation, S.T.H.G., Human resource is the key to the textile industry--development plan of Shanghai textile human resources from 1999 to 2001 (in Chinese). Shanghai Textile, 1999, p.4.
5. Anon, World cotton production.2000, CTEI.
6. Gu, Q., Effects on Chinese textile industry after its WTO entry and corresponding measures (II), (in Chinese).2000, China Textile Industry Association, China Textile Information Center.
7. Singleton, J., The world textile industry.1997,London and New Yord, Routledge.
8. Association, C.G., Self-sufficiency rate of domestic fabrics in apparel production (in Chinese).1999.
9. Lu, Q., Some reflections on China's garment processing trade (in Chinese). CHINA TEXTILE & APPAREL, 1999, p.20-23.
10. Xiang, X., Analysis of the cause of loss and deficit of national textile industry (in Chinese). China Textile Economy, 1999, p.18.
11. Wang, C., Textile Industry Gets Slimmer, in, China Daily.1999.
12. Liu, B., China's textile industry to blaze state-owned enterprises reform trail (in Chinese). CHINA TEXTILE & APPAREL, 1998, p.16-21.
13. Ye, Y., Imports of Machinery in Textile Sector Brake, in, China Daily.1999.
14. Lu, Q., A challenge for China. ATA Journal, 1996, p.22-23.
15. Gu, Q., Effects on Chinese textile industry after its WTO entry and corresponding measures (III), (in Chinese).2000, China Textile Industry Association, China Textile Information Center.
16. HKTDC, R.D., Trade Guide--Guide for HK business people to expand apparel exports (in Chinese).1999, HKTDC, HK.p.77-83.
17. HKTDC, R.D., Wholesale and retail distribution of garments in China.1995, HKTDC, HK.p.14-22.
18. HKTDC, R.D., Brand competition in China apparel market.1999, HKTDC, HK.p.1-14.
19. Anon, What is the future for the silk garment industry, (in Chinese).1999.
20. Finnie, Profile of Levi Strauss. Textile Outlook International, 1996, p.10-37.
21. Wang, L., The development and prospect of textile product processing and trade (in Chinese). China Textile Economy, 1999, p.5-9.
22. Dempsey, E., China: CITME reflects a changing world (In Chinese). ITB Yarn and Fabric Forming, 1996,p.50-57.
23. Chen, Y., Chinese Textile Machinery: global-scale production and fast-growing exports. Textile Month, 1998,p.42-45.
24. Anon, Strategy for China-made textile machinery. Textile Asia, 1999,p.4-6.
25. Anon, Machines booming. Textile Asia, 2000, p.69.
26. Anon, Machine sales reduced. Textile Asia, 2000 ,p.9.

27. Anon, German textile machinery producers remain top in difficult year. CHINA TEXTILE & APPAREL, 2000,p.24.
28. Leung, P., China--not good enough. Textile Asia, 2000, p.62.
29. Anon, German machine makers. Textile Asia, 1998, p.107.
30. Charlin, J.-C., A dynamic sector of France. Textile Asia, 1999, p.31.
31. Coster, J.d., United they stand. Textile Asia, 1999, p.99-100.
32. Steinfeld, E.S., Forging reform in China: the fate of state-owned industry.1998, Cambridge University Press.3.
33. Anon, Special report. ATA Journal, 1998, p.40.
34. Anon, China market situation. ATA Journal, 1999, p.30.
35. HKTDC, Textiles and clothing markets brace for major changes.2000, HKTDC.
36. Leung, P., More quota allotted. Textile Asia, 2000, p.48.
37. Leung, P., Quota wasters debarred. Textile Asia, 2000, p.69.
38. Hughes, J.K., Import issues and textile policy in the US. Textile Asia, 1999, p.69-72.
39. Gong, J., Development trend of current world textile technology (in Chinese). 2000, CTEI.
40. Qian, J., Review on the US textile and apparel trade in 1997 (in Chinese). China Textile Economy, 1998, p.32-34.
41. Messura, M., Steady silk. Textile Asia, 1999, p.94.
42. Wang, L., Study on the "changes of two No. 1" (in Chinese). China Textile Economy, 1999, p.13.
43. Luke, J.E., Redefining an industry--US fiber manufacturers search for new end-uses in an effort to shield themselves from a sea of imports. ATI, 1999, p.104-105.
44. Anon, Highlights of the EU-China Agreement on WTO.2000, EU.
45. HKTDC, R.D., Service liberalization in China.1999, HKTDC, HK.p.29-32.
46. Franklin Dehousse, K.G., Philippe Vincent, Market access study to identify trade barriers affecting the EU textiles industry in certain third country market.1999, EU.
47. USITC, Assessment of the Economic effects on the US of China's Accession to the WTO.1999, USITC.
48. Leung, P., Machine imports depressed. Textile Asia, 2000, p.62-63.
49. Diekmann, H.E., Insight into the EU market. Textile Asia, 1999, p.80.
50. Anon, Effects on the textile and apparel trade due to WTO entry (II), (in Chinese).2000, The State Information Center, together with provincial and municipal information centers.
51. Su, J., Implement green management and develop green textiles--new hot point of Chinese textile industry in the 21st century (in Chinese). China Textile Economy, 1999, p.18.
52. Porter, M.E., Competitive strategy.1980, , London, Toronto, Sydney, Singapore, The Free Press.34-46.
53. Anon, Import and export trade.1999, HKTDC.
54. Anon, Market profile on Chinese Mainland.1999, HKTDC.
55. HKTDC, China's WTO entry and its impact upon HK (in Chinese).2000, TDC, HK.
56. Anon, Economic and trade information on Hong Kong.1999, HKTDC.
57. Anon, HK's manufacturing sector: textiles and clothing industries.1999.

58. Anon, Profiles of HK major manufacturing industries--HK's textiles industry.1999, HKTDC.
59. Anon, Who will benefit from China's WTO accession, .2000, HKTDC.
60. HKTDC, R.D., Impact of China's WTO accession: current status and implications for Hong Kong.1999, HKTDC, HK.
61. HKTDC, R.D., HK's vital role in China's reforms and world integration.1998, HKTDC, HK.
62. Research Department, H., China's WTO Accession and Implications for Hong Kong.2000, HKTDC, Hong Kong.p.22.
63. HKTDC, R.D., Economic development in the Northwest of China: Business opportunities for HK (in Chinese).2000, HKTDC, HK.p.i-8, 16-18.
64. Agency, E.I., ed. HK Economy Yearbook., Economic Information & Agency, HK.
65. Census and Statistics Department, H., ed. HK External Trade., HK.
66. Sri Ram Khanna, a.t.I.R.T., Trends in EU textile and clothing imports. Textile Outlook International, 2000, p.55-96.
67. Sri Ram Khanna, a.t.I.R.T., Trends in US textile and clothing imports. Textile Outlook International, 1998, p.100-129.
68. Wang, L., ed. Report on foreign direct investment in China--Industrial distribution of FDI (in Chinese). 1997, Economic Management Press.
69. Group, B.o.C., Realizing our "positioning" objective through China's WTO impetus.2000, HKTDC.
70. Yongzheng Yang, W.M., Koji Yanagishima, Evaluating the benefits of abolishing the MFA in the Uruguay Round Package, in, Global trade analysis: modeling and application, T.W. Hertel, Editor.1997, Cambridge University Press, Cambridge, New York.p.253-279.
71. World Bank., China Engaged: Integration with the Global Economy,.1997, World Bank.p.15.

3 US Denim Products Trade[1]

OVERVIEW

If Levi Strauss, who was born in Bavaria in 1829 and landed in the US in 1847, had failed to notice the miners' potential demand for strong and sturdy pants, then the denim would, probably, not have developed into a historic legend. In 1873, Levi Strauss made up the first jeans in San Francisco for California miners from a heavy brown canvas. The 1960s and 70s witnessed the rapid development of denim production, rising from 270 million square yards in 1968 to approximately 700 million in 1977 [1]. With continuous technological innovation and product development since the 1980s, the denim family has grown larger, covering tops and bottoms as well. And denim apparel is now viewed less as workwear than as a fashionable item by current consumers.

Today, the US, which is the home of denim and jeans, is the most important denim market in the world. In 1999, the entire US denim market amounted to US$ 17 billion, with jeans accounted for 63% of the total market. Denim is the biggest consumer of US cotton with about 25% of the share [4]. According to a survey on world denim production [2], the US accounted for 41.7% of the global denim fabric consumption, estimated to be 3.0 million linear meters in 1996. Brands such as Levi's, Wrangler, and Lee have dominated the world denim market. The average American has seven pairs of denim jeans and nine other denim items [3]. Jeans are the largest segment of the apparel industry with annual unit sales of approximately 500 million units, or average two pairs of jeans per US citizen [5]. In the view of Cotton Inc., US retail sales of denim has enjoyed continuous growth since 1990 at the average annual rate of 10.8% [3].

The agreement between the Levi Strauss company and the American Ministry of Defense to manufacture the off-duty uniform of the US Navy Marines has, to some extent, promoted the spread of Jeans worldwide. American surplus stores began to sell jeans unused by American soldiers after WWII, starting the expansion of American brands in other continents and direct exports as well.

With the entry of more and more new competitors, the US began to import more denim products due to the comparative advantages in most developing countries. The trading volume has increased in the past few years, creating an enlarging trading deficit which was US$ 886 million in 1995 and up to US$ 1.1 billion in 1997, as shown in Figure 3.1.

Men's or boy's denim apparel has been the most important category of US imported or exported denim products, and occupied one half of US total imported

---

[1] All the data in this section was from STAT-USA, US Department of Commerce, the access time was December 1998 to March 1999.

denim products and 46.0% of US total exports. Female's denim apparel was not as important as men's or boy's denim apparel in both US imports or exports. Denim fabric had a small share in US imports but was a major category of US exports. US imported a lot of denim yarns, but exported no denim yarns in recent years.

Denim product imports were mainly from American and Asian countries. The major export destinations for US denim products were American and European countries. Mexico was the most important supplier and consumer of US denim products. American countries have emerged as very important trading partners in US denim products trading during recent years.

3.1 US denim products trading gap

3.2 USA total denim products

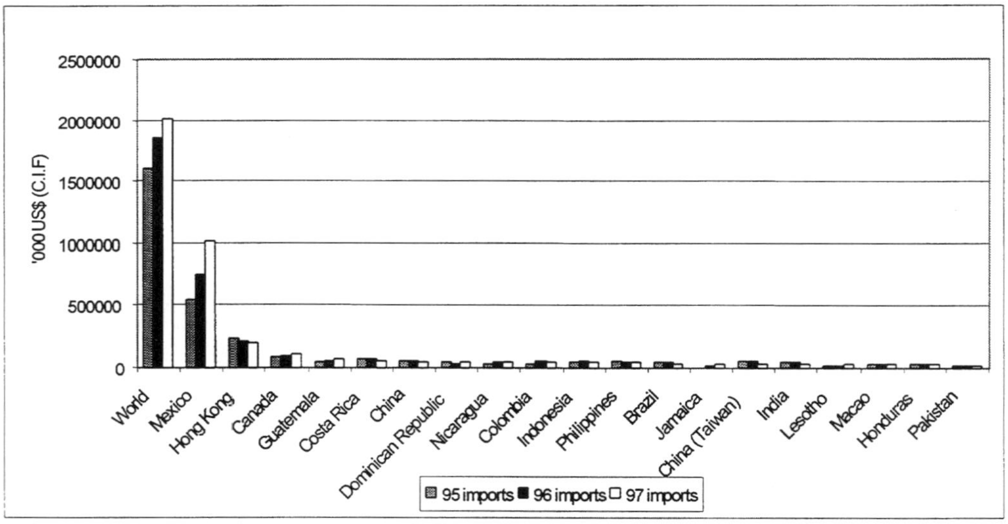

Table 3.1 USA denim products imports by categories

	USA imports of denim products (C.I.F., '000$)									
	1995	1996	1997	Jan-Nov 97	Jan-Nov 98	1995%	1996%	1997%	Jan-Nov 97%	Jan-Nov 98%
Denim Yarn	321011	351294	313557	289885	261284	20.00%	18.95%	15.65%	15.61%	11.25%
Denim Fabric	6	7157	9137	7830	31109	0.00%	0.39%	0.46%	0.42%	1.34%
Men's or boy's denim	791747	928218	993132	923747	1109327	49.33%	50.07%	49.56%	49.75%	47.76%
Women's or girl's denim	492203	567005	687965	635193	920943	30.67%	30.59%	34.33%	34.21%	39.65%
Total denim products	1604967	1853674	2003791	1856655	2322663	100.00%	100.00%	100.00%	100.00%	100.00%

3.3 US denim products imported by region

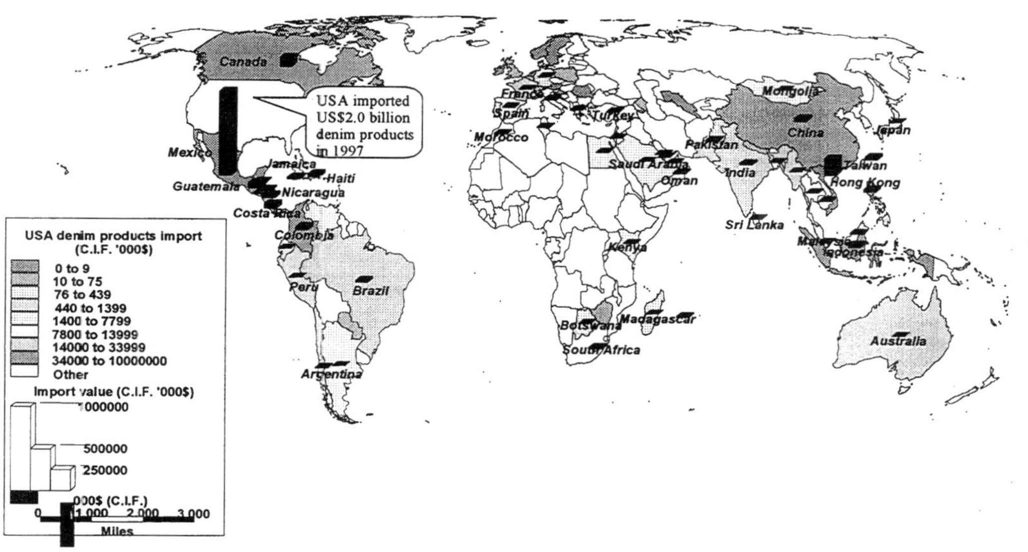

3.1 IMPORTS

3.1.1 Denim products imports

The import value of denim products has been on a continuous rise from 1995 to 1997 (Figure 3.2), growing from US$ 1.6 billion in 1995[1] to US$ 1.8 billion in 1996 and US$ 2.0 billion in 1997. The average annual growth rate was 12.4% from 1995 to 1997. The year 1998 registered a sharper increase; In the first eleven months of that year, the import value reached US$ 2.3 billion, up by 25.1% comparing with the same period in 1997.

As shown in Figure 3.3, US denim products imports were mainly from American and Asian countries or areas. The most important supplier was Mexico, which supplied one third of the total imports in 1995. Between 1995 and 1997, Mexico's denim products exports to the US increased by 42.8% per year, mainly due to NAFTA and currency devaluation. Mexico contributed 52.3% of the US denim products import market from January to November 1998.

Hong Kong, one of the world leading textile and apparel exporters, was the second most important supplier after Mexico. Hong Kong accounted for 11.6% of US imports in 1995. However, Hong Kong's share has decreased in the past few years. Its market share became just less than 10% in 1997. The market size of Hong Kong

[1] The import value is calculated on a CIF basis.

denim products in the US shrank by 6.8% per year between 1995 and 1997. From January to November in 1998, Hong Kong accounted for 9.4% of US denim products imports.

Canada was the third most important supplier. Canada exported US$ 78.7 million worth of denim products to the US in 1995. In 1997, the value increased to US$ 104.3 million, the average growth rate being 16.3% per year. Canada held 5.2% of US denim products imports in 1997.

Among other countries, Guatemala provided 2.8% of US denim product imports in 1997, Costa Rica (2.7%), China (1.9%), Dominican Republic (1.9%), Nicaragua (1.9%), Colombia (1.9%), Indonesia (1.7%), Philippines (1.7%), Brazil (1.3%), Jamaica (1.2%), China (Taiwan) (1.2%), India (1.2%), Lesotho (1.1%), Macao (1.1%), Honduras (1.0%) and Pakistan (1.0%). These nineteen countries or areas (including Mexico, Hong Kong and Canada) together accounted for 91.2% of US denim products imports.

The suppliers of US denim products can be divided into two groups. One group consists of American countries, such as Mexico, Canada, Guatemala, Costa Rica, Dominican Republic, Nicaragua, Colombia, Brazil, Jamaica, Lesotho and Honduras. Another group is comprised of non-American countries or regions, such as Hong Kong, China, Indonesia, Philippines, China (Taiwan), India, Macao and Pakistan. In recent years, the market share of the first group has increased quickly at the expense of that of the non-American countries except for Pakistan and Indonesia.

Table 3.1 lists US denim products import categories and their value from 1995 to November 1998. About one half were men's or boy's denim apparel and about one third were women's or girl's denim apparel. In 1995, one fifth of denim products imports were denim yarn, while this figure dropped to one sixth in 1997. The US did not import much denim fabric between 1995 and 1997.

3.1.2 Denim yarn imports

As shown in Fig 2.5, the US imported US$ 321 million worth of denim yarn in 1995, peaking in 1996 with US$ 351 million and dropping to US$ 314 million in 1997. Mexico has been the most important supplier since 1996.

In 1995, Mexico contributed 10.2% of US denim yarn imports as the fifth most important denim yarn supplier. Increasing by 172.3% each year, its export value topped US$ 146 million, accounting for 46.5% of US denim yarn imports in 1997.

Canada was the second most important foreign supplier to the US. It exported US$ 38.8 million worth of denim yarns to US in 1995. With an annual growth rate of 9.1%, the export value reached US$ 45.9 (14.6% of US imported denim yarn) in 1997.

Hong Kong used to be the most important overseas supplier for US denim yarn imports before 1995. It exported US$ 82.6 million worth of denim yarn to US in 1995, accounting for 25.7% of US denim yarn imports. However, with a reduction rate of 23.0% per year, Hong Kong exported US$ only 44.6 million worth of denim yarn to the US in 1997, or 14.2% of US denim yarn imports. It is now the No. 3 foreign supplier to the US market.

India exported US$ 33.3 million of denim yarn to US in 1995, accounting for 10.4% of US imports. However, with an annual decreasing rate of 21.1%, India just held 4.4% of US imports in 1997.

3.4 US denim yarn imports by region

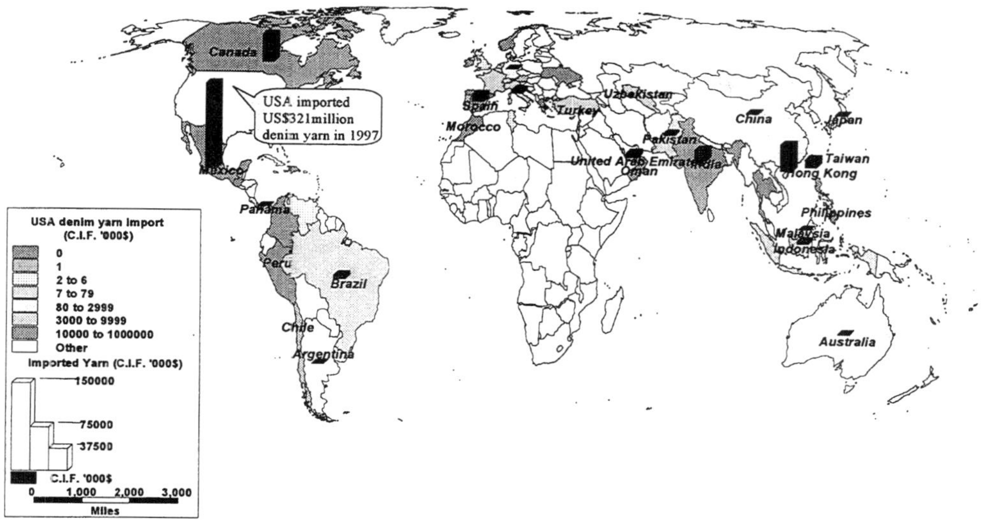

3.5 US denim yarn imports

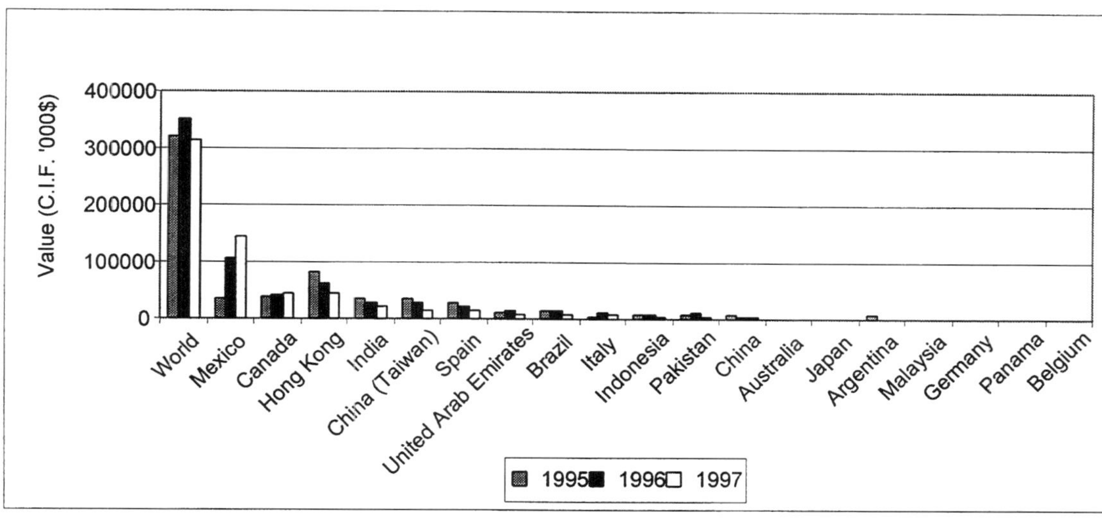

China (Taiwan) and Spain each accounted for approximately 4% of US imports in 1997, a sharp contrast with their 1995 figures, which were 10.3% and 8.6% respectively.

United Arab Emirates and Brazil exported US$ 7.8 million and US$ 6.5 million worth of denim yarns to US in 1997. Their market shares have also experienced a downward trend in the last few years. United Arab Emirates' market share fell from 3.6% in 1995 to 2.5% in 1997, while that of Brazil reduced from 4.7% in 1995 to 2.1% in 1997.

Italy provided 1.7% of US denim yarn imports in 1997. Indonesia accounted for 1.4%. Other countries, such as Pakistan, China, Australia, Japan, Argentina, Malaysia, Germany, Panama and Belgium, each contributed less than 1% of the US denim yarn imports.

Among these suppliers, Mexico experienced the fastest growth, followed by Canada. Other countries' market shares reduced sharply. Their average depreciation rates reached two digits annually.

The higher value added denim yarns imported into US were mainly from developed countries, such as Canada (US$ 2.6/ton), Italy (US$ 3.3/ton), Japan (US$ 3.7/ton), and Germany (US$ 3.1/ton). The low priced denim yarns came from Hong Kong (US$ 1.7/ton), China Taiwan (USD 1.3/ton), Brazil (US$ 1.8/ton), Indonesia (US$ 1.4/ton), Pakistan (US$ 1.3/ton) and China (US$ 1.4/ton).

3.1.3 Denim fabric imports

In 1996, the US imported US$ 7.2 million of denim fabric, and this increased to US$ 9.1 million in 1997. Before 1996, US denim fabric imports were negligible. However, in the first eleven months of 1998, US imported US$ 31.1 million worth of denim fabrics, an increase of 297% on a year-to-year basis.

Mexico was the most important exporter to the US denim fabric market (Figures 3.6 and 3.7). Mexico exported US$ 1.2 million and US$ 4.3 million worth of denim fabric to US in 1995 and 1997, hitting US$ 22.4 million from January to November in 1998, an increase of 578% on a year-to-year basis. Its market share grew from 16.8% in 1996 to 46.5% in 1997 with an average export price of US$ 2.0 per square meter.

Hong Kong was the second most important denim fabric supplier to the US. It exported US$ 3.4 million worth of denim fabric to US in 1996 and US$ 2.1 million in 1997, a reduction of 19.6%. As a result, its market share slumped from 47.9% in 1996 to 22.8% in 1997 and further dropped to 7.7% in the first eleven months of 1998.

India was the third most important denim fabric supplier. It exported US$ 0.5 million worth of denim fabric in 1996, accounting for 7.4% of US total denim fabric imports. In 1997, it enjoyed a favorable growth with an export value of US$ 0.9 million, occupying 10.2% of US total denim fabric imports.

Mexico, Hong Kong and India together contributed 79.6% of US total denim fabric imports.

The average US import price of denim fabric was around US$ 2 per square meter. The higher priced denim fabrics were from Italy (US$ 4~5 per square meter), Australia (US$ 4~6 per square meter), Spain (US$ 8~10 per square meter), Germany (US$ 4 per square meter) and Belgium (US$ 6~7 per square meter).

3.1.4 Men's or boys' denim apparel imports

Men's or boy's denim apparel was the most important product category of US denim products imports. Figures 3.8 and 3.9 show that US$ 792 million worth of men's or boy's denim apparel was imported into the US in 1995. With an annual growth rate of 12.7%, the import value reached US$ 993 million in 1997. Compared with the same period of 1997 between January and November, US imports of men's or boy's denim apparel increased by 16.4% in value, reaching US$ 1.1 billion during the first eleven months of 1998.

Mexico again was the most important supplier of men's or boy's denim apparel to the US. It exported US$ 328 worth of men's or boy's denim apparel to the US in 1995. With a growth rate of 23.6% per year, the export value reached US$ 483 in 1997. Its market share increased from 41.4% in 1995 to 51.5% in the first eleven months of 1998. Mexico has become the most important source of US men's or boy's denim apparel imports.

3.6 US denim fabric imports by region

3.7 US denim fabric imports

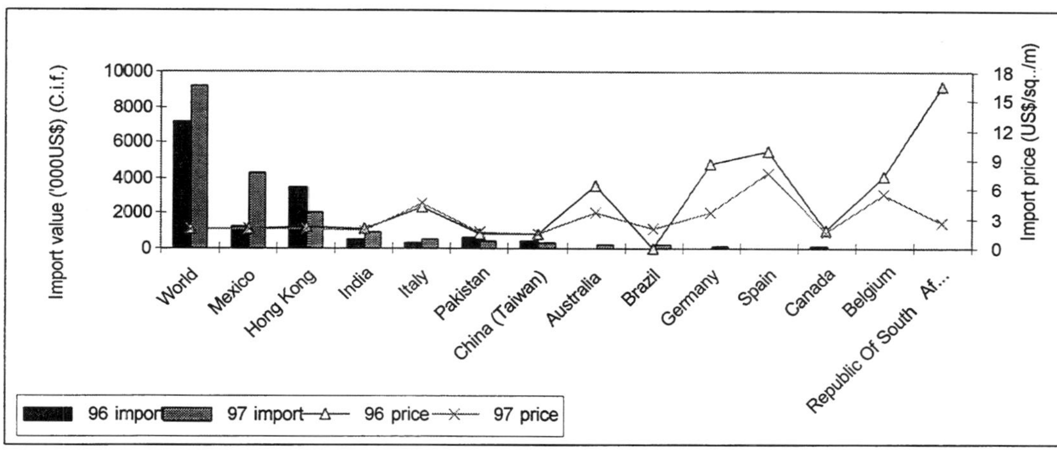

Hong Kong was the second most important supplier of US men's or boy's denim apparel imports. Based upon an average growth rate of 15.6% per year, its export value grew from US$ 49.6 million in 1995 to US$ 65.1 million in 1997, with an increasing market share from 6.3% to 6.6%.

Another important supplier was Guatemala, who exported US$ 57.0 worth of men's or boy's denim apparel to the US in 1997, accounting for 5.7% of US total imports. The average growth rate was 34.9% each year between 1995 and 1997.

In 1997, Colombia and Canada contributed 4.7% each to US man's or boy's denim apparel imports. The Colombian export value was US$ 20.8 million in 1995, peaking at US$ 50.7 million in 1996 and falling back to US$ 36.6 million in 1997. Canadian exports peaked in 1996 with US$ 43.7 million, accounting for 4.7% of US men's or boy's denim apparel imports. They reduced to US$ 36.2 million in 1997. In the first eleven months of 1998, they picked up some growth momentum, reaching US$ 38.8 million and accounting for 3.5% of US imports.

3.8 US men's or boy's denim apparel import

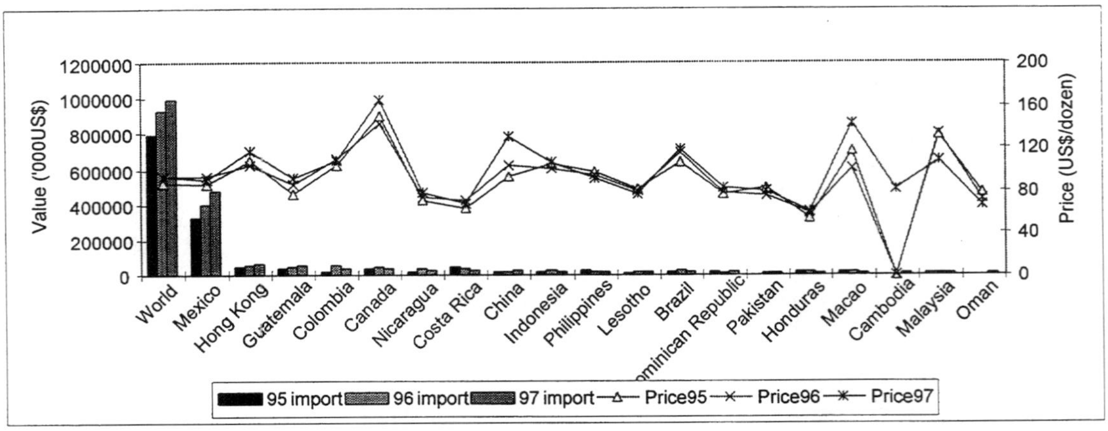

3.9 US men's or boy's denim apparel imports by region

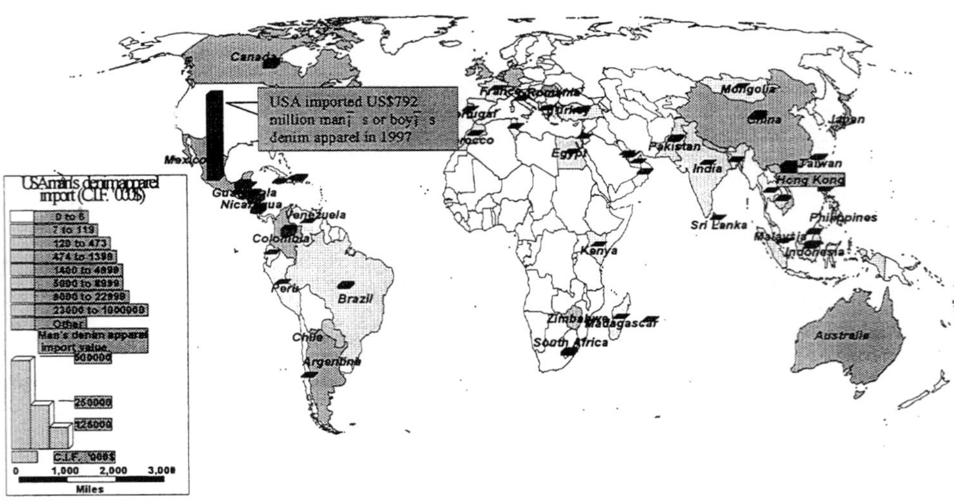

Nicaragua, Costa Rica and China exported US$ 28.4, 27.2 and 26.0 million of men's or boy's denim apparel, accounting for 2.5-2.9% of US imports in 1997. Nicaragua enjoyed a growth rate of 17.8% each year, and that of China was 7.2%. Costa Rica's exports, on the contrary, decreased at a rate of 14.2%.

Indonesia, Philippines, Lesotho and Brazil exported US$ 21.6, 20.0, 19.5 and 19.1 million of men's or boy's denim apparel respectively, accounting for 2.2-1.9% of US imports.

Dominican Republic, Pakistan, Honduras and Macao accounted for 1.5%, 1.3%, 1.3% and 1.2% of US imports in 1997 respectively. The market shares of the rest suppliers were less than 1% in 1997.

Men's denim apparel exports from Colombia experienced the fastest growth among the major suppliers (38.0% per year on average between 1995 and 1997), followed by Lesotho and Guatemala whose annual exports increased by 34.9% and 34.1% respectively. Except for Costa Rica, Philippines, Brazil and Canada whose export value fell by 21.4%, 5.8%, 3.4% and 0.1% per year on the average from 1995 to 1997, the rest of the major suppliers registered a positive growth.

3.10 US women's or girl's denim apparel imports

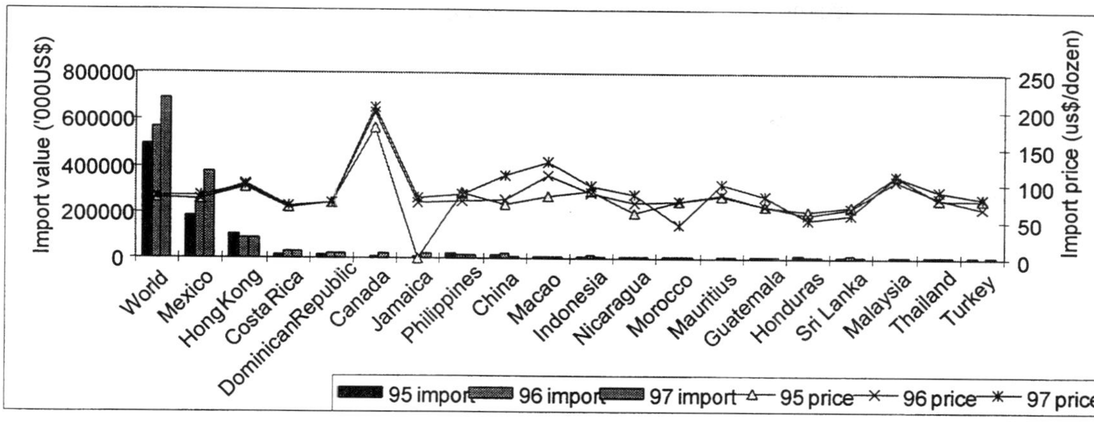

The average import price of man's or boy's denim apparel was US$ 87.9 per dozen in 1995, US$ 93.3 per dozen in 1996 and US$ 92.6 per dozen in 1997. Among the major suppliers, the average import price was close to Mexico's export price, which was US$ 84.2 per dozen in 1997, because of its significant market share in this import market. The export prices of Hong Kong, Colombia, Canada, China and Indonesia were higher than US$ 100 per dozen. Among these countries or regions, the Canadian export price was the highest (US$ 165.9 per dozen) in 1997, followed by China (US$ 131.0 per dozen) and Hong Kong (US$ 117.0 per dozen). Colombian and Indonesian export prices were US$ 107 per dozen. Other major suppliers' export prices were lower than the US average import price.

3.1.5 Women's or girls' denim apparel imports

Figure 3.10 shows that US women's or girl's denim apparel imports increased from US$ 492 million in 1995 to US$ 688 million in 1997. In the first eleven months of 1998, the import value reached US$ 921million.

Again, Mexico was the most important supplier of US women's or girl's denim apparel, exporting US$ 181 million worth to the US in 1995. With a growth rate of 53.1% per year, the export value topped US$ 373 million in 1997. In the first eleven months of 1998, Mexico exported US$ 512 million worth of women's or girl's denim apparel to US. It accounted for 36.8% and 54.3% of US imports in 1995 and 1997 respectively.

The second most important supplier was Hong Kong. In 1995, Hong Kong exported US$ 99.6 million worth of women's or girl's denim apparel, accounting for 20.2% of US imports. By sharp contrast to the rapid penetration of Mexican exports in the US market, Hong Kong's exports reduced to US$ 88.6 million in 1997, declining at an annual average rate of 5.6% with a share of 12.9% of the US denim apparel import market.

Costa Rican and the Dominican Republic's exports increased in the past few years. In 1997, Costa Rica contributed 3.8% to US imports and the Dominican Republic 3.3%.

Canadian women's or girls denim apparel exports to US grew fast from 1995 to 1997. With the growth rate of 256.3% per year, the exports increased from US$3.6 million in 1995 to US$22.2 million in 1997, accounting for 3.2% of US imports in 1997.

3.11 US women's or girl's denim apparel imports by region

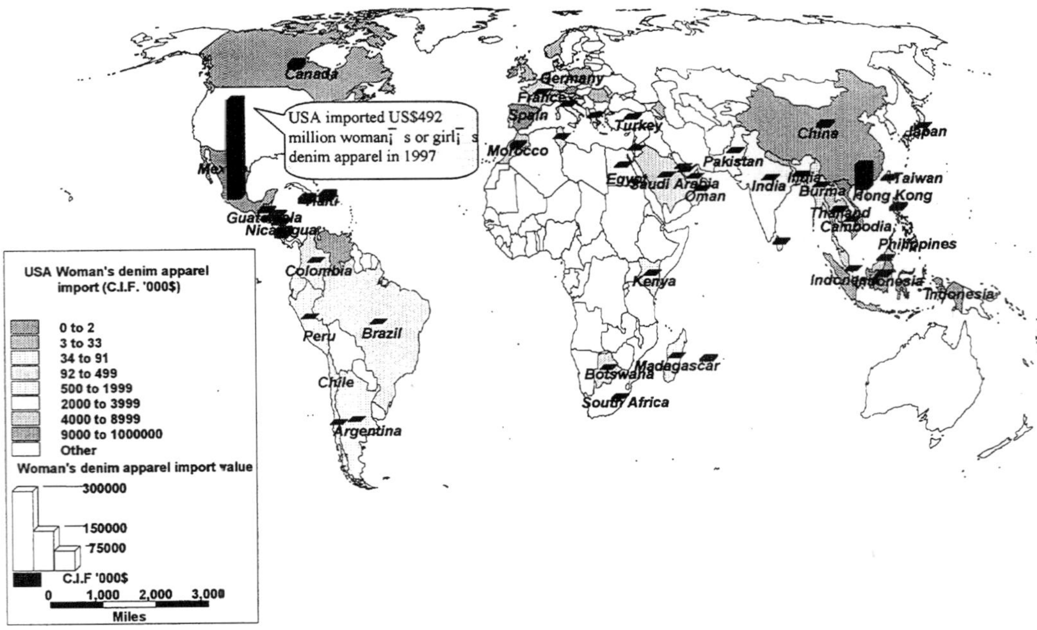

Jamaica, Philippines and China exported more than US$ 10 million of women's or girl's denim apparel to the US in 1997. Jamaica accounted for 2.9% of US imports in 1997, Philippines 2.0% and China 1.5%. While Jamaica's exports increased from 1996 to 1997, Philippines' and China's exports reduced by 20.0% and 18.0% per year from 1995 to 1997 respectively. Other countries' exports were lower than US$ 10 million, contributing less than 1.5% to the total US imports.

The average import price was US$ 82.7 per dozen in 1995, US$82.0 per dozen in 1996 and US$ 85.4 per dozen in 1997. The Mexican export price was US$ 80.7 per dozen in 1995, US$79.2 per dozen in 1996 and US$84.2 per dozen in 1997. The price of exports from Hong Kong was about US$ 96~101 per dozen, which was higher than the average.

Among other major exporters, the Canadian export price ranked the highest, and was US$176 per dozen in 1995, US$196 per dozen in 1996 and US$204 per dozen in 1997. Canada focused on the high value added market niche and its market size has grown quickly. As regards other important suppliers, the export prices of China, Macao, Philippines and Indonesia were US$111.5 per dozen, US$130.1 per dozen, US$86.5 per dozen and US$97.8 per dozen respectively in 1997. Costa Rica, Dominican Republic and Jamaica's export prices were lower than the average.

3.2 EXPORTS

3.2.1 Denim products exports

Table 3.2 shows that the US exported much less denim products than she imported. She exported US$ 591 million worth of denim products in 1993 and, in 1997, the figure reached US$ 856 million, increasing by an annual average rate of 11.2%.

3.12 US denim product exports

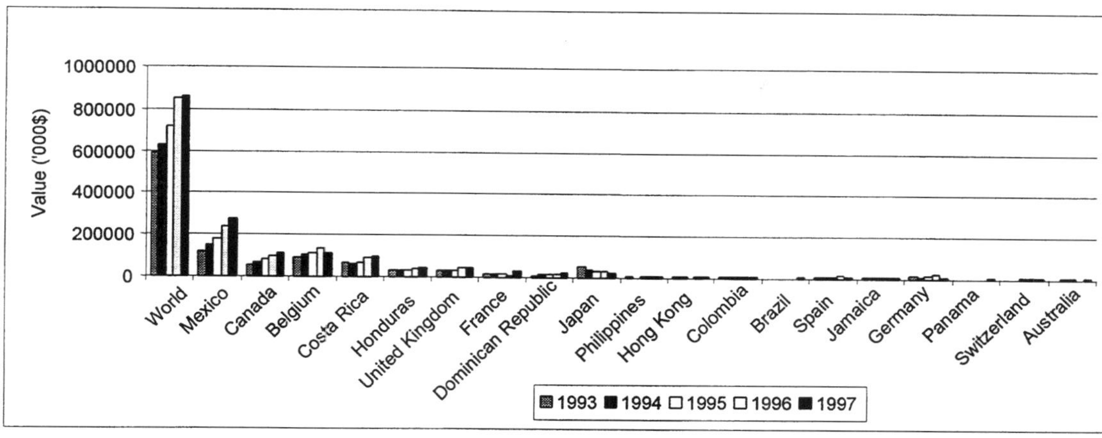

3.13 US denim product exports by region

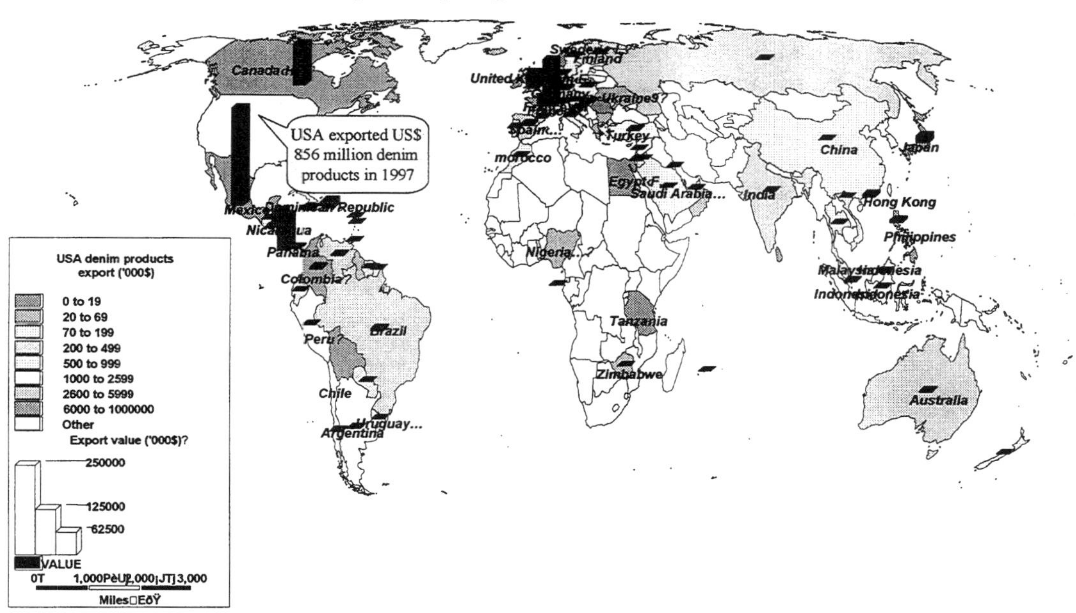

3.2 US denim product exports by categories

	USA denim product exports ('000$)							
	1997	1996	1995	1994	1993	Growth %	1-8 98	1-8 97
Denim Fabric	287205	262245	228501	212574	182379	14.37%	217104	184706
Man's or boy's denim apparel	393327	432939	355568	313787	322868	5.46%	273516	269012
Woman's or girl's denim apparel	175235	158718	135296	99317	86428	25.69%	114822	116188
Denim Products	855767	853902	719365	625678	591675	11.16%	605442	569906

In 1993, about 30.8% of US denim product exports were denim fabrics, and the ratio increased to 33.6% in 1997. Women's or girl's denim apparel also enjoyed a favorable growth, from 15.9% in 1993 to 20.5% in 1997. In contrast, the share of men's or boy's denim apparel exports reduced from 50.2% in 1993 to 46.0% in 1997. Crucial foreign markets for US denim products were American and European countries as shown in Figures 3.12 and 3.13. To be more specific, the major

3.14 US denim fabric exports

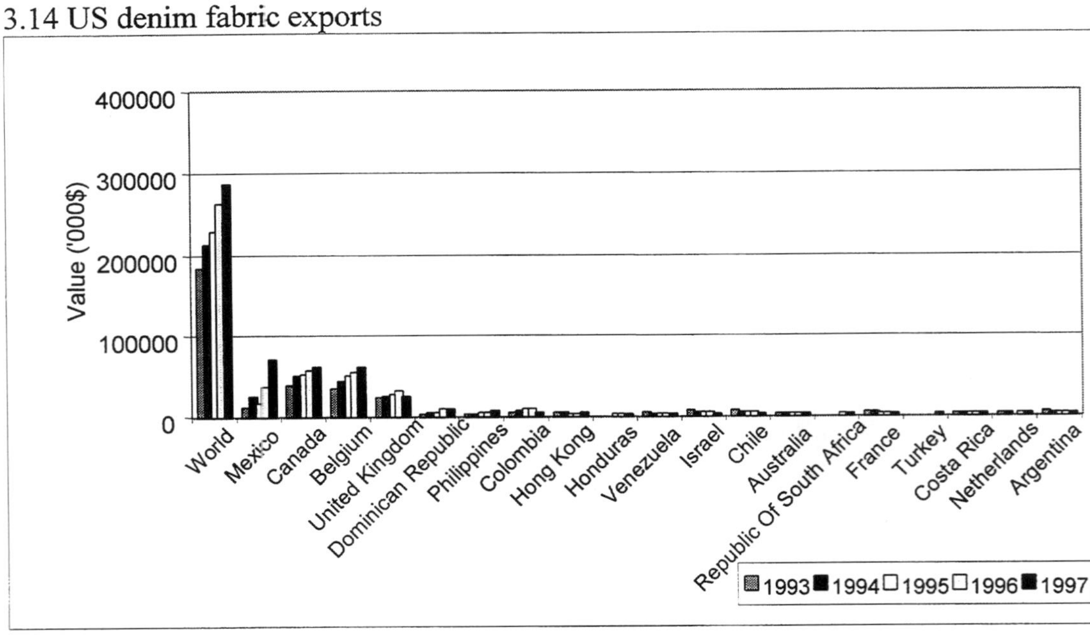

importers were Mexico, Canada, Belgium, Costa Rica, Honduras, United Kingdom, France, Dominican Republic, and Japan. Of these countries, Mexico was the most important destination for US denim products with an average annual growth rate of. 31.0%. The export value to Mexico increased from US$ 122.5 million in 1993 to US$ 274.4 million in 1997, accounting for 32.1% of total US exports. Canada was the second most important importer of US denim products. It imported US$ 115.4 million (13.5% of US exports) of denim products in 1997 with a similar growth rate to that of Mexico. Following Mexico and Canada, Belgium was the third most important consumer of US denim products and imported US$ 110.0 million of (12.9% of US total exports) denim products in 1997. Costa Rica imported US$ 97.3 million worth of denim products in 1997, accounting for 11.4% of US exports. Honduras and the United Kingdom each imported about 5% of US exports respectively in 1997. About 2.0% of US exported denim products went to France and 1.7% to Dominican Republic. US denim product exports to Japan continuously declined from 1993 to 1997. Japan imported US$ 48.9 million of denim products in 1993 and the figure dropped more than a half to US$ 20.5 million in 1997.

3.2.2 Denim fabric exports

In contrast to the import product mix, denim fabric is a major category of US denim product exports. About one third of the exports were denim fabrics. The US exported US$ 182 million worth of denim fabrics in 1993 and US$ 287 million in 1997, increasing by 14.4% per year on the average.

Figure 3.15 shows that US denim fabrics were mainly exported to Mexico, Canada, Belgium, United Kingdom, Dominican Republic, Philippines, Colombia, Hong Kong and Honduras. Mexico was the most important US denim fabric consumer in 1997. It imported US$ 70.7 million worth of denim fabric, accounting for 24.6% of US exports. However, US denim fabric exports to Mexico in 1993 were only US$12.2 million with a market share of 11.2%. It merits special attention that the annual growth rate for US denim products to the Mexican market was as high as 119.5%, which far exceeded its overall 11.5% export growth rate.

3.15 US denim fabric exports by region

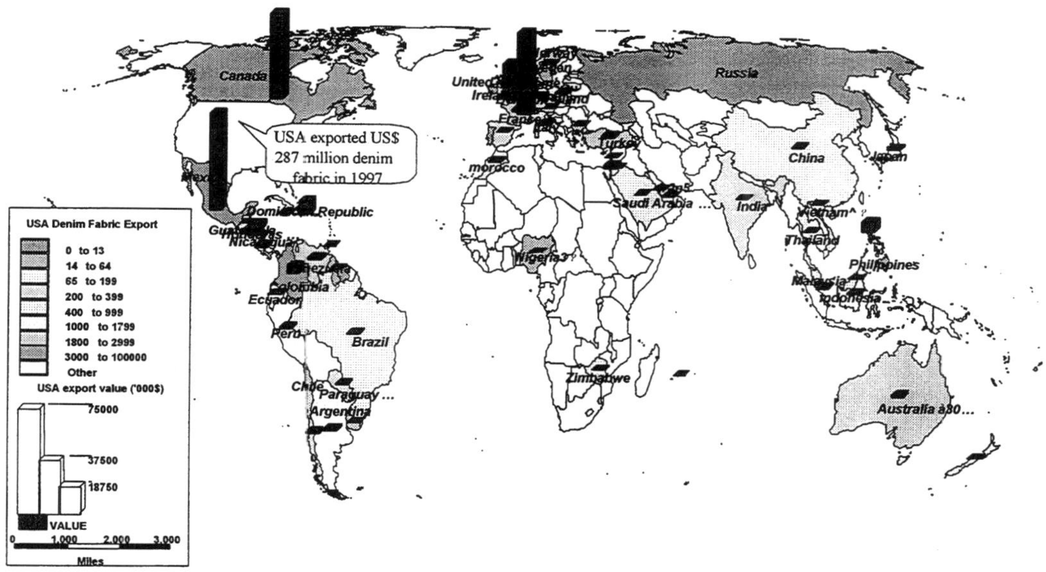

Canada and Belgium were the other two important foreign markets for US denim fabric. More than 21% of US denim fabric exports went to Canada in 1997, which represented over US$ 62 million in value. Another 20.9% went to Belgium in 1997. Both Canada and Belgium witnessed a relatively strong import penetration for US denim fabrics with annual growth rates of 16.5% and 19.5% respectively from 1993 to 1997.

The United Kingdom imported US$ 25.5 million worth of denim fabric from the US in 1997, accounting for 8.9% of US denim fabric exports. Other countries imported less than US$ 9 million.

3.2.3 Men's or boys' denim apparel exports

Figure 3.16 shows that the US men's or boy's (MB) denim apparel exports increased slowly in the past few years. They were US$ 322.8 and US$ 393 million in 1993 and 1997 with the annual average growth rate of 5.5%. The growth rate was much less significant than that of denim fabric and woman's or girl's denim apparel exports.

In 1993, Mexico, Belgium and Japan were the three most important destinations for US men's and boy's denim apparel. The former two imported US$ 77 million (17.6% of US exports) and US$ 57 million (24.0% of US exports) respectively in 1993. Japan imported another US$ 45.8 million (14.2% of US exports). However, in 1997, Canada replaced Japan as the No. 3 importer for US men's or boy's denim apparel. The top three accounted for 34.0%, 12.52% and 10.2% of US exports respectively. Canada was the fastest growing market for US man's or boy's denim apparel, and grew by 83.7% per year between 1993 and 1997. Exports to Mexico were also up by 18.1% per year on the average. But those to Belgium were down by 3.4% each year in the same period.

3.16 US men's or boy's denim apparel exports

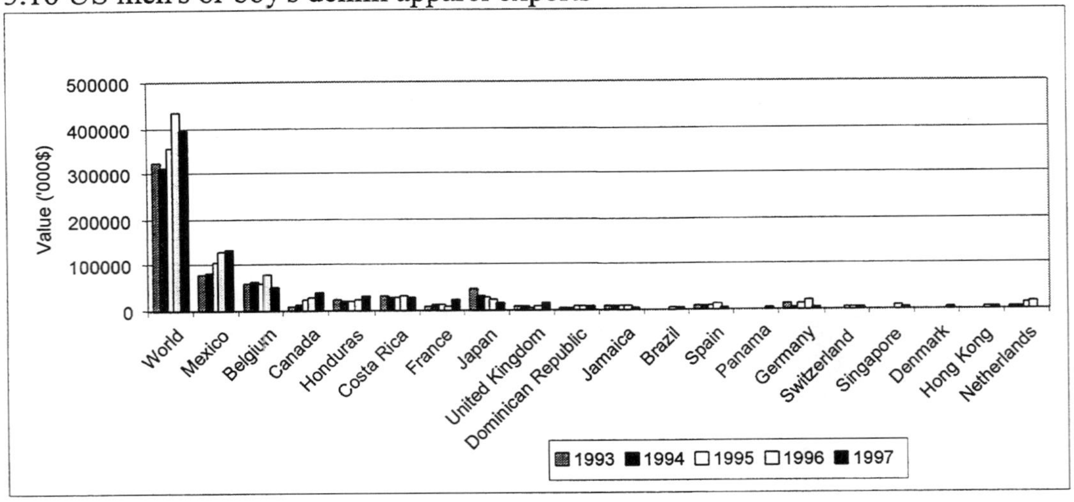

3.17 US men's or boy's denim apparel exports by region

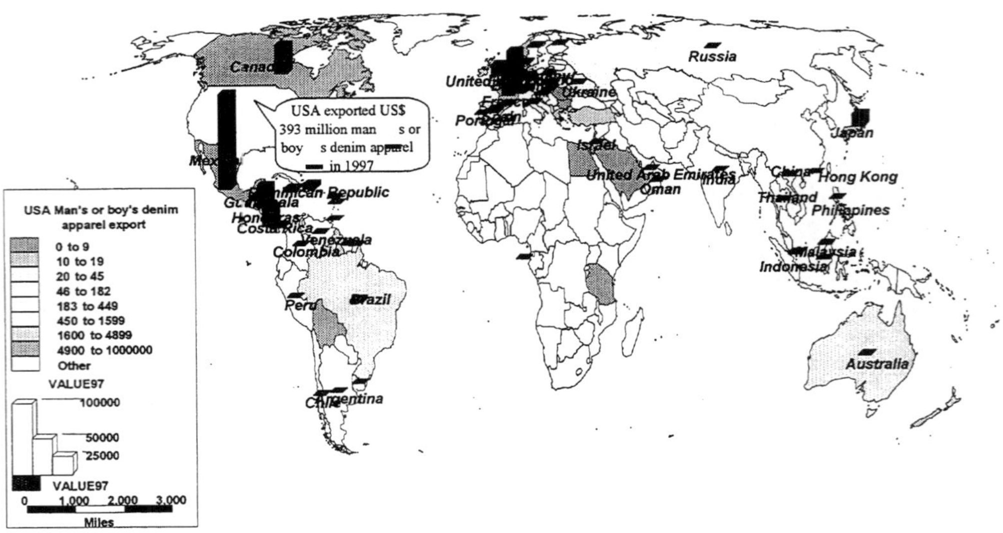

Honduras imported US$ 31.8 million worth of men's or boy's denim apparel in 1997, accounting for 8.1% of US exports. Costa Rica accounted for 6.9%, France 5.9%, Japan 4.4% and the United Kingdom 4.0%.

In view of the above analysis, it is clear that US men's or boy's denim apparel exports were concentrated on American countries, while the non-American markets became less attractive for the US men's or boy's denim apparel exports.

3.2.4 Women's or girls' denim apparel exports

US women's or girl's denim apparel exports grew significantly in recent years. According to Figures 3.18 and 3.19, the export value amounted to US$ 86.4 million in 1993 and it doubled by 1997, recording US$ 175 million at an annual average growth rate of 25.7%.

Parallel with the situation of men's or boy's denim apparel exports, American countries took a remarkable share of US women's or girl's denim apparel, notably Mexico, Costa Rica, Canada and Honduras. The first two accounted for 78.9% of US exports.

3.18 US women's or girl's denim apparel exports

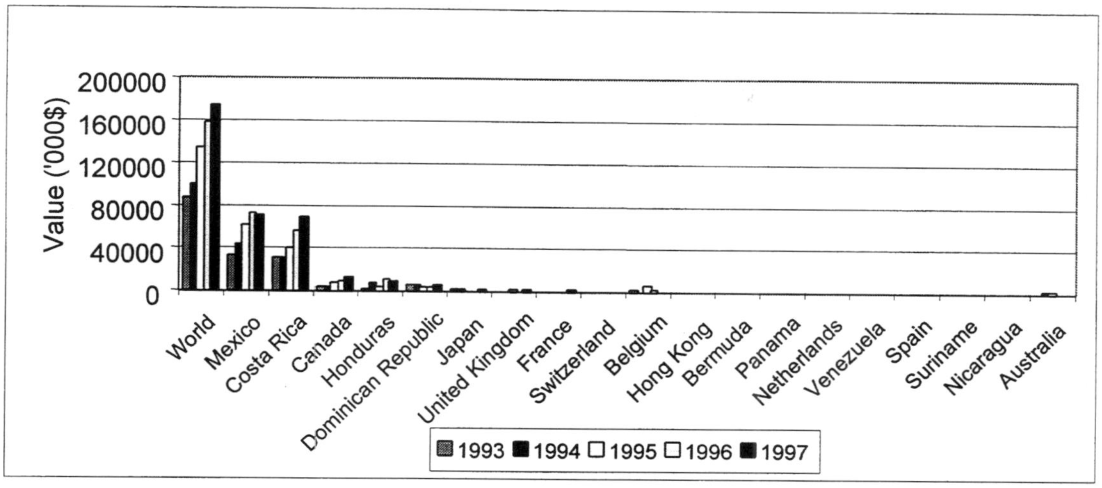

3.19 US women's or girl's denim apparel exports by region

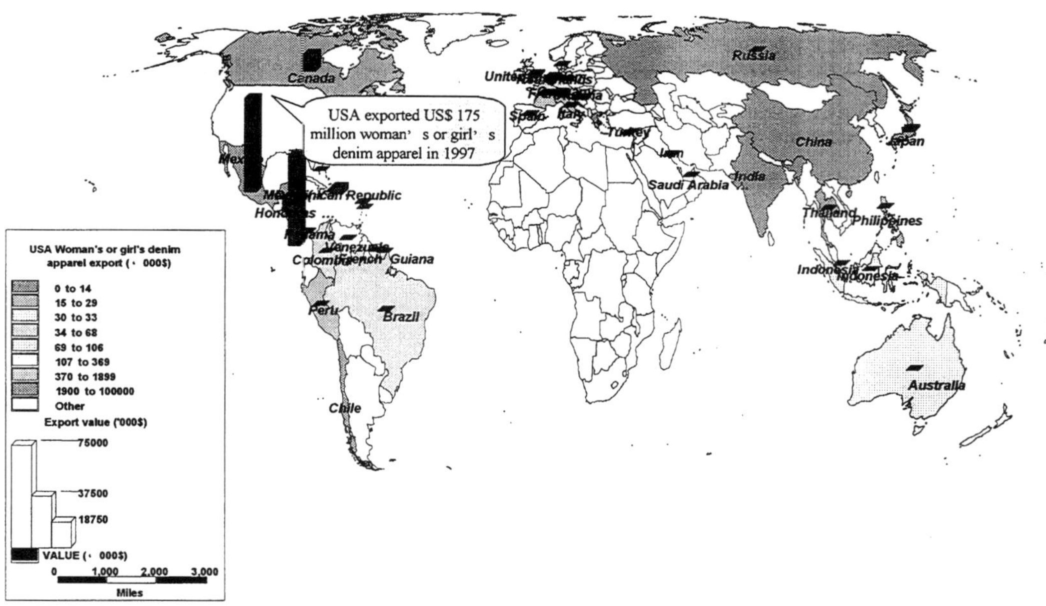

Mexico was the most important importer. It imported US$ 32.9 million worth of women's or girl's denim apparel from the US in 1993. The value jumped to US$ 70.1 million in 1997, or 40.0% of US total exports. The growth rate was 28.4% per year on average from 1993 to 1997.

Costa Rica was the second most important consumer, and imported US$ 68.1 million of goods from US in 1997, accounting for 38.9% of US exports. Its imports grew at an average rate of 29.1% each year from 1993 to 1997.

As regards other important importers, Canada imported 7.6% of US women's or girl's denim apparel exports, Honduras 5.1%, Dominican Republic 3.1% and Japan 1.1%) in 1997.

REFERENCES

1. Roshan Paul, S.R.N., Jegadeesh Thampi, Denim: the evergreen favorite. Textile Dyer & Printer, 1996, p.15-19.
2. Zimmermann, R., Denim: what might the future hold for the world's hottest textile sector ITS Textile Leader, 1998, p.34-48.
3. Anon, The denim dilemma.2000, www.cottoninc.com.
4. Jurg Rupp, Andrea Bohringer., Denim: jack in a box of the fashion industry, ITB International Textile Bulletin, March 2000, p 10
5. Pushpa Bajaj, Rohit Agarwal., Innovations in denim production, American Dyestuff Reporter, p 26-38

4 EU Denim Products Trade[1]

4.1 EU DENIM PRODUCTS TRADING

OVERVIEW

Denim products in Europe have a history of only about half a century, though denim itself, according to many different sources, originated in Nimes--a French city.

The end of WWII started the jeans presence on the European continent. Official imports from the US are unavailable due to the disorder of the post-war period. Sefranek's (founder of Mustang) first 300 pairs of jeans were produced in Germany in 1949, imitating the pattern of American jeans with a coarse blue drill instead of indigo denim. "Made in America" denim was widely supplied since the late 1950s and European mills established specialist mills for local denim production in the early 1970s [4].

Having recognized the importance of this new growing denim market, American brands rushed to open European offices in the early 1960s, with Levi's and Wrangler in 1961 and Lee in 1963 [4]. The mid-1970s witnessed rapid development of denim and jeans, growing into a defined force in the textile and jeans sector. Total jeans sales in ten European countries[2] jumped from 47 million units in 1970 to 170 million in 1977 [1, p12]. However, denim jeans suffered a sudden decline in the first half of the 1980s due to over production and the emergence of other casual wear items. Sales slumped from 250 million units in 1980 to 150 million in 1985 [2]. During the second half of the 80s, denim consumption began to pick up. In 1991, denim production in Europe was 228.7 million square meters and it rose to 241 million square meters in 1995 [1, p16].

Nowadays, the EU is the second largest denim market in the world, accounting for 23.3% of global denim fabric consumption [3]. Some European labels are widely welcomed in the world market, such as Mustang, Ge Pa Fin, and Pepe. Figure 4-1 shows the jeans production in some European Countries. It indicates that jeans production experienced a healthy growth in the Italian and British markets.

Denim product imports from the world (not including intra-EU trade) increased from 1.1 billion Euro in 1994 to 1.5 billion Euro in 1997 with a growth rate of about 6.1% per year. When compared with the import market, the export market is relatively small and experiencing a downward trend. There were 537 million Euro worth of denim product exports to the world in 1994 but it went down to 481 million Euro in 1997. The average dropping rate was about 2.9% per year. The trade gap

---

[1] Data in this part is from EUROSTAR, accessed around July 1998.

[2] The ten countries include France, Germany, Italy, Austria, Switzerland, Sweden, the Netherlands, Belgium, the UK and Spain.

4.1 Jeans production by some European countries in 1990 and 1994[3]

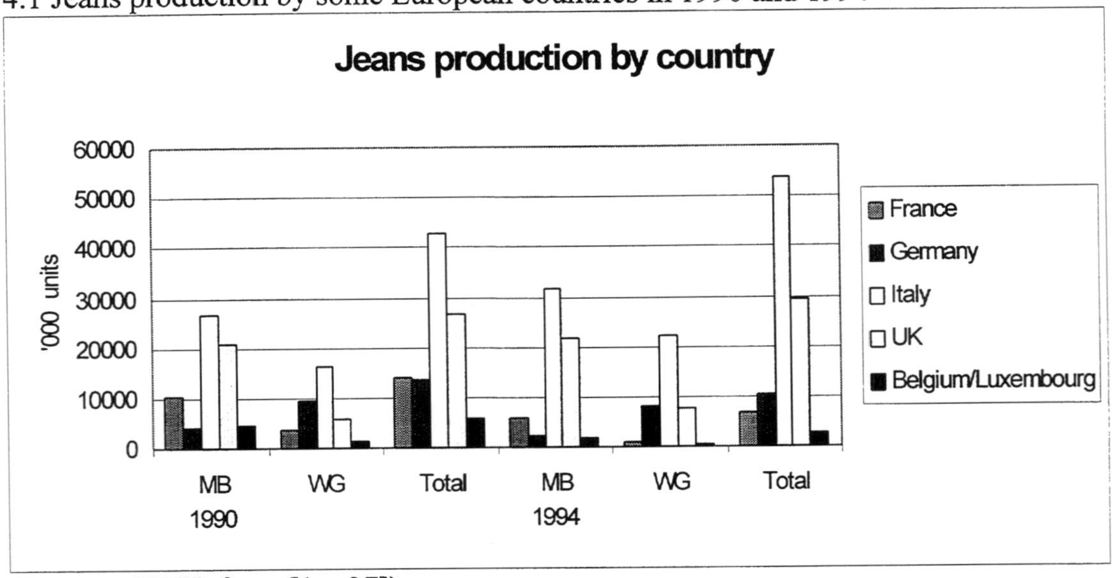

(Source: CITH, from [1, p27])

(imports - exports) was 613.8 million Euro in 1994 but in 1997, the figure became 1 billion Euro, increasing by 10.3% per year.

The trade gap for denim fabric was -18.2 million Euro in 1994 and -79.7 million Euro in 1997, growing at an annual rate of 19.3%. The jeans trade gaps were 632 million Euro and 1.1 billion Euro during the same period and the growth rate was about 11% per year. These figures indicate that the EU was a net denim fabric exporter and denim jeans importer, a trend becoming more significant during recent years.

4.1.1 Imports

Almost a half (51.8%) of the EU imported denim products came from other EU members. Figure 4.2 shows different countries' (including the EU itself) contributions toward EU imported denim in a geographic picture.

Apart from internal trading within EU itself, the Union also has extensive trading connections with other European and North African countries, such as Tunisia, Turkey, Morocco, Malta, Poland, and Croatia. For example, EU denim product imports from Tunisia, Turkey, Morocco, Malta, Poland and Hungary represented 20.8% of the total EU imports as well as 43.1% of the total imports from non-EU countries. Denim product exports from EU to Tunisia, Switzerland, Morocco, Turkey, Poland, Malta, Hungary, Croatia and Israel accounted for 14.6% of the EU's total exports. If viewed in relation to exports to the non EU countries alone, the ratio jumps 77.1%.

The US and Hong Kong were relatively important to the EU denim market as their market shares were 7.3% and 4.5% respectively.

Figure 4.3 shows that high-price denim jeans came from the United States and Japan; while less expensive goods were mainly from Hong Kong, Morocco, Pakistan and Bangladesh.

[3] MB = Men's or boy's; WG = Women's or girl's

4.2 Different countries' contributions toward EU denim product imports in value.

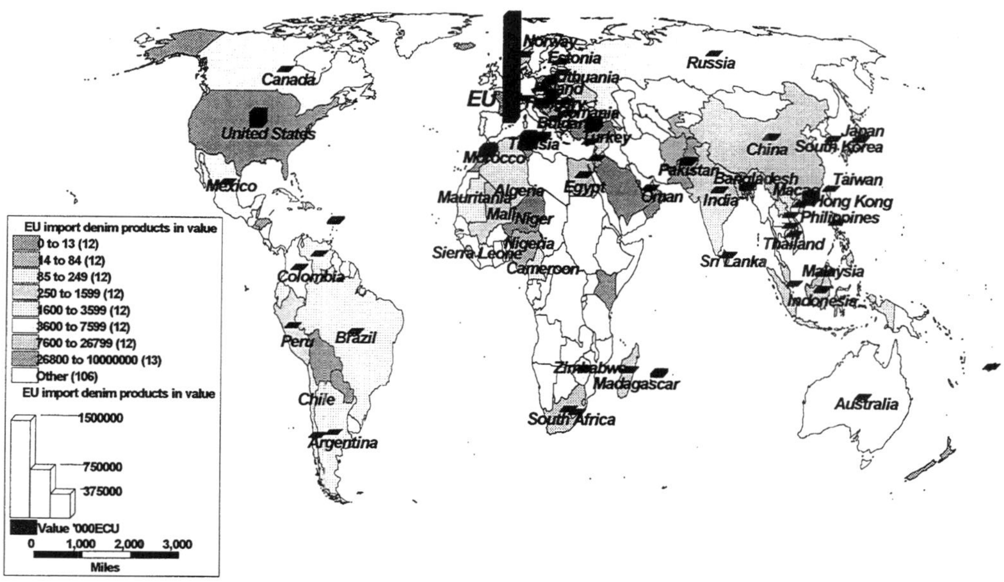

4.3 EU imported denim jeans market share

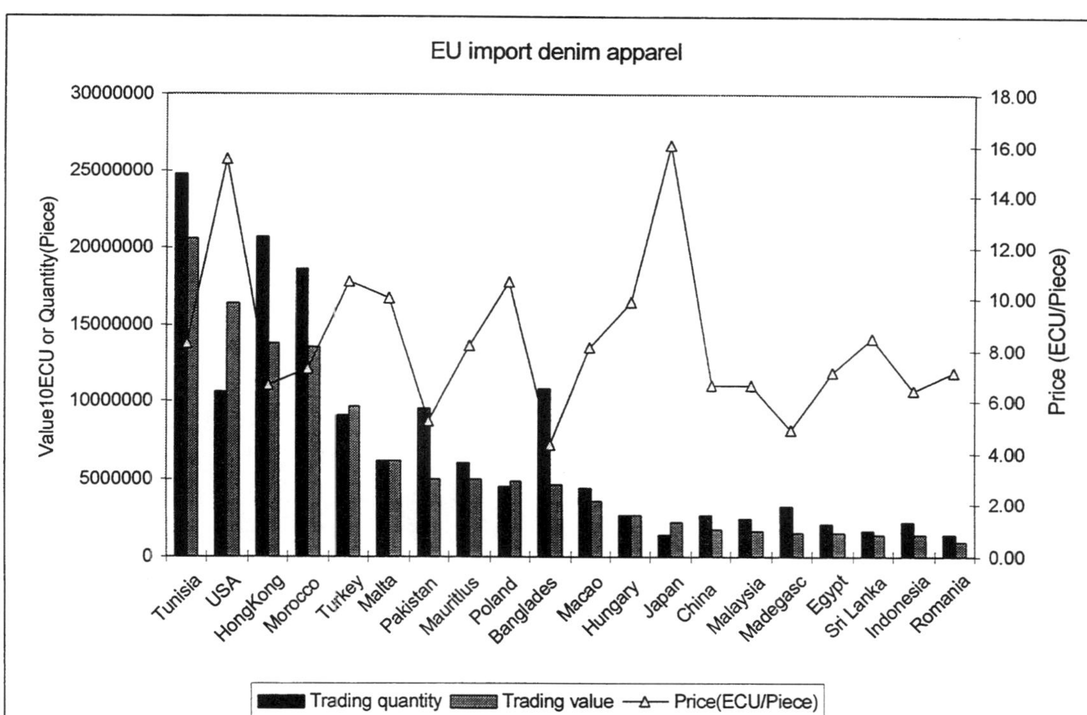

4.4 EU Import Denim Fabric

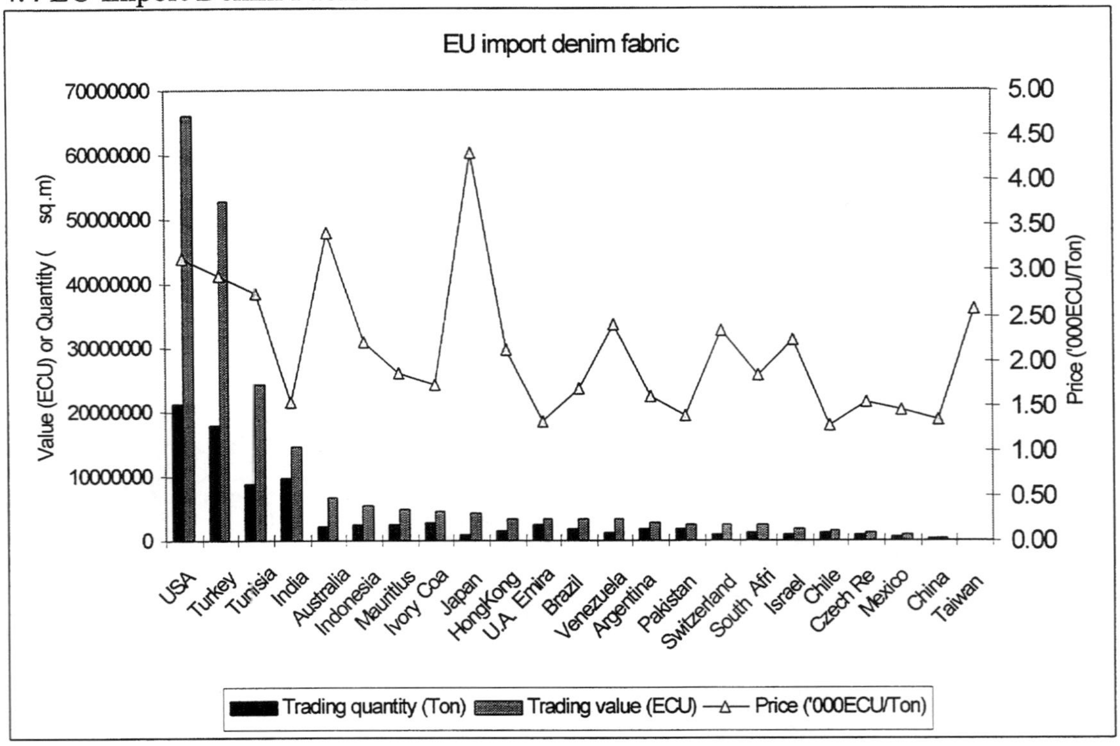

Figure 4.4 shows that higher price denim fabrics were from the United States, Australia, Turkey, Tunisia and Japan while lower price denim fabrics were imported from India, U.A. Emirates, Pakistan, Chile, Mexico and China.

4.1.2 Exports

As with the import market, EU members dominated the denim product export market, as illustrated in Figure 4.8. In 1997, the EU exported about 81.1% of its denim products to its members; while non-EU countries imported 481 million Euro worth of denim products from the EU. Among these non-EU importers, Tunisia had the largest share of about 18.5% of the total exports to the non-EU countries, with an import value amounting to 81.7 million Euro. Morocco, with 60.0 million Euro and a market share of 13.6%, was the No. 2 importer. Switzerland and Turkey came third and forth with import values of 50.3 million and 39.0 million Euro respectively. Hong Kong was in 14th place, with an import value of 5.6 million Euro. Fashion leaders the US and Japan came 6th and 15th. Figure 4.5 shows the market share of individual countries based on the export value from the EU.

Countries with an import value lower than 3.0 million Euro are defined as 'Others' and there are 121 countries within this group. China, ranking 90th, was in this group with an insignificant import value of 71 thousand Euro. Three countries didn't import EU denim products into their countries. They were Bolivia, Ecuador and E. Guinea. There were 22 non-EU countries with an import value over 3.0 million Euro in 1997, ranging from 3.14 million Euro to 81.7 million Euro with a mean of 17.8 million Euro.

4.5 Pie Chart of EU denim product exports to individual non-EU countries

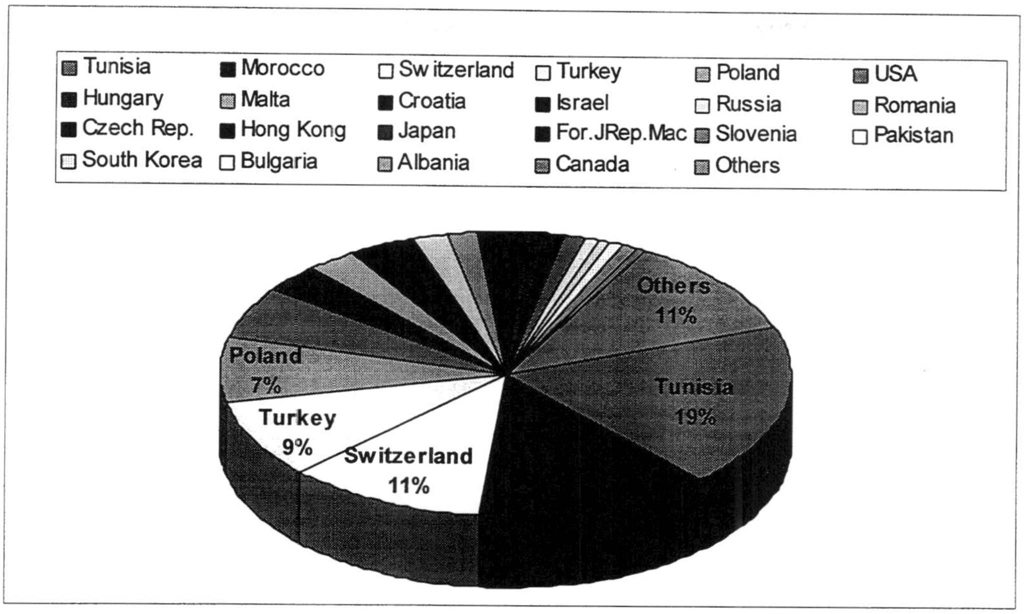

Figure 4.6 shows how EU countries exported various grades of denim products to the non-EU countries. The higher price denim jeans were destined mainly for the US, Russia, Japan, Hong Kong and Andorra, while lower price ones were exported to Hungary, Romania and Bulgaria.

Figure 4.7 shows that the more expensive denim fabrics, priced around 7 to 23 thousand Euro per ton, were destined for Romania, South Korea, Hong Kong and China. Lower price denim fabrics, about 3.5 thousand Euro per ton, were exported to the United States.

4.6 EU denim jeans exports to non-EU countries

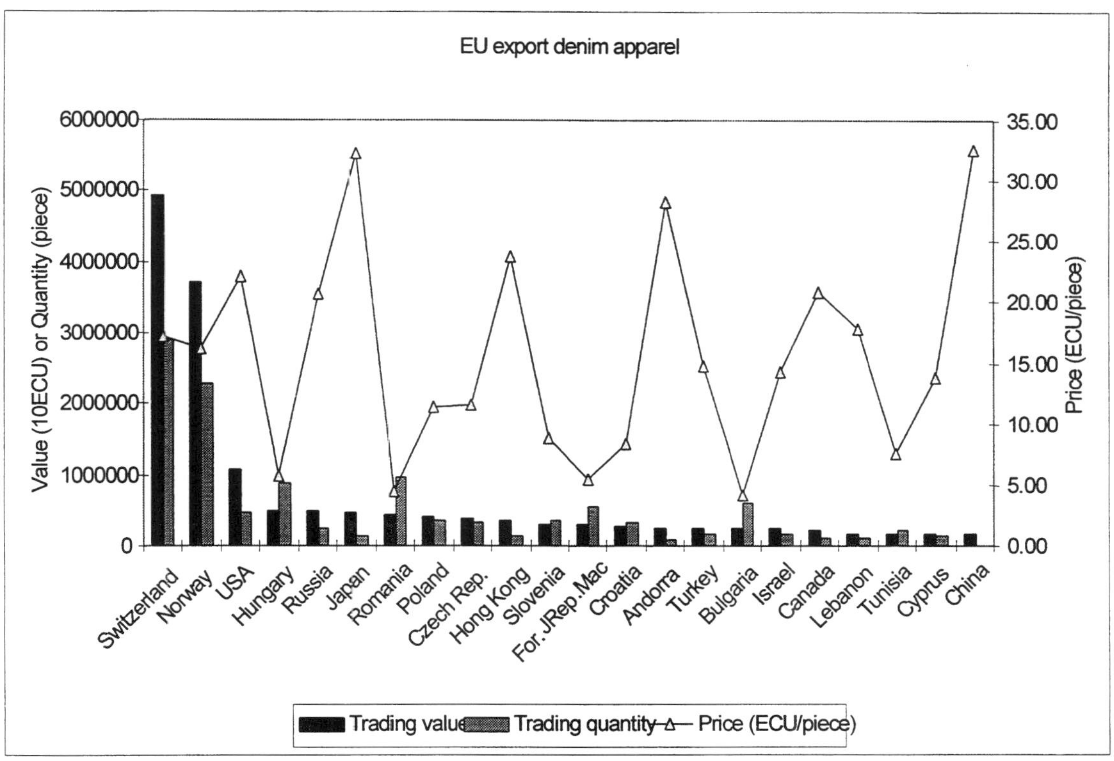

4.7 EU denim fabric exports to non-EU countries

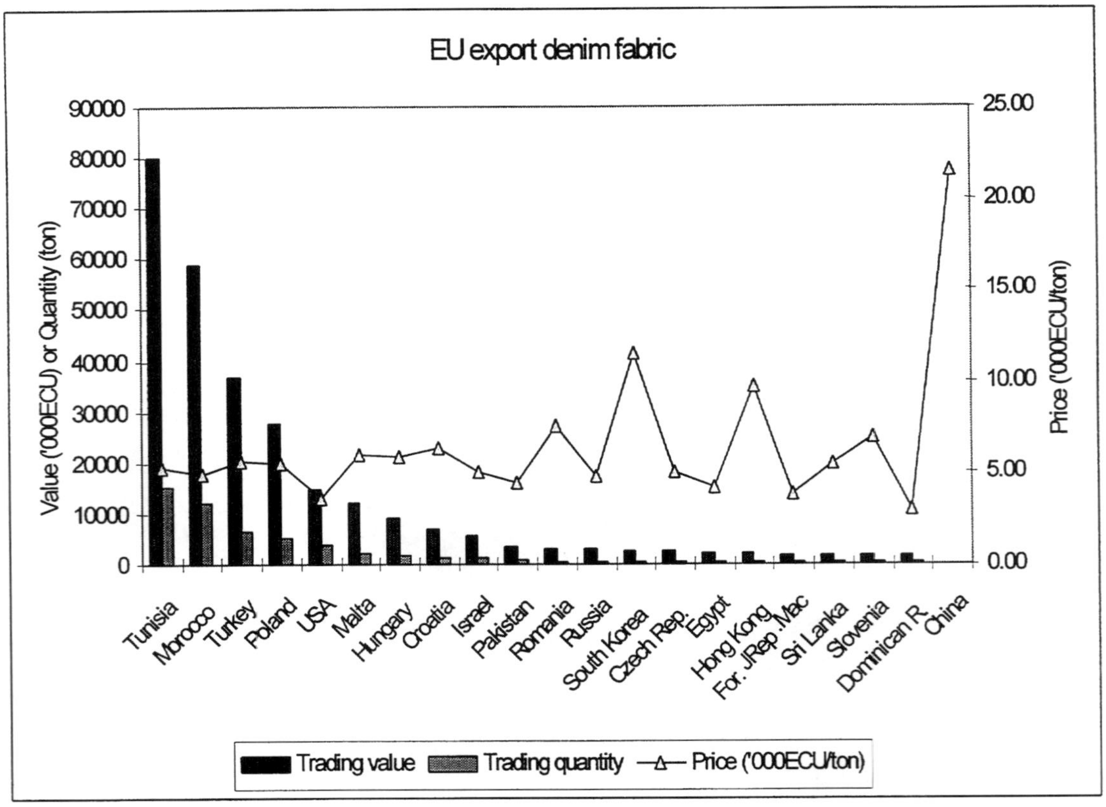

4.8 Different countries' contributions toward EU denim product exports in value

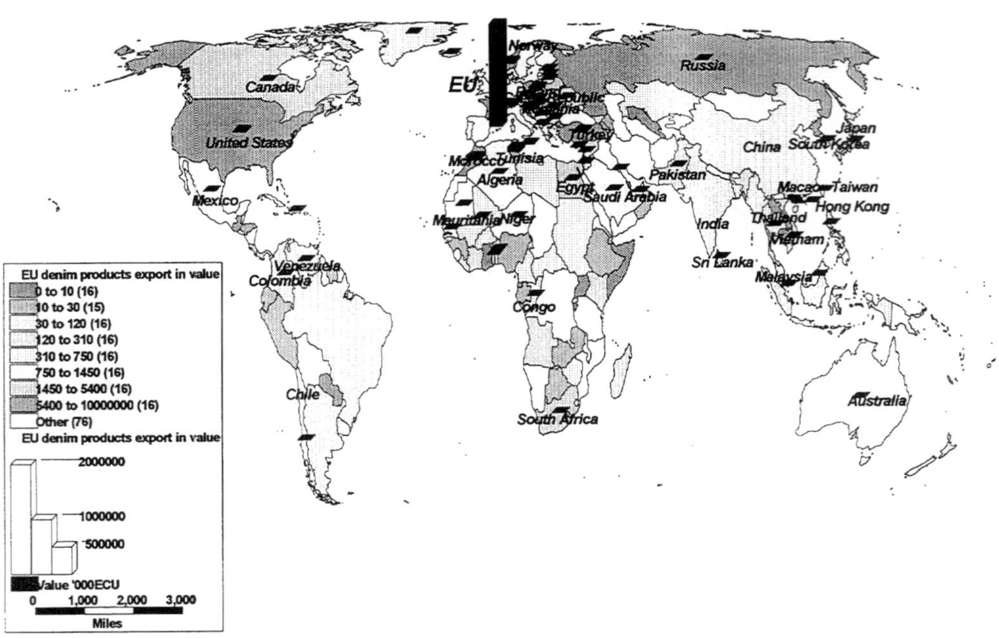

4.1.3 Comparison of the trade prices between intra-EU and extra-EU

As an important world denim market, the EU denim market has its own unique characteristics. For example, denim jeans in Europe has always been perceived more as a fashion item with a higher price and status, but lower consumption rate, as distinct from the US market [3]. The EU denim trading market partly embodied this feature. According to Figure 4.9, we can see that the prices of denim items were different in the internal and external markets. The price of internal denim trading within EU is higher than that of the external trading with non-EU countries. Figure 4.9 shows the price differences of various kinds of denim fabric and jeans between EU internal and external trade.

The above analysis clearly shows that EU denim product trading is concentrated on the Europe continent, especially between EU members. Such a trading pattern is attributed not only to their similar culture and geographic proximity, but also to the aim of the EU and the new concept of EU, which is 'Make Europe Together' and 'Enlarge EU' [5]. As a result, trading within European countries is promoted and developed into a new level. Denim trading well reflects this trend, which will be explained in detail in the following pages.

The EU is composed of 15 member states. It has been a Single Market since 1992, with free movement of production factors. However, each country has its own features and market situation despite economic integration, which makes the EU denim product market much more complicated and diversified. The next section will deal with imports and exports on a country-to-country basis.

4.9 Price comparison for different types of denim fabric and jeans

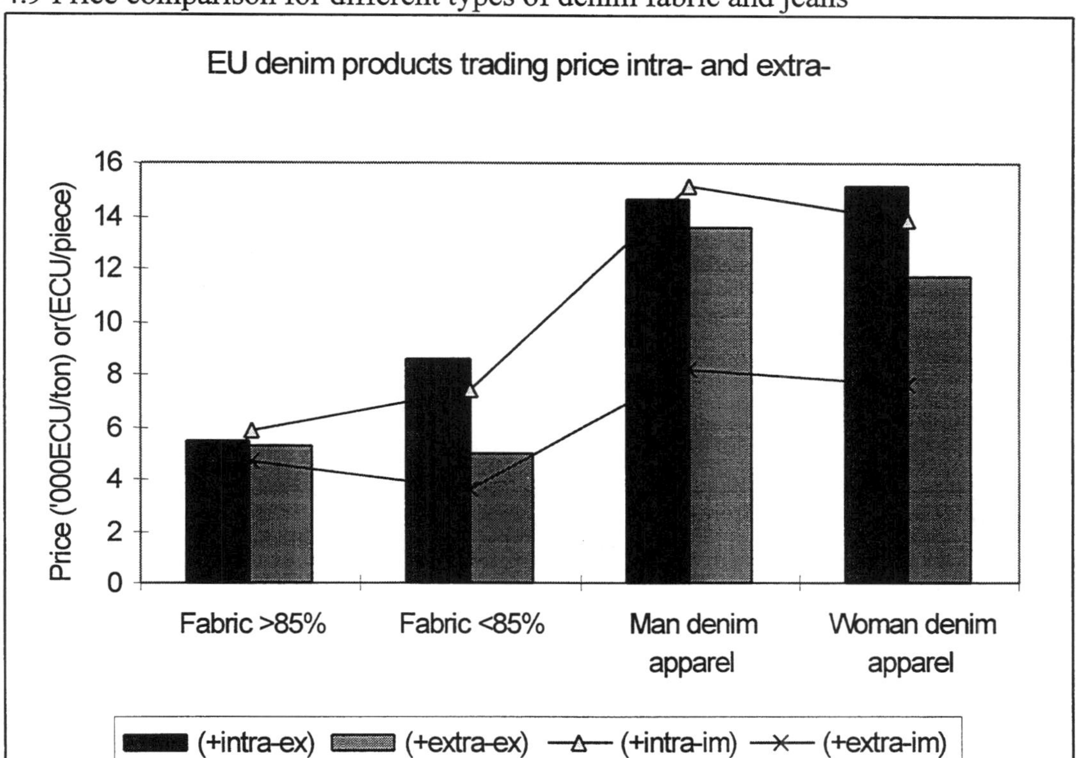

4.10 Denim products trade gap

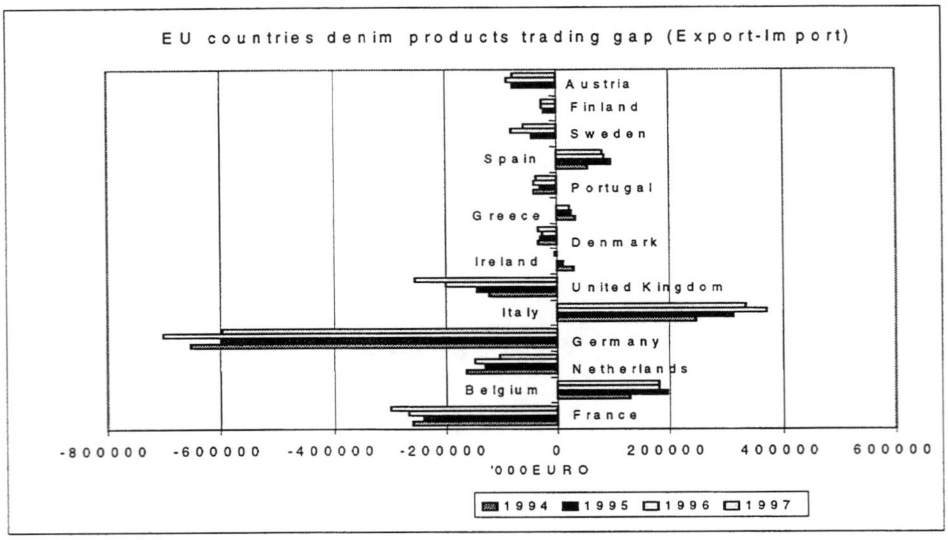

4.2 EU DENIM PRODUCTS TRADING (WITHIN EU)

OVERVIEW

In general, more than a half of the traded denim products were jeans; denim fabrics only represented a small proportion for all EU countries. The market was concentrated on other EU countries, and Belgium, Italy and Germany were the three most powerful EU countries in the denim-related industry.

As regards imports, Germany absorbed one quarter of the EU imported denim product imports in 1997, with an import value of 813 million Euro. The United Kingdom, Belgium and France had similar market sizes, each importing 439, 437 and 430 million Euro worth of denim products respectively, about 13%-14% of the total import market.

Among all EU countries, Italy and Belgium were the market leaders in the denim product export market. The total export value of Italy was 651 million Euro and that of Belgium was 613 million Euro. Germany ranked third with 216 million Euro worth of exports.

Figure 4.10 shows that nine EU members--Austria, Denmark, Finland, France, Germany, Netherlands, Portugal, Sweden and United Kingdom--had trade deficits, indicating that they relied more on denim product imports.

Germany had the largest trade deficit, followed by France, United Kingdom and Netherlands. The British trading deficit increased sharply at an annual rate of 27.3% between 1994 and 1997, while that of Belgium increased by 9.7 % per year. In contrast, the trade gaps of Germany and Netherlands narrowed at the annual rate of 2.1% and 9.1% respectively between 1994 and 1997.

Italy enjoyed the largest positive trade balance, followed by Belgium, Spain, Greece, and Ireland. The Italian trade gap increased at the rate of 8.9% each year from 1994 to 1997, and those of Belgium and Spain increased by 9.7% and 11.3% respectively.

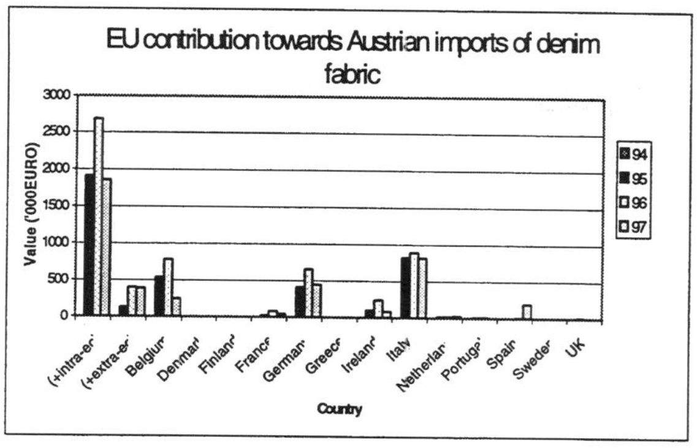

4.11

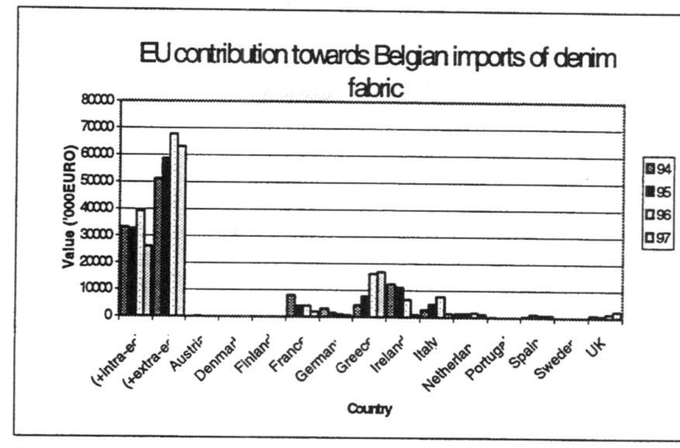

4.12

The remaining countries had relatively insignificant trading gaps, indicating that there was not much difference between their denim product importation and exportation.

4.2.1 Imports

4.2.1.1 Denim fabric

<u>Austria (See Fig. 4.11)</u>

- Austria was not admitted into the EU until the year 1995. Therefore, trade data for 1994 was not available.
- In the Austrian denim fabric import market, an average of 88% of denim fabric imports came from other EU countries in the last 3 years.
- Denim fabric imports from EU members reached a peak in 1996 with turnover about 2.68 million Euro.
- The main suppliers were Belgium, Italy and Germany. Other suppliers included Ireland, France and Netherlands. All these countries registered their record sales in 1996.
- Spain has the potential to grow into one of the important suppliers in the future.

<u>Belgium (See Fig. 4.12)</u>

- Belgium merits special attention in that imports from non-EU countries played an important role in the Belgian denim fabric import market. Between 1994 and 1996, the average market share of non-EU imports was 63%, but in 1997 the ratio increased to 70%.
- Within the EU market, there were about 8 major suppliers, including Greece, Ireland, Italy, Netherlands, France, Germany, Spain and the United Kingdom. Members such as Austria, Denmark and Finland also exported their denim fabric to Belgium but the value was relatively small in comparison with the major suppliers.
- Most of the suppliers experienced a downward trend since 1995, except for Greece and the United Kingdom. Although the market share of France underwent a significant increase from 1995 to 1996, it dropped again in 1997.
- In 1994 and 1995, Ireland had the biggest market share of 37% and 34% respectively. In 1996 and 1997, Greek denim fabric dominated the Belgium import market, accounting for 41% and 65% respectively.

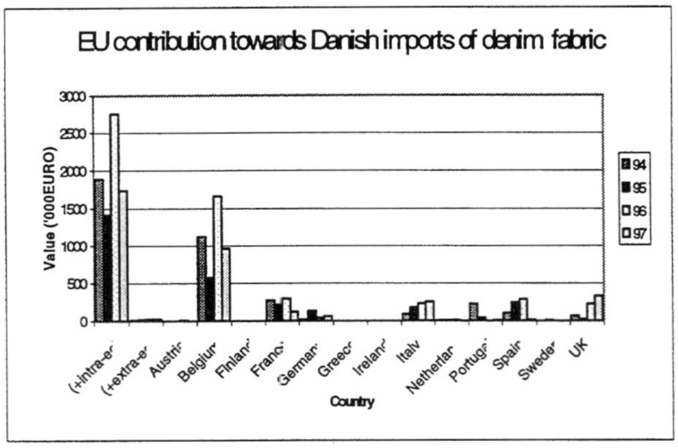

4.13

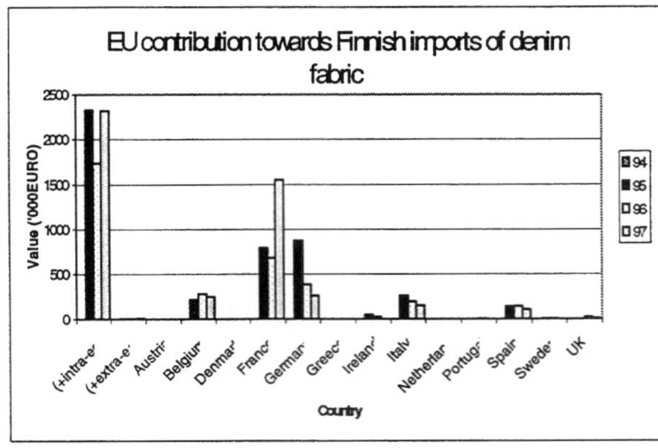

4.14

Denmark (See Fig. 4.13)

- The denim fabric import market in Denmark was very dependent on other EU countries. On average, 99.2% of the imported denim fabric came from the EU members between 1994 and 1997.
- The market for denim fabric imported from the EU reached its peak in 1996 with a turnover around 2.8 million Euro. In 1997, it dropped back to 1.7 million Euro.
- Belgium was the leading supplier. Other exporters included France, Germany, Italy, Spain, United Kingdom and Portugal.
- Most of the suppliers had no clear increasing or decreasing trend except Italy and the UK. The former enjoyed favorable increases since 1994 and the latter has grown since 1995.

Finland (See Fig. 4.14)

- Finland did not join the EU until 1995. Therefore, only 3 years data are available.
- The import situation is the same as Denmark. About 99.5% of denim fabric imports were from other EU members in 1997. Except for the year 1996 when the import value from EU suppliers hit the lowest level of 1.7 million Euro, EU denim fabric imports into Finland in 1995 and 1997 were almost the same, reaching 2.3 million Euro.
- The major suppliers included Belgium, France, Germany, Italy and Spain.
- Germany held the biggest market share of 37.5% in 1995. France was fairly close with 33.9%. In 1996 and 1997, France replaced Germany as the leading supplier with a share of 39.1% and 66.7% in the two years.
- Germany, Italy and Ireland had a clearly decreasing trend in recent years.
- Though most of the suppliers exported less in 1997 than in 1996, the overall Finland import value in 1997 was still higher than that in 1996, mainly due to the sharp increase from the French market.

France (See Fig. 4.15)

- Between 1994 and 1997, the average market share from the EU was 75.9%, hitting a low in 1996 with 71.1% and a high in 1997 with 80.8%.
- Corresponding with these import ratio indications, imports from the EU reached its highest level of 39.1 million Euro in 1997, while imports from non-EU markets registered record sales in 1996 with 14.5 million Euro.
- The leading supplier was Belgium with approximately 60% of the total denim fabric imports. Other suppliers included Italy, Spain Germany, Greece, Ireland

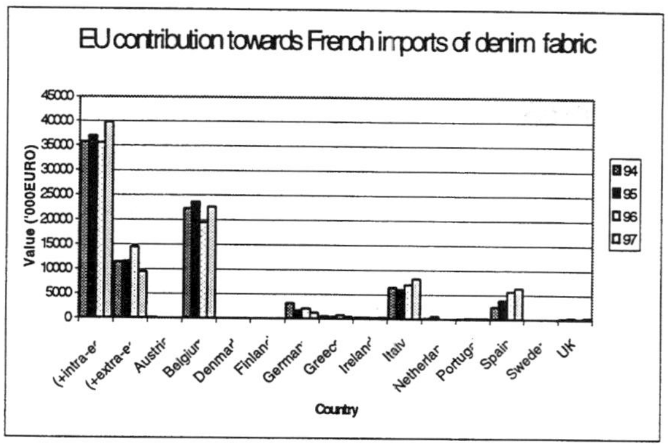

4.15

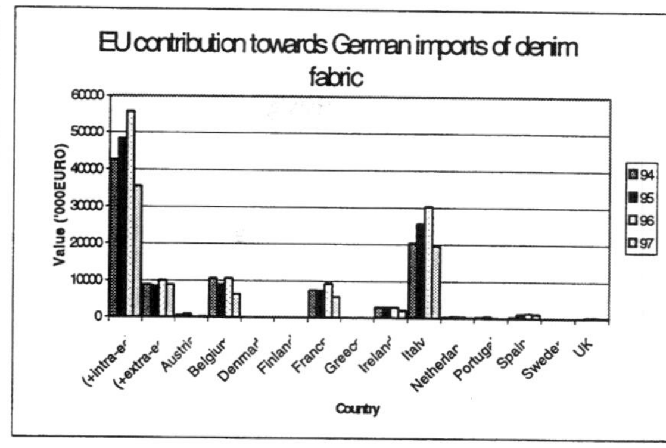

4.16

and the United Kingdom. Imports from Denmark, Finland and Sweden were relatively small compared with other EU countries.

- Germany had a decreasing trend except in 1996. By contrast, Spain had a very clearly increasing trend. Trends in other countries are unclear with up-and -down variations.

Germany (See Fig. 4.16)

- The EU members played the key role in Germany's denim fabric import market. About 84.8% (55.7 million Euro) of the denim fabric was imported from the EU in 1996 but this share declined to 80.0% (35.5 million Euro) in 1997. The largest market share from EU members appeared in 1995, accounting for 85.2%.
- During the investigation period, total imports from the EU and non-EU countries as well as that from most individual members reached their record level in 1996.
- Within the EU market, Italy contributed the most, followed by Belgium and France. The other EU suppliers included Ireland, Spain, Netherlands, Portugal and the United Kingdom.
- Finland made no contribution to the German denim fabric import market. Exports from Denmark, Greece and Sweden were trivial and unstable. For example, imports from Sweden only occurred in 1995 and 1997.

Greece (See Fig. 4.17)

- Imports from EU members accounted for the lion's share of its denim fabric import market. This was 93.2% in 1996 but slumped to 81.0% in 1997. Imports from the EU had a decreasing trend with a peak value of 13 million Euro in 1995.
- In 1994, 1995 and 1997, Italy was the largest supplier within the EU, accounting for more than half (average of 52.4%) of the imports from EU members. But in 1996, Germany was the leading supplier instead of Italy. It held 30.6% of the market share, 5.5 percentage higher than that of Italy.
- Ireland demonstrated a clear decreasing trend. Belgium and Spain reached their peak in 1995, and then started falling. Others showed no clear trend.
- Imports from France reached the highest level in 1996 when the value was much larger than in other years.
- There were no imports from Finland and Greece only began to import from Sweden in 1997.

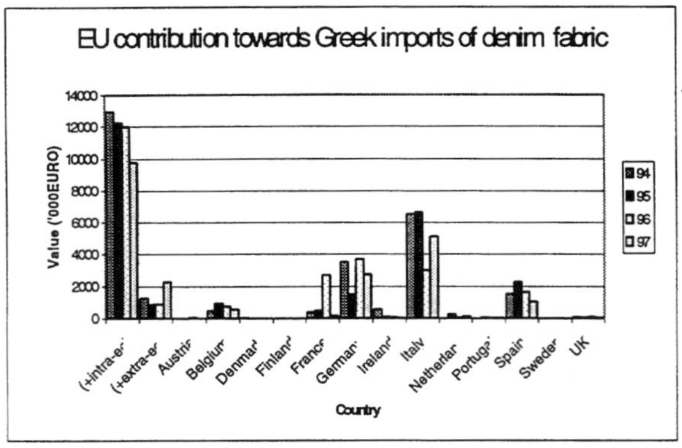

4.17

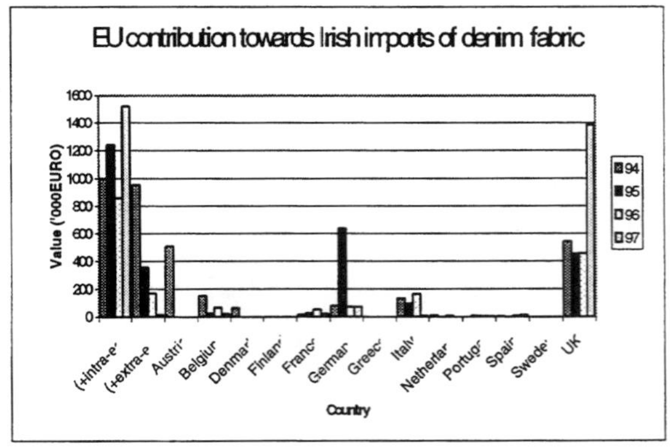

4.18

Ireland (See Fig. 4.18)

- Imports from non-EU countries decreased rapidly from 0.95 million Euro in 1994 down to 15 thousand Euro in 1997. No trend could be identified as regards the intra-EU trade. Imports from the EU reached a peak in 1997 amounting to 1.5 million Euro with 99% of the total denim fabric imports.
- The top supplier was the United Kingdom. Its exports to Ireland increased from 0.46 million Euro in 1996 to 1.4 million Euro in 1997, a three fold increase. This amounted to 91.2% of the total Ireland denim fabric imports. Other suppliers of consequence included Austria (1994 only), Belgium, France, Germany (especially in 1995) and Italy.
- Finland, Sweden and Greece made no contribution to Ireland's denim fabric imports.

Italy (See Fig. 4.19)

- Imports from the EU and the non-EU countries were of comparable sizes, especially in 1996 and 1997. In 1996, 50.5% and 49.5% of denim fabric imports were from the EU and non-EU countries respectively. In 1997, the difference between the two was even closer, 49.97% for the EU imports and 50.03 for the non-EU imports.
- Imports from both the EU and non-EU countries underwent a decreasing trend.
- Main suppliers included Belgium, France, Germany, Greece, Netherlands and Spain, of which none had an outstanding performance in the Italian import market.
- In 1996, the top four suppliers were Belgium, Greece, Germany and Spain. Their share of the total EU denim fabric import market in Italy were 30.2%, 25.1%,

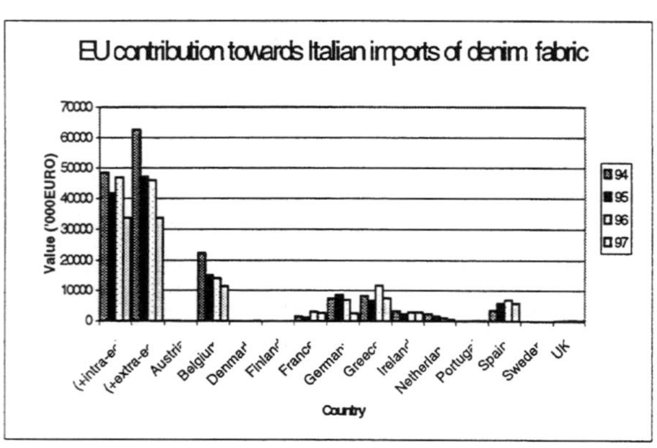

4.19

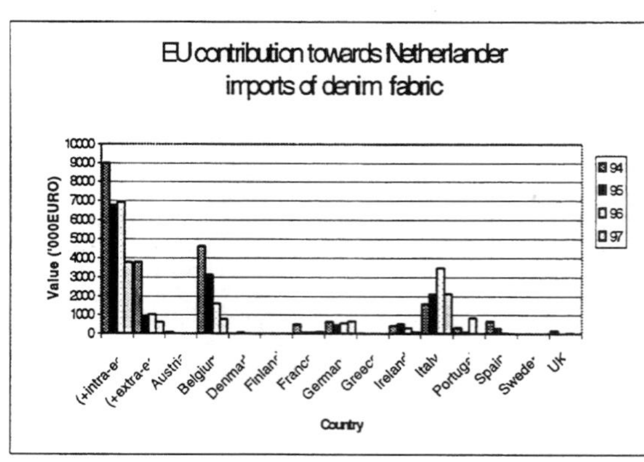

4.20

14.4% and 14.5% respectively. In 1997, the top four changed to Belgium, Greece, Spain and Ireland, accounting for 34.1%, 22.2%, 17.1% and 8.5% respectively.

- Imports from Belgium and Netherlands showed a clearly decreasing trend in terms of import value.
- There were no imports from Sweden and those from Finland were almost negligible.

<u>Netherlands (See Fig. 4.20)</u>

- As with most of the EU countries, imports from the EU were much more significant than those from the non-EU countries. In 1997, 86.0% of the denim fabric imports were sourced in the EU.
- In 1997, the leading supplier was Italy, which held about 56.2% of the EU market share. It was Belgium that ranked first in 1994 and 1995.
- Belgium, Spain and Ireland had a clear decreasing trend, while Germany had an increasing trend at a relatively slow pace.

<u>Portugal (See Fig. 4.21)</u>

- In 1996 and 1997, 86.6% and 88.7% of the denim fabrics were imported from the EU. Although the EU market share in 1997 was larger than that in 1996, the actual import value dropped drastically from 33 million Euro (38 million in total value) in 1996 to 24 million Euro (27 million Euro in total value).
- Spain was the leading supplier with a similar trade pattern to the overall EU imports. Other suppliers included Belgium, France, Germany, Greece, Italy, Netherlands and the United Kingdom.
- In 1997, Spain had 44% of the total 'intra-EU' market share, followed by Italy with 24% and Belgium with 14%.
- Italy had a very clear increasing trend. In 1994, its exports to Portugal were 2.8 million Euro and doubled to 5.7 million Euro in 1997. By sharp contrast, the exports from the United Kingdom to Portugal shrank from 1.4 million Euro in 1994 to 0.48 million Euro in 1997.

<u>Spain (See Fig. 4.22)</u>

- It is immediately obvious from the graph that the 1994 and 1995 intra-EU import value were very different from the 1996 and 1997 figures. Between 1994 and 1995, the average import value from the EU was 3.5 million but it jumped to 27 million Euro between 1996 and 1997.

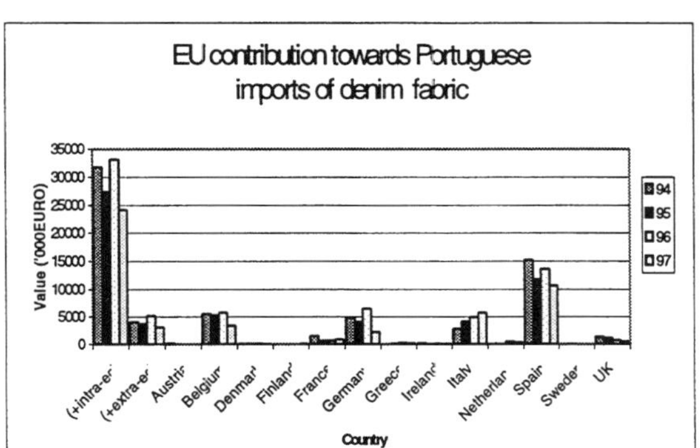

4.21

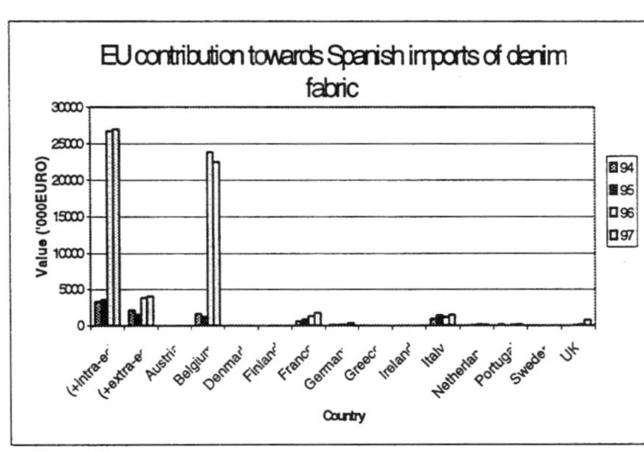

4.22

- As a result, the market share of the EU imports rose from 70% (3.6 million Euro) in 1995 to 87% (27 million Euro) in 1997.
- Such a significant change was mainly attributed to the sudden increase in imports from Belgium in 1996. It is now the single most important supplier to Spain, accounting for 83% of the EU market share in 1997.
- France had the second largest market share of 6.5%. The Italian market share was 5.5%, occupying the third place.
- Other suppliers included Germany, Netherlands, Portugal and the United Kingdom.
- Imports from France grew at a fairly constant rate and those from the UK increased significantly in 1997 though their value remains small--from 0.12 million Euro to 0.76 million Euro in 1997.

Sweden (See Fig. 4.23)

- Sweden was not an EU member until the year 1995, so only three years data were available.
- Imports from the EU had a very clear declining trend, from 0.16 million Euro in 1995 down to 97 thousand Euro in 1997. On the contrary, imports from non-EU countries enjoyed a strong growth, reaching 40 thousand Euro in 1997. Therefore, the ratio of EU imports dropped from 87% in 1995 to 71% in 1997.
- The leading supplier was Italy in 1995 and 1997. Although imports from France were larger than those from Italy in 1996, the difference between the two was only 1 thousand Euro.
- Imports from Germany slumped rapidly and significantly in 1996 and 1997. Those from Belgium decreased at a slower rate in the same period. By contrast, imports from the UK increased slowly.
- Other suppliers included Denmark, Netherlands, Portugal and Spain.

The United Kingdom (See Fig. 4.24)

- The UK was one of the few EU members which relied more on non-EU denim fabric imports.
- In 1997, imports from non-EU countries accounted for 70% of the total with a value of 89.7 million Euro. This peaked in 1996 with 91.6 million Euro (71% of the total imports). Imports from EU members peaked in 1994 with 50 million Euro, accounting for 42% of the total imports.

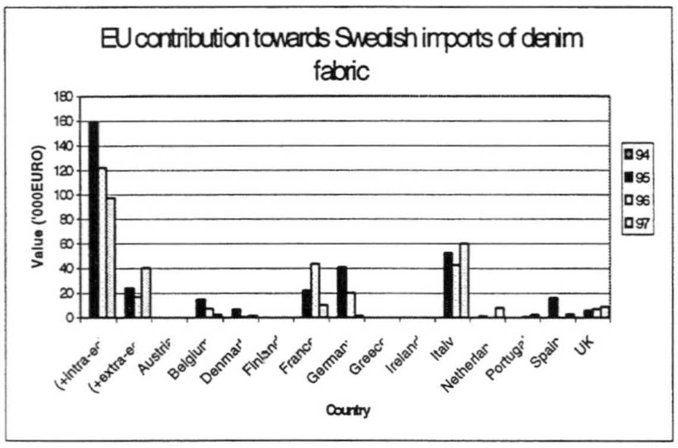

4.23

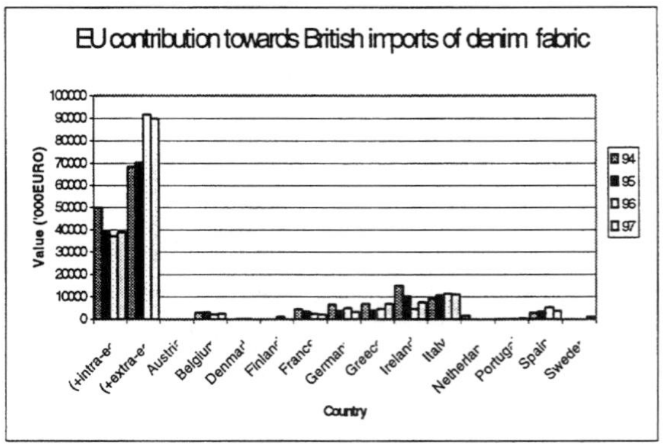

4.24

- Suppliers included Belgium, France, Germany, Greece, Ireland, Italy and Spain. None of the above countries had very outstanding performance.
- The top three suppliers in 1997 were Italy (29% of the EU imports), Ireland (19%) and Greece (18%). Ireland was the leading supplier in 1994 with 30% of the EU imports, reaching 15 million Euro.
- France had a decreasing trend while Italy had an increasing trend though the value dropped a bit in 1997.

4.2.1.2 Denim jeans

Austria (See Fig. 4.25)

- On the average, 83 million Euro of denim jeans were imported from all sources and about 87.1% (71 million Euro) of them were from the EU between 1995 and 1997.
- A peak occurred in 1996 for imports from both EU and non-EU countries, amounting to 77.2 million Euro and 14.8 million Euro respectively.
- During the observed period, the total import value was between 82.9 million Euro (1995) and 92.1 million Euro (1996).
- The leading supplier was Germany. In 1997, 54.8% of the total EU-originated imports came from Germany. Another key supplier was Italy with 23.9% of the market share. Other suppliers included Belgium, Netherlands, the United Kingdom and France.
- As regards exports from the UK and France, both had a very clear decreasing trend. In contrast, those from Netherlands enjoyed a positive growth.

Belgium (See Fig. 4.26)

- Similar to its import pattern for denim fabrics, Belgium imported more denim jeans from non-EU countries. During the observed period, 54.7%, 66.1%, 65.7% and 62.3% of denim jeans imports were from non-EU countries.
- For imports from the EU, a peak appeared in 1994 with a value of 161 million Euro; imports from non-EU countries reached the highest level in 1996 with a value of 238 million Euro.
- EU Suppliers included France, Germany, Italy, Netherlands, Portugal, Spain and United Kingdom. France had the biggest market share in 1994 but for the other years, Netherlands was the leading supplier with a positive growth in 1996 and 1997.

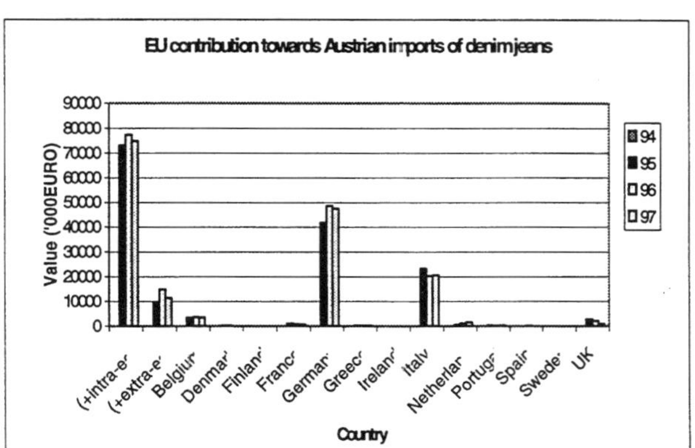

4.25

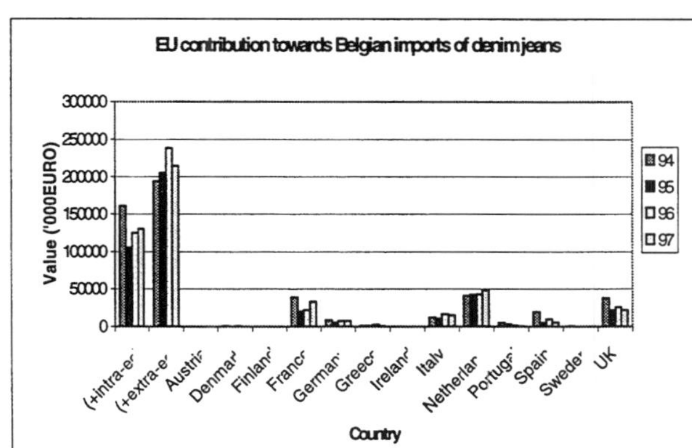

4.26

- Imports from Portugal underwent a downward trend during these years.

Denmark (See Fig. 4.27)

- Imports from the EU countries had a clear fast-growing trend, with the import value up from 34 million Euro in 1994 to 60 million Euro in 1997. The biggest 'jump' appeared between 1994 and 1995, increasing by 44.7%. About 80% of the denim jeans imports came from other EU countries in 1997.
- Denmark imported from a wide range of EU countries, including Italy, Netherlands, Sweden, the United Kingdom, Belgium, France, Germany and Greece. Imports from Italy and Netherlands were on the rise while those from Greece and the United Kingdom were decreasing.
- In 1997, Italy had the biggest share of 37% of the total imports from the EU, followed by Sweden with 20%, Netherlands with 15% and the United Kingdom with 11%.

Finland (See Fig. 4.28)

- Imports from the EU reached 21.3 million Euro in 1997, accounting for 86% of the total denim jeans imports. The market share of the EU imports was growing at the expense of the non-EU imports.
- Main contributors included Sweden, Denmark, Italy, Germany, Portugal and the United Kingdom. Sweden was the leader with the largest market share in the past few years. In 1997, for example, 55% of imported EU denim jeans came from Sweden. Italy came second with 18%.
- Imports from Italy and Sweden peaked in 1997, and those from Germany, Denmark and the United Kingdom peaked in 1996.

France (See Fig. 4.29)

- French denim jeans imports showed a different perspective from its denim fabric imports, with the non-EU suppliers having the larger share. They accounted for 66% of the total denim jeans imports in 1994, dropping to 60% in 1996 and 1997.
- In terms of import value, imports from the EU peaked in 1997 with 152 million Euro and those from non-EU suppliers peaked in 1994 with 237 million Euro.
- Belgium was the leading exporter, taking an overwhelming share in the French market. It held about 62% of the imports from all the EU suppliers. Except in 1995, it enjoyed a healthy growth in the French import market. Other major suppliers included Germany, Italy, Netherlands, Spain and the United Kingdom.

Germany (See Fig. 4.30)

- Imports from EU suppliers took a downturn despite the good performance in 1996. Conversely, those from outside suppliers enjoyed a favorable growth except for the year 1995. As a result, the gap (imports from the EU-imports from the non-EU countries) narrowed from 172 million Euro in 1994 to 50 million Euro in 1997.
- The two significant EU suppliers were Belgium and Italy, accounting for 39% and 38% respectively in 1997, followed by Netherlands and the United Kingdom with 6%, Denmark and France with 3.3%.
- Imports from Belgium peaked in 1994 and then started to decrease. Those from Italy peaked in 1996 but dropped significantly in 1997.

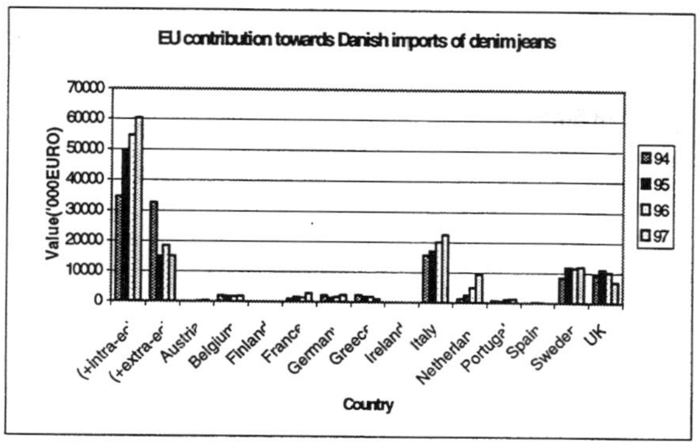

4.27

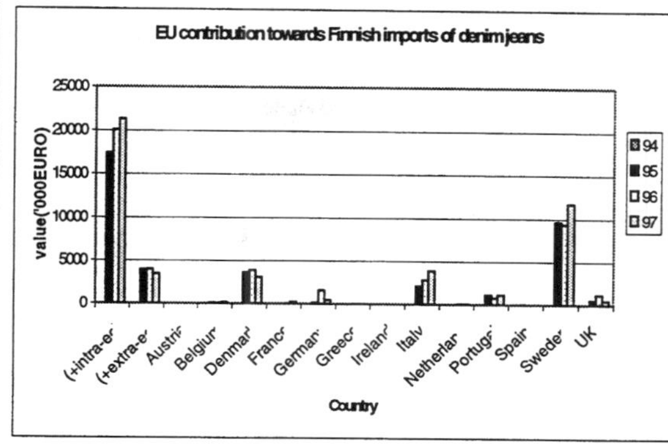

4.28

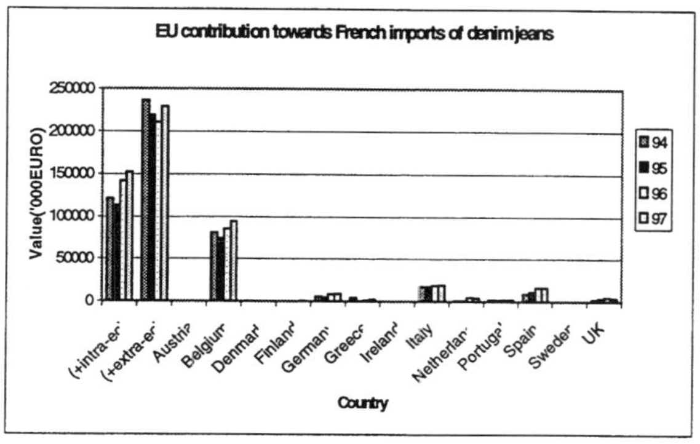

4.29

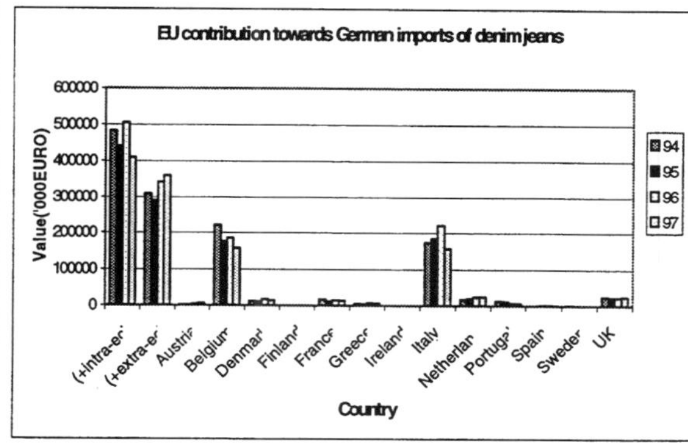

4.30

Greece (See Fig. 4.31)

- Imports from EU suppliers remained at almost the same level between 1994 and 1997 except for a slump in 1995. The average import value from the EU in these three years was 20 million Euro whereas it was only 15 million Euro in 1995.
- Although non-EU suppliers' export value underwent a remarkable increase, its market size was still much lower than that of the EU suppliers, only accounting for 29% of the total denim jeans imports in 1997.
- Belgium and Italy were of great importance to the Greek denim jeans import market. In 1997, about 58% (11.4 million Euro) of denim jeans imports came from Belgium and 25% (4.9 million Euro) came from Italy. Imports from Belgium slumped sharply in 1995 and then climbed at a relatively fast speed.
- Other EU exporters included France, Germany, Netherlands, Spain and the United Kingdom each with a similar export performance. The average market share of these countries was 3.3%.

Ireland (See Fig. 4.32)

- Denim jeans imports concentrated upon EU members, which accounted for 94% of the total imports. Both EU and non-EU exporters have increased their supplies to Ireland since 1994, though the latter's value was much less significant.
- The total import value was 66.2 million Euro, of which 90.6% (60 million Euro) were imported from the United Kingdom. The average annual growth rate of UK imports was 5.8% between 1994 and 1996, jumping to 26.7% between 1996 and 1997.

- Other EU countries including Denmark, France, Germany, Italy and Netherlands share the remaining 9.4%.

Netherlands (See Fig. 4.33)

- Imports from EU suppliers shrank continuously whereas those from non-EU countries increased. Consequently, the EU suppliers' market share dropped from 65% (158 million Euro) in 1994 to 51% (116 million Euro) in 1997.
- Belgium was the leading supplier, accounting for 64% of the total imports from the EU. Its export value peaked in 1996 with 87 million Euro.
- Other notable EU suppliers included France, Germany, Italy and the United Kingdom. Their export pattern had something in common--all of their exports to the Netherlands peaked in 1994.
- Another point worth mentioning is that imports from France dropped significantly from 20 million Euro in 1995 to 2 million Euro in 1996.
- Imports from Italy also experienced a downturn while those from the United Kingdom have picked up since 1995.

Italy (See Fig. 4.34)

- Denim jeans imports into Italy amounted to 248 million Euro in 1997. About 52.2% (129 million Euro) came from EU suppliers, decreasing from 62.6% (139 million Euro) in 1994.
- Though Belgium still had the largest share, which was 46.2% in 1997, its exports to Italy shrank very quickly. It once accounted for 80.2% in 1994.

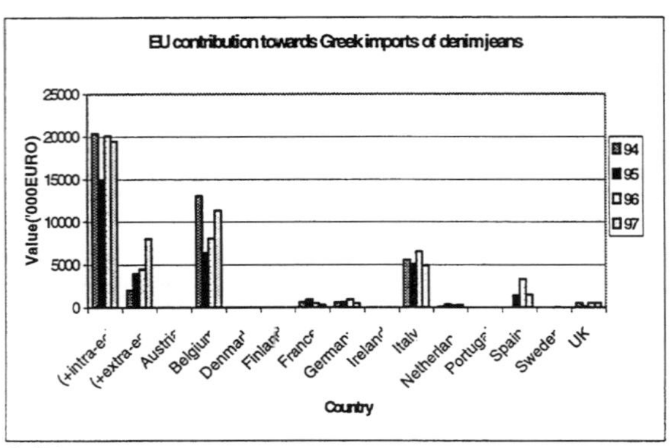

4.31

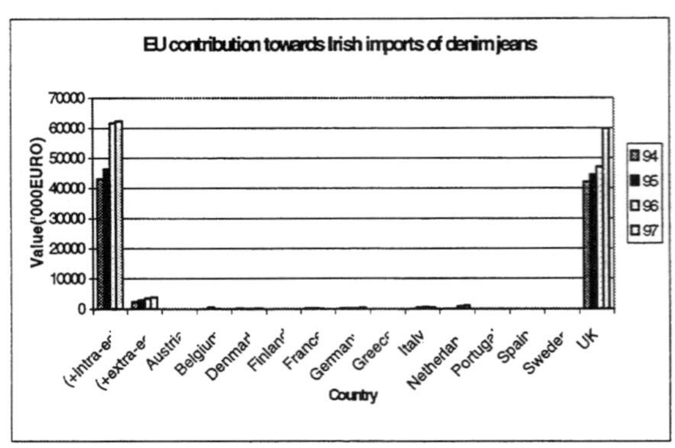

4.32

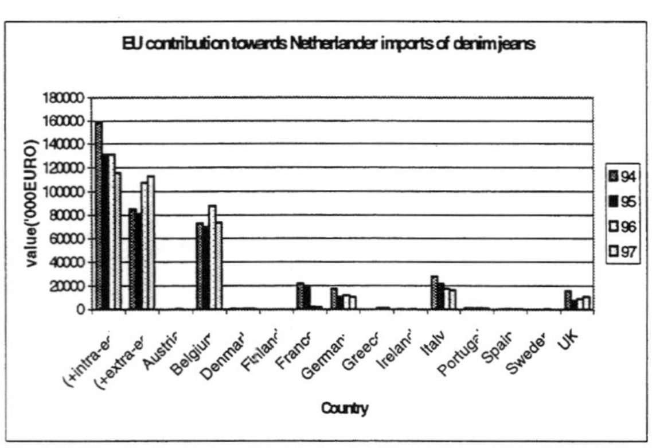

4.33

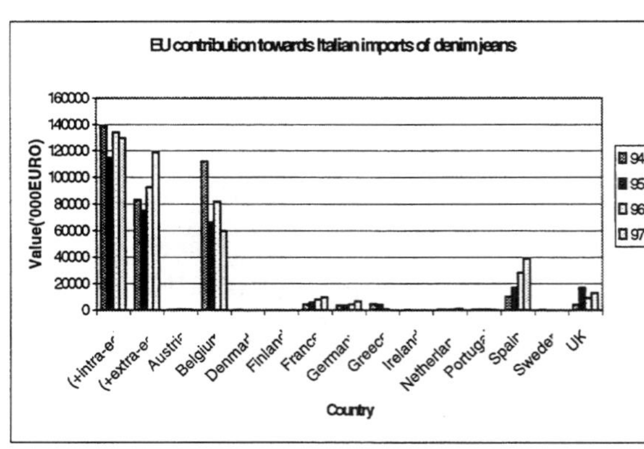

4.34

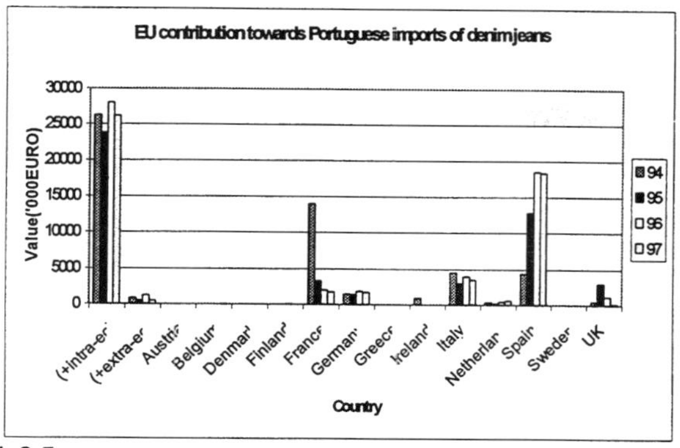

4.35

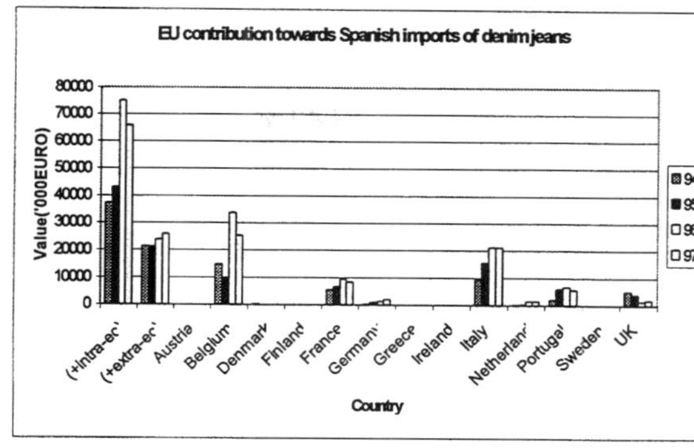

4.36

- Italy has imported more and more denim fabrics from the United Kingdom (peaked in 1995), France, Germany and Spain. Spain was the second most important supplier with 30% of the total imports from the EU, followed by the United Kingdom, France and Germany with 9.9%, 7.5% and 5.2% respectively.
- Other EU countries included Austria, Denmark, Netherlands and Portugal, holding 0.27% of the market share.

Portugal (See Fig. 4.35)

- Imports from EU suppliers dominated the domestic import market, though import value change patterns were identical between the EU and non-EU imports. On the average, 97% of the imported denim jeans came from EU countries and in 1997, the EU suppliers' share was up to 98%.
- France was the leading supplier in 1994. However, it was replaced by Spain as the most important exporter to the Portuguese market since 1995. In 1996 and 1997, Spain exported 28 million Euro (66% of the total EU supplies) and 26 million Euro (70%) worth of denim jeans to Portugal respectively. Under such heavy competition from its Spanish counterparts, imports from France shrank significantly from 14 million Euro in 1994 to 3.3 million Euro in 1995.
- Portugual also imported from other EU countries, including Germany, Italy Netherlands and the United Kingdom.

Spain (See Fig. 4.36)

- As with most of the EU countries, Spanish importers focused more on EU suppliers, reaching 75 million Euro (76% of total denim jeans imports) and 66 million Euro (72%) in 1996 and 1997 respectively.
- As regards imports from the EU members alone, Belgium and Italy were the top two suppliers, capturing 39% and 32% of the import market in 1997. Other exporters included France, Germany Netherlands, Portugal and the United Kingdom, together holding 29%.
- Italy, Netherlands and Germany increased their exports to the Spanish market while the United Kingdom found its market share dropping very quickly.

Sweden (See Fig. 4.37)

- In 1995, 53% of the imported denim jeans came from EU countries but in 1996 and 1997, this figure jumped to 77%.
- Italy was the leading EU exporter in 1995 with a market share of 68%. However, in 1996 and 1997, the United Kingdom suddenly overtook Italy, capturing 61%

and 52% of the total EU supplies. During the same period, the Italian share dropped to 30% and 26%.

- Other notable EU suppliers included Belgium, Denmark, Germany and Portugal.

The United Kingdom (See Fig. 4.38)

- As with its denim fabric import pattern, the UK imported more from the outsiders of the EU, which accounted for 59% of the total 310 million Euro imports in 1997. Imports from both EU and non-EU suppliers demonstrated strong growth since 1994 except for a relatively lackluster performance in 1995.
- The market share was quite evenly spread over Belgium, France, Ireland, Italy and the Netherlands, most of which experienced an upturn since 1994, except Italy. Imports from Netherlands increased sharply from 1 million Euro (1.7% of the total EU supplies) in 1994 to 33 million Euro (26%) in 1997. Other suppliers included Denmark, Germany, Greece, Portugal, Spain and Sweden.

4.2.1.3 Denim products

The previous sections discussed the import market of individual EU countries for denim fabrics and denim jeans in separate pictures. This section will combine the two to get some insight into the overall denim product importation in each EU member's market, including the import pattern, the leading partner and the product mix.

Austria (See Fig. 4.39)

- Although denim product imports from EU and non-EU suppliers had similar development patterns in Austria, their figures were very different. 77 million Euro (86% of the imported denim products) worth of denim products came from other EU suppliers whereas only 12 million Euro (14%) worth of denim products came from non-EU countries in 1997.
- 98% and 97% of the denim products imported from the EU and non-EU countries in 1997 were jeans.
- Most of the denim products came from Germany, accounting for about 63% of the total EU supplies, followed by Italy with 30%.
- As regards the small suppliers, France and the United Kingdom exported less since 1995 while Netherlands increased their exports to Austria.

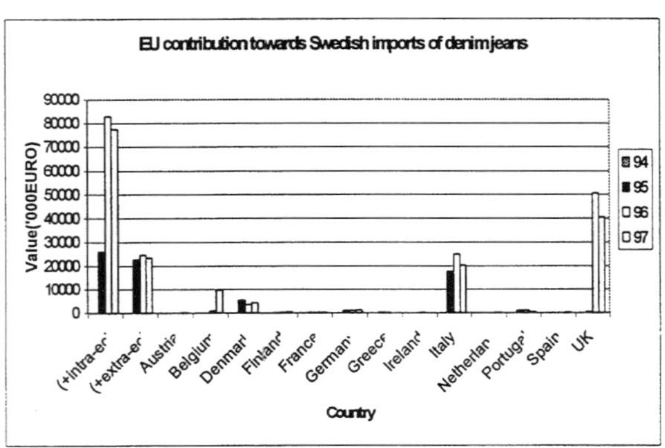

4.37

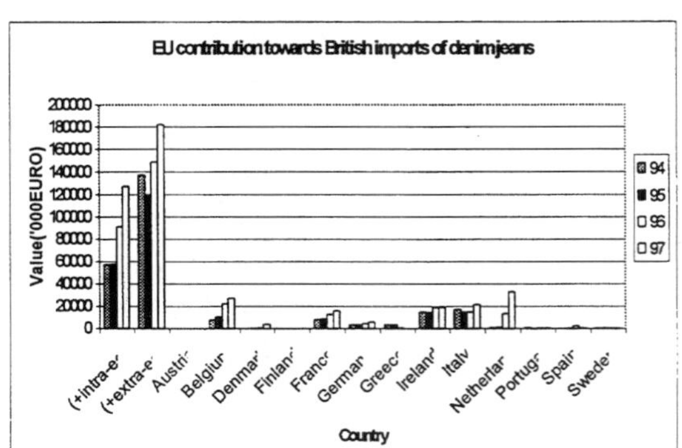

4.38

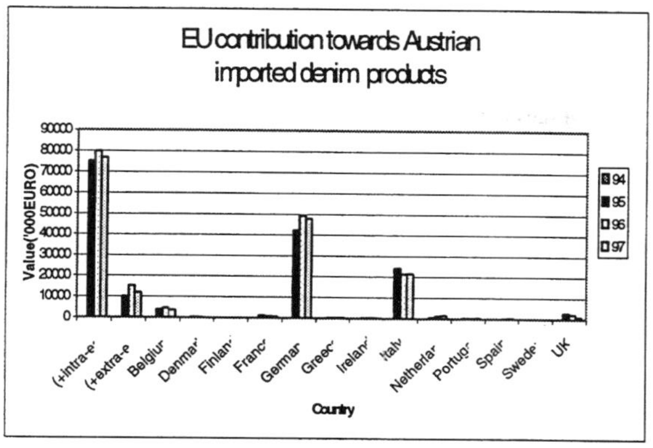

4.39

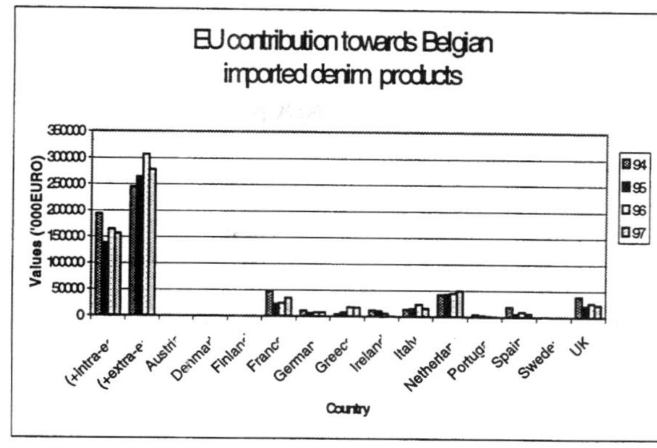

4.40

Belgium (See Fig. 4.40)

- The import pattern in Belgium is very different from that of most of the EU members. Between 1994 and 1997, its denim product imports from non-EU countries took a larger share than those from the EU, increasing from 56% in 1994 to about 65%.
- Imports from the non-EU markets reached their peak in 1996 with 306 million Euro and then dropped to 278 million Euro in 1997, a similar market share to that of 1996.
- 83% an 77% of the denim products from the EU and non-EU suppliers in 1997 were denim jeans.
- As regards EU suppliers, the French had the biggest market share in 1994 with 24% (46 million Euro), followed by Netherlands with 22% (42 million Euro). But in the next three years, the French share dropped to 17% and 15% and 22% while Netherland's shares increased to 31%, 27% and 32%, replacing France as the leading exporter to Belgium.
- Other EU suppliers included the United Kingdom, Spain, Italy, Ireland, Greece, Germany and Portugal.

Denmark (See Fig. 4.41)

- 53% (36 million Euro) of the imported denim products came from other EU members in 1994 but this share increased to 81% (62 million Euro) in 1997.
- In 1997, 97% of the denim products imported from the EU were jeans, while imports from non-EU countries were almost exclusively denim jeans.

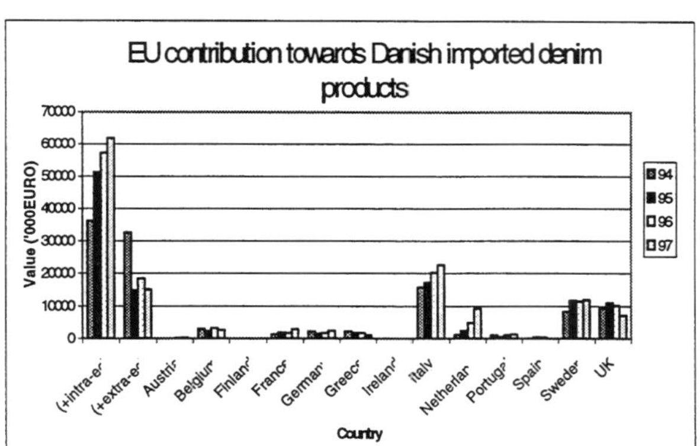

4.41

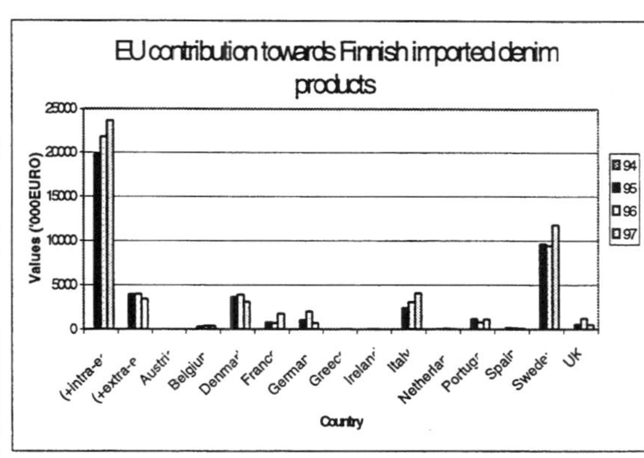

4.42

- Italy, Sweden and Netherlands were the top EU suppliers with market shares of 37%, 19% and 15% respectively in 1997. Austria, Belgium, France, Germany, Greece, Portugal and Spain together shared another 17% of the EU market.
- Imports from Italy and Netherlands increased rapidly, especially those from Netherlands, with a growth rate of 89% from 1996 to 1997.

<u>Finland (See Fig. 4.42)</u>

- EU suppliers held an overwhelming share of 87% of the total denim product imports, amounting to 24 million Euro in 1997, the result of also took an upward trend since 1995.
- 90% and 100% of the denim products imported from EU and non-EU suppliers in 1997 were denim jeans.
- Half of the imported denim products, valued at about 12 million Euro, came from Sweden in 1997. Denmark and Italy had shares of 13% and 17% respectively. Although imports from Italy have gone up continuously since 1995, it still has a long way to catch up with Sweden.
- Imports from the United Kingdom and Germany peaked in 1996 with 5.6% and 9.3% of the total imports from the EU countries, then dropping to 2% and 3% respectively in 1997.
- French market size increased significantly from 0.7 million Euro in 1996 to 1.8 million Euro in 1997, or 8% of the market.
- Other EU suppliers included Belgium, Portugal and Spain, together capturing another 7%.

<u>France (See Fig. 4.43)</u>

- Though non-EU suppliers exported more denim products to France than EU suppliers, their share dropped from 61% in 1994 to 55% in 1997.
- 79% of the imported EU denim products and 96% of the imported non-EU denim products were denim jeans.
- Belgian was the leading exporter, accounting for 61% of the imported EU denim products. Italy came second with 14%.
- Denim product imports from Spain, the third leading supplier, increased from 11 million Euro in 1994 to 23 million Euro in 1997.
- EU suppliers holding much smaller market shares included Germany, Greece, Netherlands, Portugal and the United Kingdom.

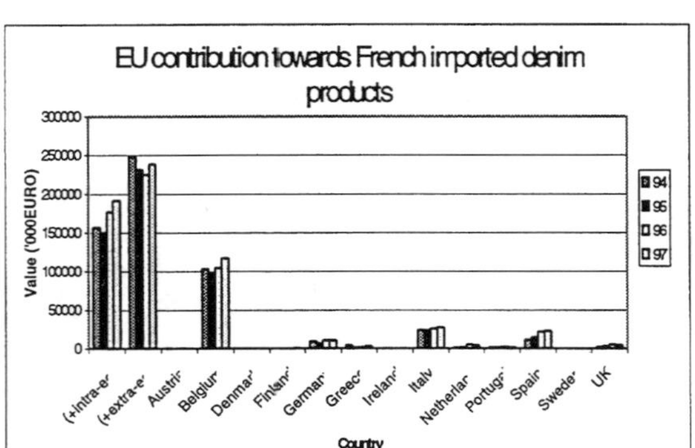

4.43

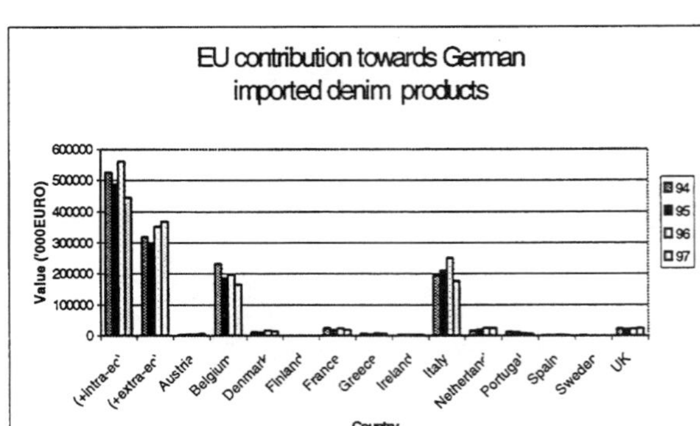

4.44

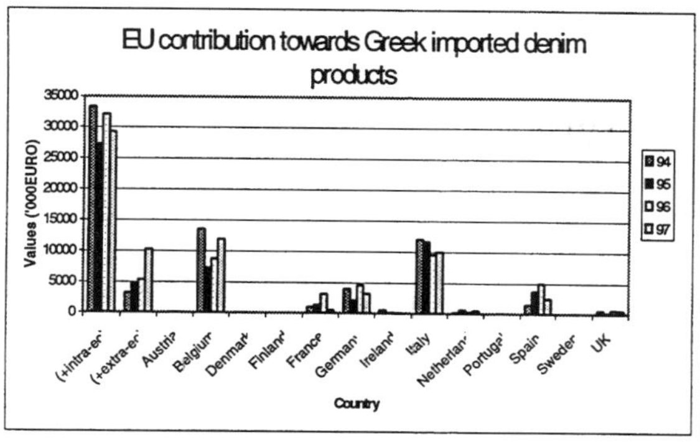

4.45

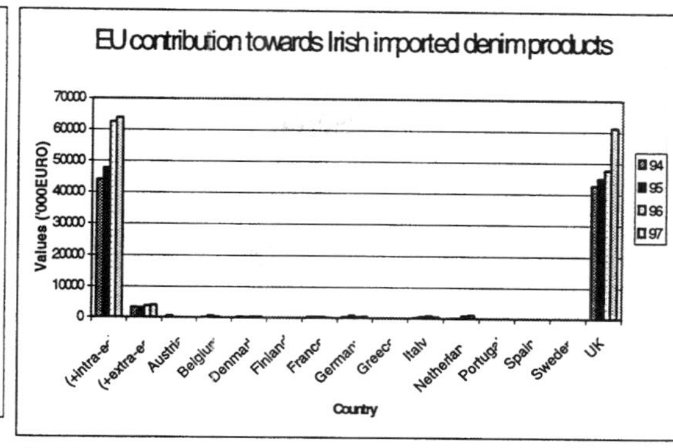

4.46

Germany (See Fig. 4.44)

- Though Germany still imported more denim products from other EU countries, the gap between the two markets (EU suppliers and non-EU suppliers) became smaller. The EU market share had decreased from 62% in 1994 to 55% in 1997.
- Imports from EU suppliers peaked in 1996 with 560 million Euro and those from non-EU suppliers peaked in 1997 with 368 million Euro.
- 92% and 98% of the denim products imported from EU countries and non-EU countries were denim jeans.
- Italy and Belgium were the two key EU suppliers, accounting for 77% of the total EU supplies in 1997, Italy having 40% and Belgium 37%.
- Other EU exporters included Austria, Denmark, France, Greece, Ireland, Netherlands, Portugal, Spain and the United Kingdom. Netherlands and the United Kingdom each had 5.5%, followed by France with 4.3% of the market.

Greece (See Fig. 4.45)

- The EU countries' market share dropped from 91% in 1994 to 74% in 1997. Although non-EU suppliers thus gained a more notable presence in the Greek import market, it will be difficult for them to attain the same export level as their EU counterparts within the next few years.
- In 1997, 67% and 78% of the EU and non-EU supplies were denim jeans.
- As in Germany, Belgium and Italy were very important suppliers to the Greek denim product import market, accounting for 75% of the total EU supplies in 1997, of which Belgium held 41%. This share was two percentage points lower than in 1994, due to the drop of Italian exports from 36% in 1994 to 34% in 1997.

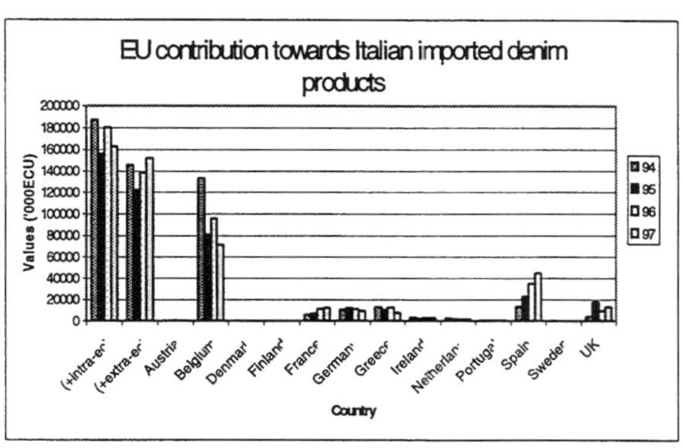

4.47

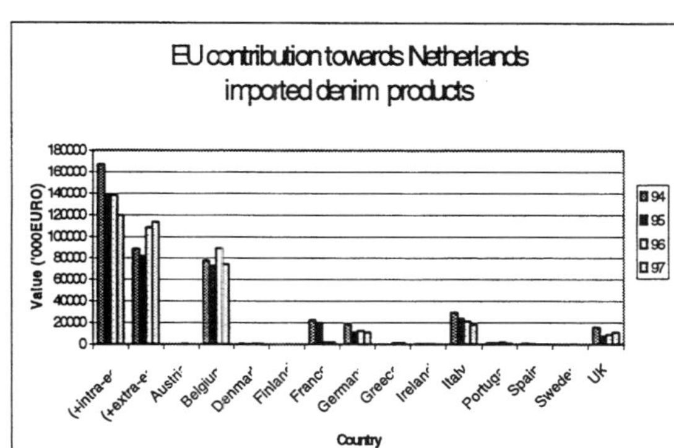

4.48

- Germany and Spain ranked third and fourth with 11% and 9% respectively.
- Other EU exporters included France, Netherlands and United Kingdom. It is worth noting that imports from France slumped from 10% in 1996 to 1.7% in 1997.

Ireland (See Fig. 4.46)

- On average, 94% of the imported denim products in Ireland came from other EU countries. The EU market size enlarged significantly in 1996, increasing from 48 million Euro in 1994 to 62 million Euro.
- In 1997, 98% of the denim products imported from the EU were denim jeans. Imports from non-EU suppliers were almost all denim jeans.
- The United Kingdom was the single most important force in the Irish import market, capturing about 96% of the total EU supplies. It enjoyed a very fat growth rate of about 30% between 1996 and 1997.
- Netherlands and Germany came second and third with 1.9% and 0.85% in 1997 respectively.

Italy (See Fig. 4.47)

- EU and non-EU supplies almost equally divided the Italian denim product import market. The market share of the EU suppliers decreased from 56% in 1994 to 52% in 1997.
- 79% and 78% of the denim products imported from EU and non-EU suppliers were denim jeans.
- Belgium was still the leading exporter despite its less significant performance since 1995. Its market share was 71% in 1994 but this dropped rapidly to 44% in 1997. On the other hand, Spain quickly emerged as the second leading supplier to Italy with a 27% market share in 1997, and the export value growing from 13 million Euro (7%) to 45 million Euro. Imports from France enjoyed a steady growth as well, from 5.9 million Euro (3%) in 1994 up to 12 million Euro (8%) in 1997.
- The market shares of German, Greek and British exports ranged from 5% to 8%. Imports from Ireland and Netherlands accounted for 2% and 1% respectively.

Netherlands (See Fig. 4.48)

- Imports from EU suppliers dropped from 167 million Euro in 1994 to 119 million Euro in 1997 while those from the non-EU suppliers increased to 113 million Euro.
- 97% of the imported denim products from EU countries and 99% from non-EU countries were denim jeans.
- Belgium was the No. 1 supplier to the Netherlands. In 1994, for example, its market size was 78 million Euro, capturing 47% of the total EU supplies. Italy came second with 29 million Euro and an 18% market share. However, in 1997, imports from Belgium, though declining to 75 million Euro, accounted for 62% of the total EU supplies, indicating the overall decreasing trend in other EU exporters.
- Imports from France experienced the most drastic decrease by 89% from 1995 to 1996. Those from Germany and the United Kingdom also declined significantly from 1994 to 1995, 40% for Germany and 53% for the UK. Nevertheless, UK

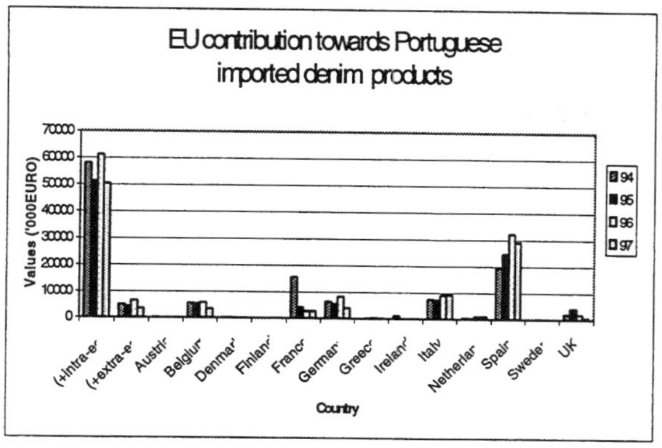

4.49

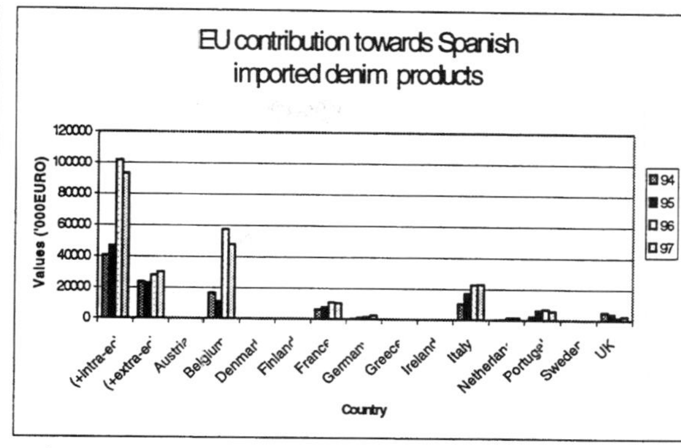

4.50

- exports recovered progressively since 1996, accounting for 19% of the total EU supplies.

<u>Portugal (See Fig. 4.49)</u>

- On average, 92% of the imported denim products came from other EU countries and a peak appeared in 1996 with 61 million Euro.
- Within the EU market, 45% to 52% of the denim product imports were denim jeans but for the non-EU market, the figure was 12% to 20%.
- Spain was the leading supplier with a market share increasing from 34% in 1994 to 58% in 1997, although the 1997 value was 3 million Euro less than that of the previous year.
- Others EU suppliers included Belgium, France, Germany, Italy, Netherlands and United Kingdom. The French market share dropped significantly between 1994 and 1995, from 16 million Euro to 4 million Euro.

<u>Spain (See Fig. 4.50)</u>

- Imports from EU suppliers increased rapidly between 1995 and 1996. In 1994, their market size was 41 million Euro, but in 1996 it jumped up to 102 million Euro and then dropped a little to 93 million Euro in 1997. As a result, their share increased from 63% to 79% in 1996 and fell to 76% in 1997.
- Unlike other EU countries, denim jeans imports from the EU countries slumped heavily from 92% to 71% while those from outside countries fell from 91% to 87%.

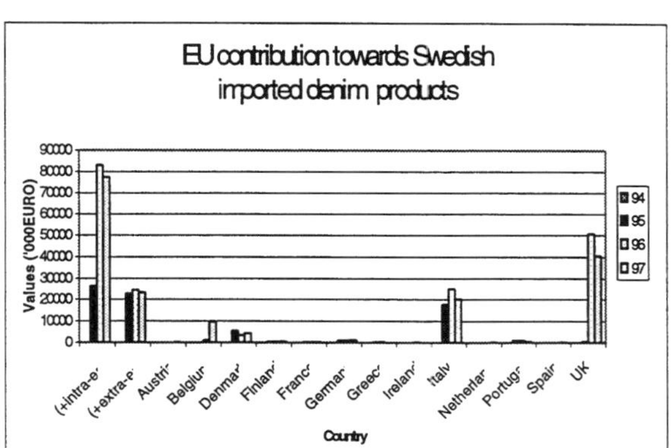

4.51

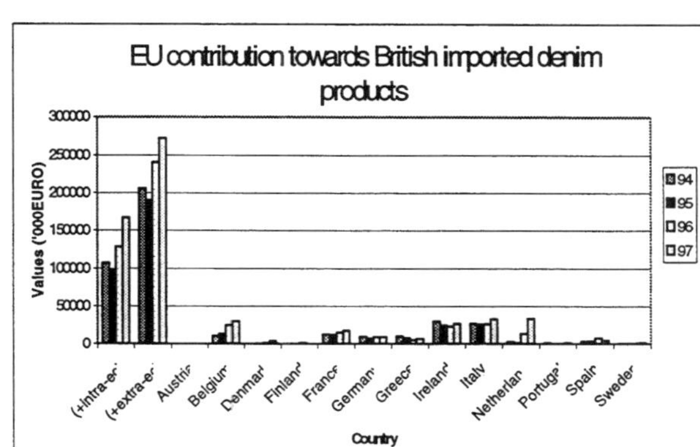

4.52

- A sharp increase from the Belgian market, which was up from 11 to 58 million Euro, helped explain the sudden growth of the total EU supplies in 1996
- Italy was the second most important exporter, accounting for 24% of the total EU supplies in 1997.
- Other EU suppliers included France, Germany, Netherlands, Portugal and the United Kingdom.

<u>Sweden (See Fig. 4.51)</u>

- As in Spain, imports from the EU suppliers experienced a sudden growth in 1996, jumping to 83 million Euro. Therefore, the EU countries' market share increased from 54% in 1995 to 77% in 1996 and 1997. The import value dropped a little to 76 million Euro in 1997.
- The denim product imports, either from EU and non-EU suppliers, were composed almost exclusively of denim jeans.
- A sudden increase from the UK market contributed a lot to the good performance of EU exports. It grew from 0.4 million Euro in 1995 to 51 million Euro in 1996 taking 61% of the total EU supplies. However, in 1997, its market share dropped to 52%.
- The other important supplier was Italy, which accounted for 30% of the EU supplies with a value of 25 million Euro in 1996. Belgian exports grew rapidly as well, from 0.9 million Euro in 1996 to 9.8 million Euro in 1997.
- 5.6% of the EU supplies were attributed to Denmark. Germany and Portugal also exported to Sweden but their value was insignificant.

<u>The United Kingdom (See Fig. 4.52)</u>

- The UK importers sourced more from non-EU suppliers, reaching 272 million Euro in 1997 comparing with the 167 million Euro of the EU suppliers. However, the EU market share increased from 34% in 1994 to 38% in 1997. Both markets had similar behavior, peaking in 1997 and hitting the lowest in 1995.
- Denim jeans played an important role in both markets--77% in the EU market and 67% in the non-EU market in 1997.
- In terms of sizes of market share, no EU country held a dominant role. Suppliers such as Belgium, Italy and Netherlands each had about 20% of the EU supplies. Ireland had 16%, France 11% and Germany 6%. Denmark, Greece and Spain shared the rest of the import market.

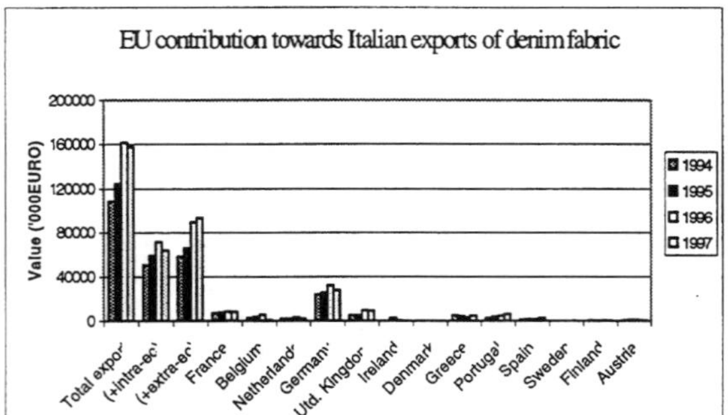

4.53

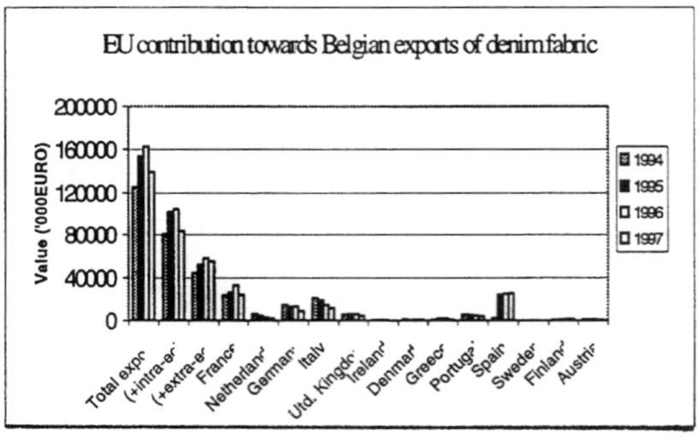

4.54

4.2.2 Exports

4.2.2.1 Denim fabric

Italy (See Fig. 4.53)

- Italy was the most important denim fabric exporter among EU countries. Its export value increased sharply in the past few years, growing from 108 million Euro in 1994 to 158 million Euro in 1997 at an average growth rate of 11.3% each year.
- Exports to non-EU countries accounted for 59.4%, increasing at an annual rate of 15.2% between 1994 to 1997. At the same time, exports to the EU market also grew from 50 million Euro in 1994 to 64 million Euro in 1997.
- Germany, the most important consumer for the Italian denim fabric industry, imported 17.8% of Italian denim fabrics in 1997 and its import value increased by 4.0% per year between 1994 to 1997.
- Other important consumers included the United Kingdom, France, Portugal and Greece with 6%, 5.3%, 3.9%, and 3.0% market share respectively in 1997.

Belgium (See Fig. 4.54)

- 139.0 million Euro worth of denim fabrics were exported in 1997 with an average annual growth rate of 2.78% between 1994 to 1997.
- More than 60% of exported denim fabric went to other EU countries although the actual value fluctuated between 1994 to 1997.
- Exports to non-EU countries grew at the rate of 6% per year between 1994 to 1997.
- The most important customer was Spain which imported 18.3% of Belgian denim fabric exports in 1997, with the average import value around 25 million Euro.
- Others importers included France (17.4%), Italy (8.4%) and Germany (6.4%) in 1997.

Spain (See Fig. 4.55)

- Spain was an important denim fabric exporter among EU members with an export value of 96.7 million Euro in 1997, 16.7 million Euro more than that in 1994.
- In 1997, Spain exported 43 million Euro worth of denim fabric to other EU countries and 53.6 million Euro of denim fabric to non-EU countries.
- The main customers among EU countries were Portugal (imported 13.4% of Spanish exported denim fabric), France (7.7%), Italy (6.6%), Netherlands (4.4%) and Germany (3.7%).
- Exports to Portugal fell from 24 million Euro in 1994 to 13 million Euro in 1997 and its correspondent market share reduced from 30.5% to 13.4%.

Germany (See Fig. 4.56)

- Germany was the fourth largest denim fabric exporter in the EU.
- German denim fabric exports peaked in 1996 and then dropped to 65.9 million Euro in 1997.
- Exports to non-EU destinations accounted for 74.5% of Germany's denim fabric exports in 1997 and the average growth rate was 4.38% per year during the observed period.

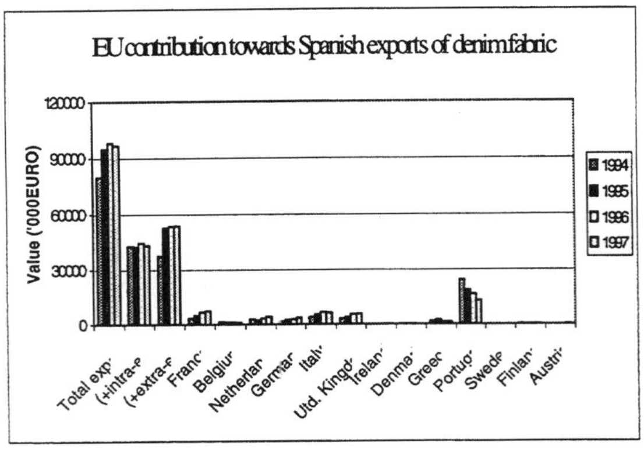

4.55

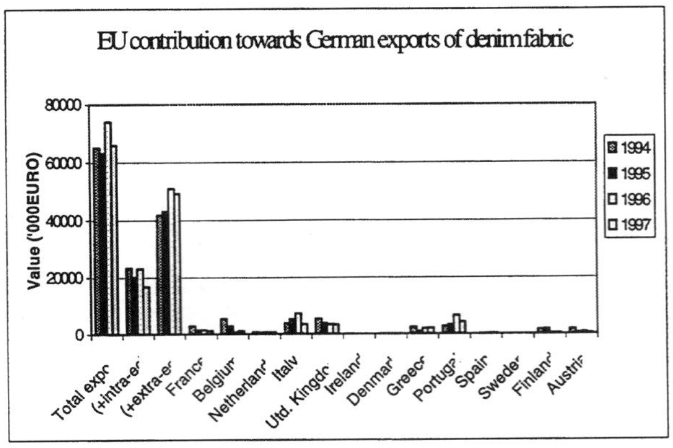

4.56

- For the EU countries, Portugal, United Kingdom and Italy were the top three importers, capturing 6.3%, 5% and 5% of Germany's denim fabric exports respectively.

Ireland (See Fig. 4.57)

- Denim fabric exports amounted to 58 million Euro in 1994 but in 1997 it went down to 42 million Euro, down by 27.5%.
- Denim fabric exports absorbed by EU countries reduced from 89.6% in 1994 to 61.5% in 1997; while those imported by non-EU countries rose from 6 million Euro in 1994 to 16 million Euro in 1997, growing at the rate of 41.7% per year.
- The United Kingdom was a remarkable importer and attracted 22.9% of Irish denim fabric exports in 1997. However, the import value dropped sharply since 1995 though it still accounted for about 37.1% of the total exports to the EU in 1997.
- Belgium was the leading importer for Irish denim fabric exports in 1997, accounting for 23.3% of the market. Other major importers included Germany and Italy, each importing about 6% of Irish denim fabric exports in 1997.

France (See Fig. 4.58)

- French denim fabric exports had a very clear downward trend, reducing from 54.9 million Euro in 1994 to 26.0 million Euro in 1997. Poor performance in both the EU and non-EU markets attributed to the quick fall.
- More than 72% of the French denim fabric exports were destined for other EU countries.

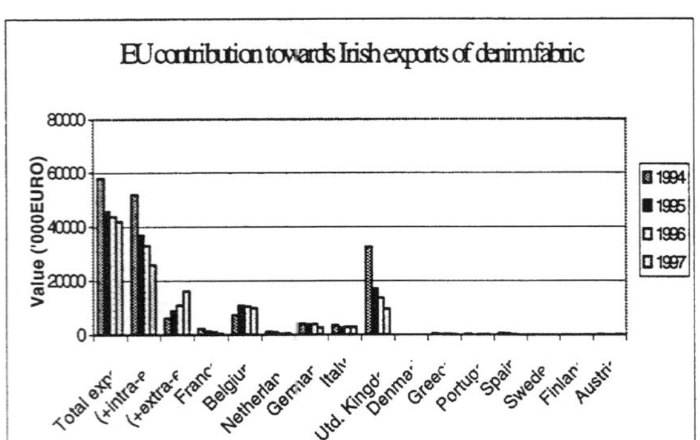

4.57

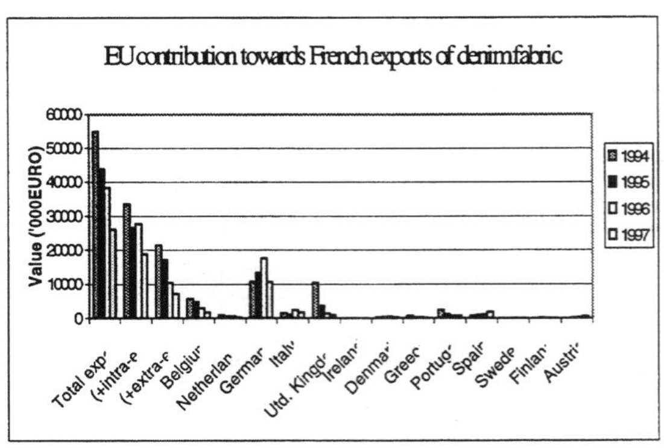

4.58

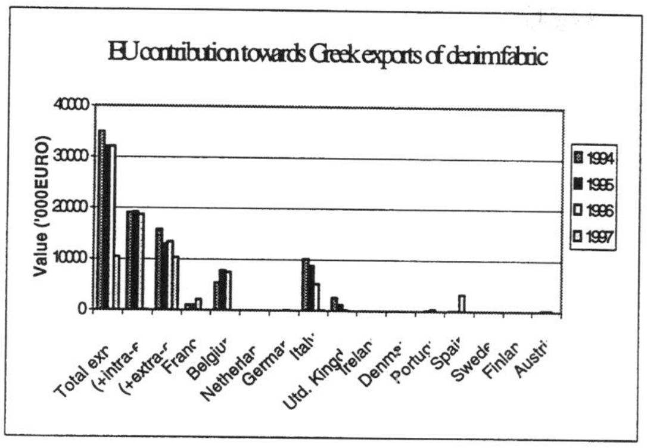

4.59

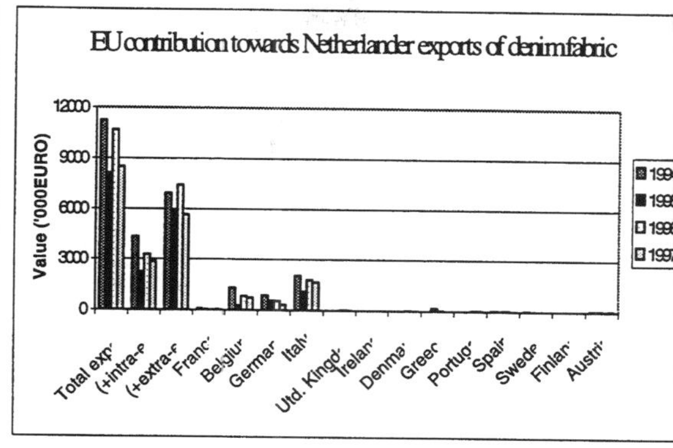

4.60

- The leading customer was Germany. The import value rose from 10.8 million Euro in 1994 to 17.5 million Euro in 1996 and fell back to 10.7 million Euro in 1997. Despite the sharp drop in 1997, Germany still accounted for more than 40% of French denim fabric exports. The UK used to be an important destination for French denim fabrics in 1994, but it was reduced to a less significant importer in 1997, having 3.6% of the market.

Greece (See Fig. 4.59)

- Greek denim fabric exports peaked in 1994 and then dropped to about 32 million Euro in 1995 and 1996. But in 1997, they suddenly slumped to 10.4 million Euro due to the sharp drop of exports to the EU market.
- Exports to other EU countries accounted for about 60% of Greek exported denim fabric before 1996. But in 1997, about 99.3% of Greek denim fabric exports were absorbed by the non-EU market.
- In 1996, the most important consumer for the Greek denim fabric industry was Belgium (imported 23.3% of exported Greek denim fabric), followed by Italy (16.4%), Spain (10.1%) and France (6.5%).
- The Italian import value decreased significantly between 1994 and 1996. In 1994, Italy was the most important consumer of Greek denim fabric, and accounted for 28.9% of Greek denim fabric export.

Netherlands (See Fig. 4.60)

- The Netherlands exported about 8.5 to 11.3 million Euro worth of denim fabric between 1994 and 1997.

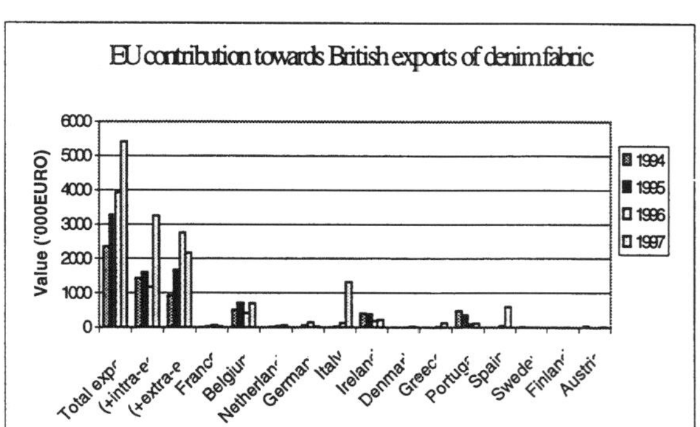

4.61

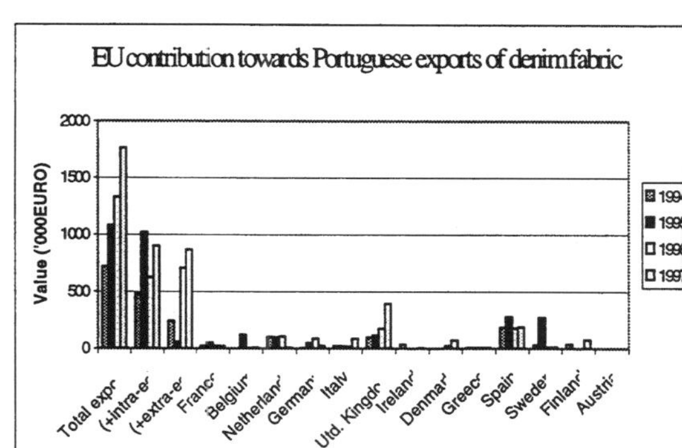

4.62

- Exports to non-EU countries were about 66.5% of its total denim fabric exports in 1997.
- Italy, Belgium and Germany were the top three leading importers, accounting for 19.8%, 8.5% and 3.9% respectively in 1997.

<u>The United Kingdom (See Fig. 4.61)</u>

- The United Kingdom exported 5.4 million Euro of denim fabric in 1997. The amount grew sharply since 1995 with a growth rate of 32.3% per year.
- Exports to EU countries increased by 31.6% per year and in 1997, 60% of the total denim fabric exports went to EU countries. At the same time, exports to non-EU countries also increased sharply with an average growth rate of 33.5% per year.
- Italy and Spain were the leading importers in 1997, though their import figures were trivial before 1997. They captured 24.5% and 11.0% of the British denim fabric exports respectively in 1997. Another important consumer was Belgium, accounting for another 13.1%.

<u>Portugal (See Fig. 4.62)</u>

- Denim fabric exports increased from 718 thousand Euro in 1994 to 1.77 million Euro in 1997 with an annual average growth rate of 36.5%.
- In 1997, about 898 thousand Euro of denim fabric were exported to EU countries, accounting for 51% of Portugal's total denim fabric exports. The non-EU countries' market share grew sharply between 1995 and 1996 and became very close to the EU level in 1997, reaching 866 thousand Euro.
- Portuguese denim fabric exports once covered many EU countries, including France, Belgium, Netherlands, Germany, Sweden, Italy, the United Kingdom, Spain, and Finland. But the first five stopped importing denim fabric from Portugal in 1997.
- In 1997, the major consumers amongst the EU members were the United Kingdom (accounting for 22.4% of Portuguese denim fabric exports), Spain (10.7%), Italy (4.9%), Finland (4.2%) and Denmark (4.1%).

<u>Austria (See Fig. 4.63)</u>

- In 1997, denim fabric exports reached 683 thousand Euro, 140 thousand Euro more than those in 1996.
- About 10 thousand Euro worth of denim fabrics (1.5%) was exported to EU members, other 673 thousand Euro denim fabrics (98.5%) went to non-EU countries.
- Only a few EU countries imported denim fabric from Austria, including Germany and Italy. Germany imported about 70 thousand Euro of denim fabric in 1996, and 10 thousand Euro in 1997. Italy imported 103 thousands Euro of denim fabric in 1995, but did not import from Austria in other years.

<u>Denmark (See Fig. 4.64)</u>

- Denmark's denim fabric exports were 357 thousand Euro in 1994, but they reduced sharply in 1995, slumping to 170 thousand Euro. The exports remained at the same level in 1997.
- Non-EU importers play a critical role in Danish denim fabric exports; about two thirds of them went to non-EU countries.

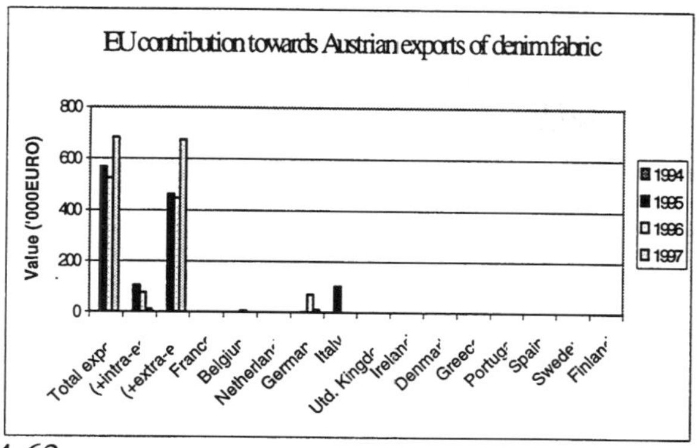

4.63

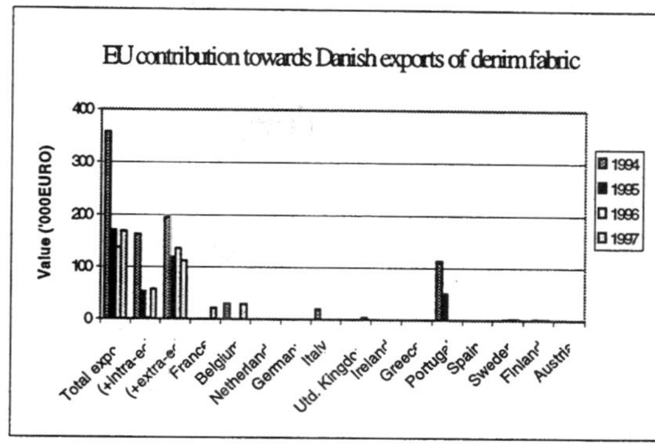

4.64

- Portugal was once the most important customer before 1996, accounting for about 30% of the total denim fabric exports. However, since then, Portugal has stopped importing denim fabric from Denmark, and Belgium and France became the most important destinations. They imported 17% and 13% of Danish denim fabric exports respectively.

Finland (See Fig. 4.65)

- Finland exported more than 150 thousand Euro worth of denim fabrics between 1994 and 1996, decreasing to 134 thousand Euro in 1997.
- Nearly all denim fabric exports were headed to non-EU countries.

Sweden (See Fig. 4.66)

- Swedish denim fabric exports grew sharply in 1997, up from 33 and 35 thousand Euro in 1995 and 1996 to 82 thousands Euro in 1997.
- Imitating the trade pattern in Finland, almost all the denim fabric exports went to non-EU markets, which accounted for 99.8% of the total Swedish denim fabric exports.

4.2.2.2 Denim jeans

Italy (See Fig. 4.67)

- Italy was the No. 1 denim jeans exporter among the EU members, growing at the annual average rate of 1.1% from 1994 to 1997. In 1997, its total denim jeans export value reached 493.1 million Euro.
- Exports to EU markets increased at 3.3% per year and peaked in 1996. In 1997, Italy exported 430.7 million Euro of denim jeans to EU members, accounting for 87.4% of Italian denim jeans exports.
- Exports to non-EU markets remained at 62 million Euro in 1996 and 1997, a drop of 30 million Euro from the export peak in 1994.
- Germany was the most important consumer, accounting for 41.0% of Italian denim jeans exports, but its market size decreased since 1994 at the average rate of around 4.9% per year. Other EU members, such as the United Kingdom, Netherlands, France, Austria and Belgium, held shares of 7.9%, 7.6%, 5.9%, 4.8% and 4.3% respectively. Except for Austria and Germany, the rest of the EU consumers increased their imports from Italy in the past few years.

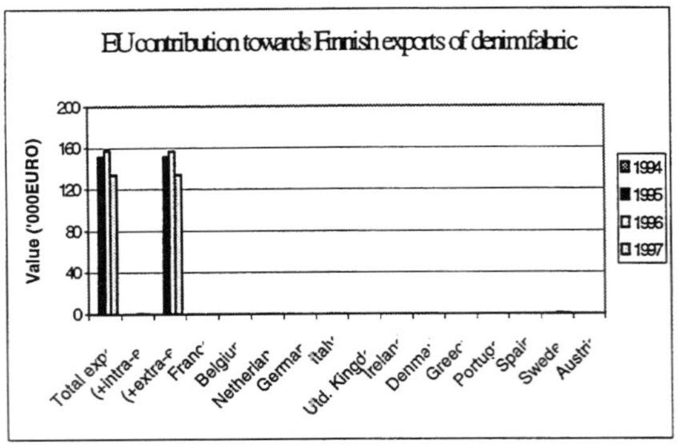

4.65

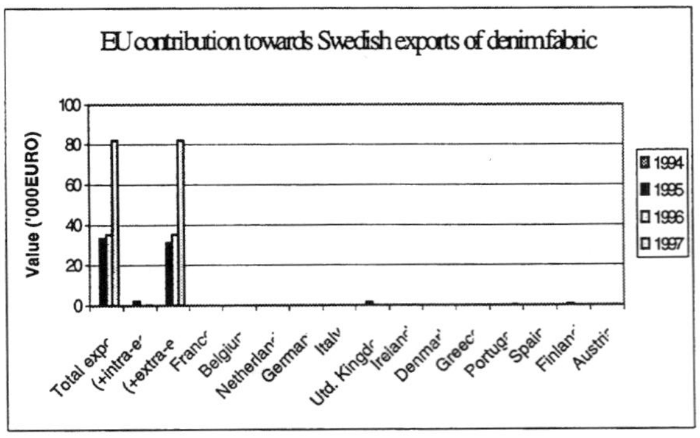

4.66

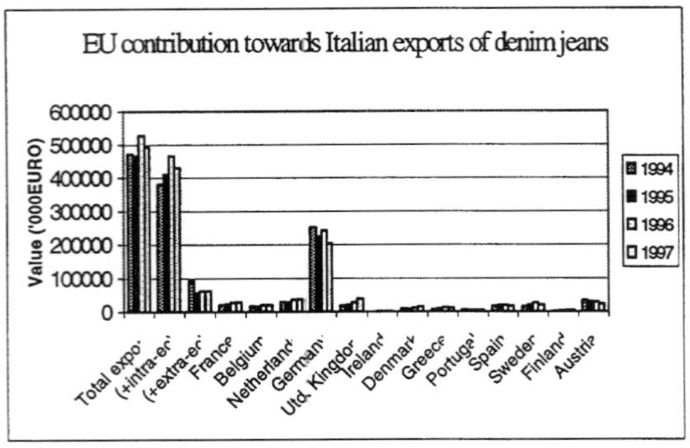

4.67

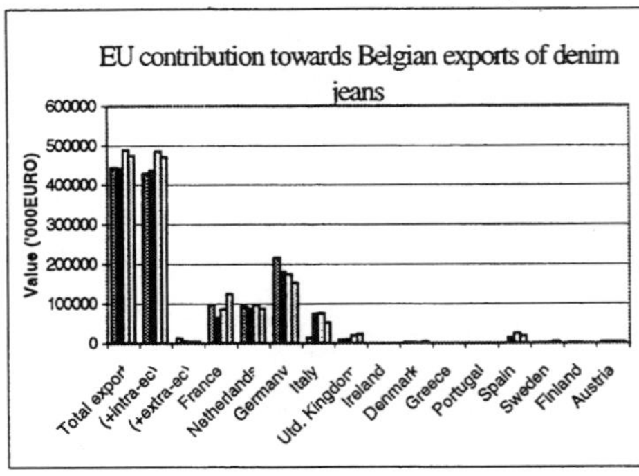

4.68

Belgium (See Fig. 4.68)

- Belgium exported more than 473.7 million denim jeans in 1997, ranking as the second most important exporter within the EU. The average growth rate of the export value was 1.73% per year between 1994 to 1997.
- Most of the Belgium exported denim jeans went to other EU countries, accounting for 97% or above since 1994.
- In 1997, Belgium's important EU consumers included Germany (32.2%), France (26.3%), Netherlands (18.6%), Italy (11.0%), the United Kingdom (4.9%) and Spain (3.9%).
- Between 1994 and 1997, exports to Germany decreased at the rate of 7.2% per year but those to France increased at an annual rate of 7.23%.

The United Kingdom (See Fig. 4.69)

- British denim jeans exports peaked in 1994 with 189.6 million Euro, slumping to 139.4 million Euro in 1995 and rising to 179.8 million Euro in 1997.
- Exports to EU members accounted for 96.5% of the total denim jeans exports, growing at 7.1% annually from 1994 to 1997. By sharp contrast, exports to non-EU markets were much smaller, decreasing fast to 6.3 million Euro in 1997.
- The top two importers were Ireland and Belgium with market shares of 34.0% and 22.7% respectively. Netherlands, Germany and Sweden each imported around 10%. Exports to Ireland, Belgium and the Netherlands, though fluctuating in some years as regards the former two countries, enjoyed an overall growth from 1994 to 1997 at the rate of 4.0%, 16.5%, and 8.3% respectively. On the contrary, exports to Germany and Sweden decreased from 1994 to 1997.

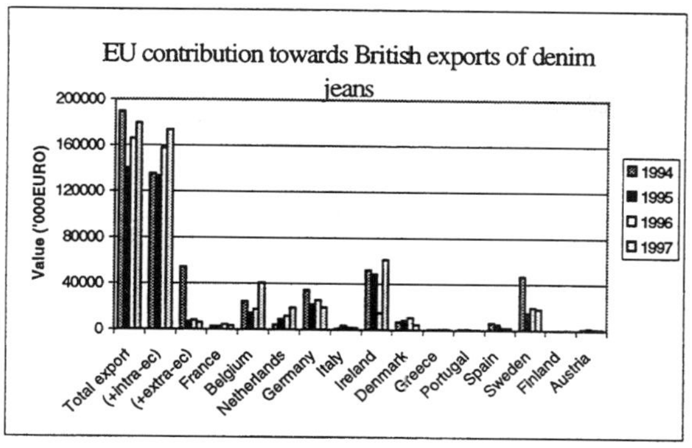

4.69

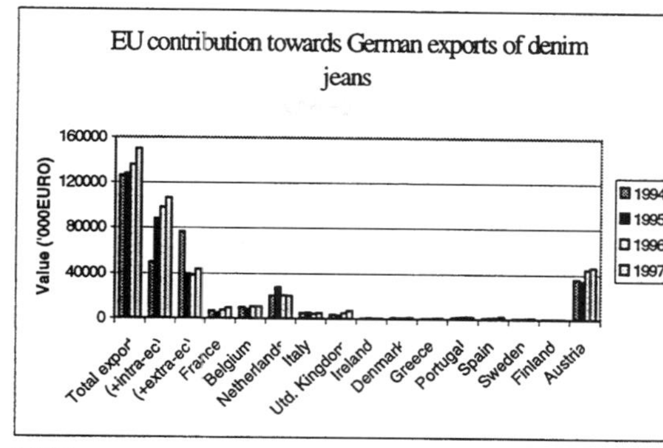

4.70

<u>Germany (See Fig. 4.70)</u>

- Germany exported 127.4 million Euro worth of denim jeans in 1997. Between 1994 and 1997, the average growing rate was 4.73% per year.
- Exports to EU destinations increased quickly, growing from 49.4 million Euro in 1994 to 106.2 million Euro in 1997 at the rate of about 28.7% per year. By sharp contrast, those to non-EU importers shrank by 10.8% each year, dropping to 43.6 million Euro in 1997.
- Austria was the most important importer, capturing 30.8% of Germany's denim jeans export with an annual growth rate of 7.3% between 1994 and 1997. Other major customers included Netherlands, Belgium, France, the United Kingdom and Italy. The former three imported 13.6%, 7.0% and 6.3% of German denim jeans respectively in 1997.

<u>Netherlands (See Fig. 4.71)</u>

- Denim jeans exports were quite steady before 1997, ranging from 81.9 million Euro to 83.8 million Euro from 1994 to 1996. The year 1997 saw a sharp increase of the value of denim jeans exports, growing to 121.2 million Euro.
- More than 97% of the denim jeans exports went to other EU countries.
- The denim jeans exports concentrated on three EU countries, namely Belgium, the United Kingdom, and Germany. They imported 37.2%, 26.5% and 21.6% of Netherlands' jeans exports respectively in 1997. The sharp rise of UK imports in 1997 was the main reason for the sudden jump in the total EU exports from the Netherlands.

<u>Spain (See Fig. 4.72)</u>

- Denim jeans exports amounted to 108 million Euro in 1997, almost a two fold increase from 1994. The annual growth rate was 41.8% between 1994 and 1997.
- About 90.5% of Spanish denim jeans exports went to other EU members. The major importers included Italy (32.8%), France (26.7%), and Portugal (20.1%) in 1997.
- Exports to Italy were not significant before 1996. However, it emerged as the leading importer in 1996 and 1997, importing 30 and 35 million Euro respectively.

<u>France (See Fig. 4.73)</u>

- Denim jeans exports grew from 90 million Euro in 1994 to 104 million Euro in 1997 at the growing rate of 3.92% per year.

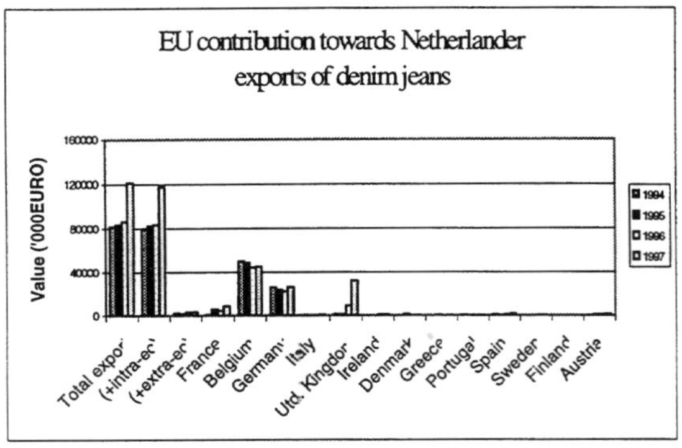

4.71

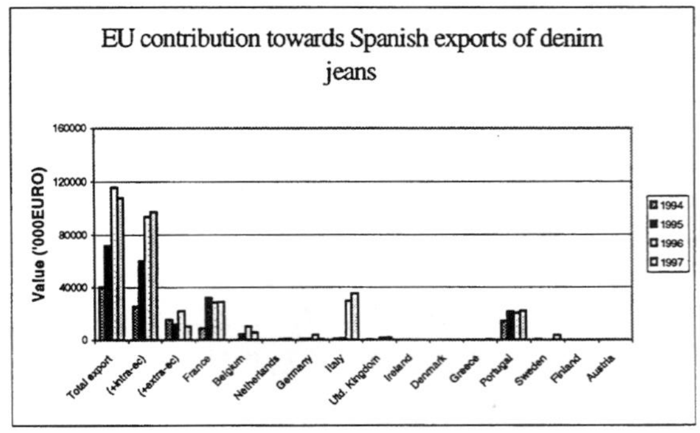

4.72

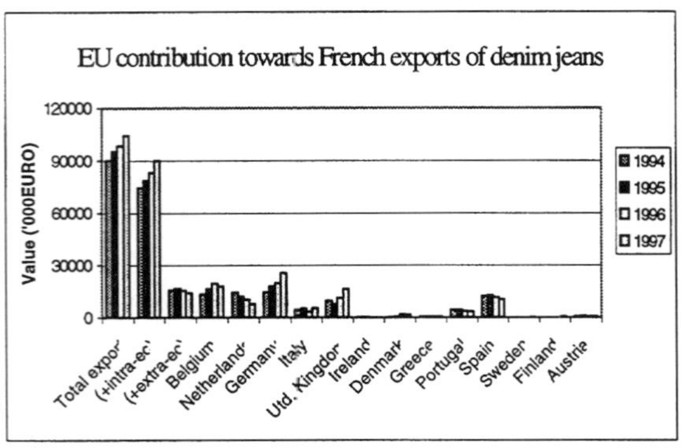

4.73

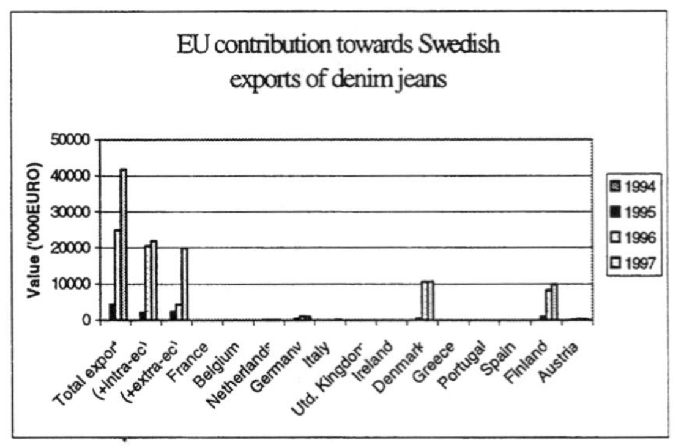

4.74

- About 86.3% of French denim jeans were exported to other EU countries.
- The important EU customers were Germany (28.2%), Belgium (19.9%), United Kingdom (18.0%), Spain (11.4%), Netherlands (8.5%) and Italy (6.1%). Exports to Germany experienced strong growth in the past few years and the rate was 18.1% per year from 1994 to 1997.

<u>Sweden (See Fig. 4.74)</u>

- When Sweden joined the EU in 1995, its denim jeans export value was very low, only about 4.2 million Euro. However, in 1996 and 1997, the export value jumped to 25 million Euro and 42 million Euro respectively.
- Exports to the EU and non-EU markets both increased by a big margin. Those to EU members accounted for 52.6% of the total denim jeans exports.
- The major EU customers were Denmark and Finland, accounting for about a half of the total denim jeans exports. In 1997, Denmark imported 10.7 million Euro worth of denim jeans from Sweden, which was 25.5% of Swedish denim jeans exports. At the same time, Finland imported 10 million Euro, capturing another 23.8%.

<u>Denmark (See Fig. 4.75)</u>

- The export value was 41.7 million Euro in 1997, dropping from the 1996 peak of 48.7 million Euro.
- Exports to EU countries played a more and more important role, accounting for 78.3% of the total denim jeans exports with an average annual growth rate of 40.7% between 1994 and 1997. The non-EU market share, once dominating in the denim jeans export market in 1994, shrank at 15.9% per year.

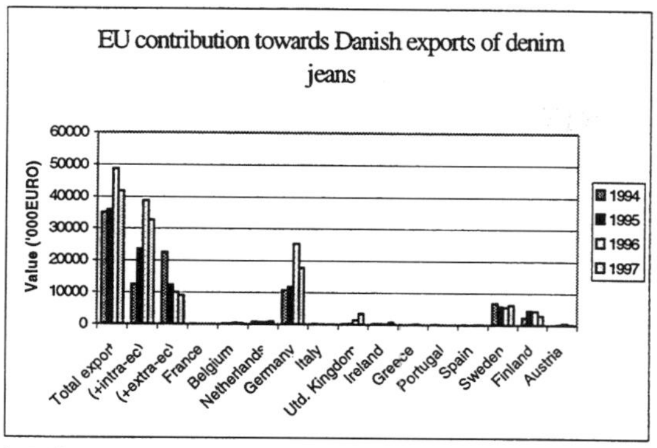

4.75

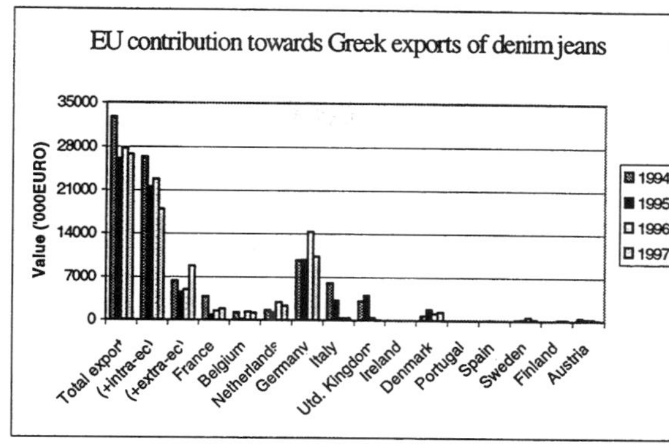

4.76

- The most important customer was Germany holding 42.8% of the total denim jeans exports, followed by Sweden (14.9%) and Finland (7.0%).

Greece (See Fig. 4.76)

- Denim jeans exports reduced from 32.7 million Euro in 1994 to 26.8 million Euro in 1997 at the annual rate of 4.6%.
- 82% of Greek denim jeans exports went to other EU countries before 1997 but the proportion fell to 67.1% in 1997. On the other hand, exports to non-EU markets kept growing at the average annual rate of 9.6% between 1994 and 1997, reaching 8.8 million Euro in 1997.
- Germany was the most important destination, importing 38.8% of Greek denim jeans in 1997. Others importers included Netherlands (8.7%), France (6.8%), Denmark (5.0%) and Belgium (4.4%).

Ireland (See Fig. 4.77)

- Denim jeans exports reached 20.5 million Euro in 1997. EU importers dominated this Irish export market (except for the year 1996), taking an overwhelming share of 97.9% of the total denim jeans exports in 1997.
- Most of the exports were absorbed by the United Kingdom, accounting for 97.4% of the total denim jeans exports in 1997.

Portugal (See Fig. 4.78)

- Denim jeans exports were over 23 million Euro in value in 1996 but suddenly dropped to 14.7 million Euro in 1997.
- The EU was the most important market for Portuguese jeans because it imported around 93% of the denim jeans exports. Amongst these members, Germany had the largest share of 38.2%. Spain came second with 27.7%. France and Finland came third and forth with 6.3% and 5.0% respectively.
- Although Germany was still the most important consumer for Portuguese jeans, its import value dropped sharply from 1994 to 1997 at the average rate of 12.9%. Sweden and Belgium also showed a clear decreasing trend.

Austria (See Fig. 4.79)

- 9.6 million Euro worth of denim jeans was exported to the world in 1997, increasing from 6.2 million Euro in 1995.
- The EU was the major market for Austrian denim jeans exports, accounting for 78.4% in 1997.

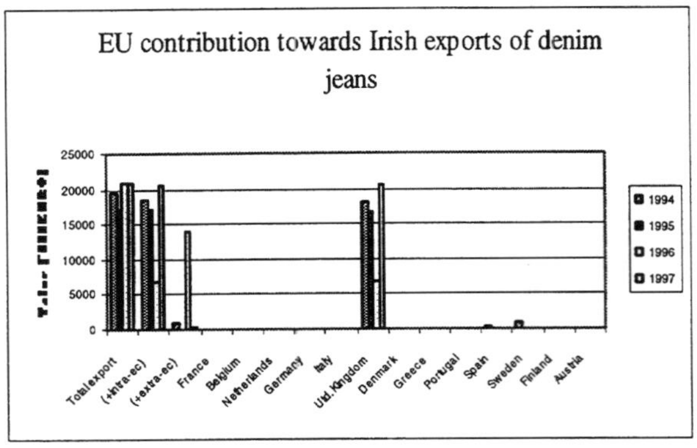

4.77

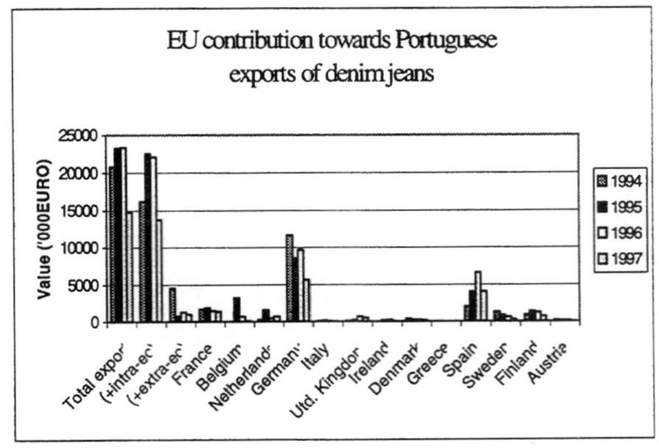

4.78

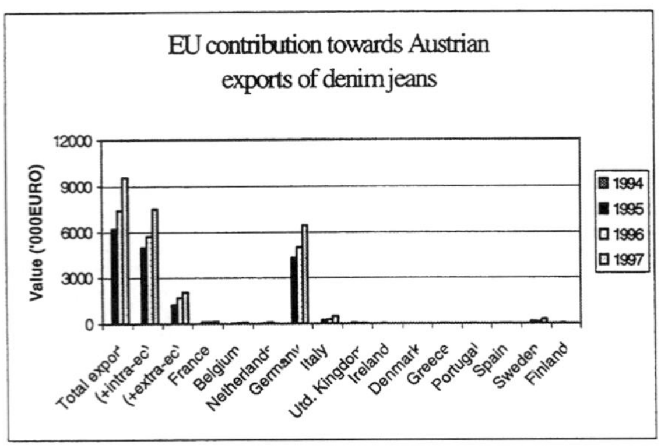

4.79

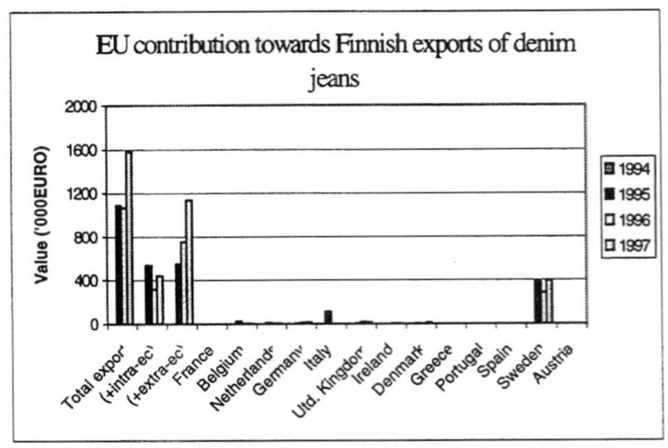

4.80

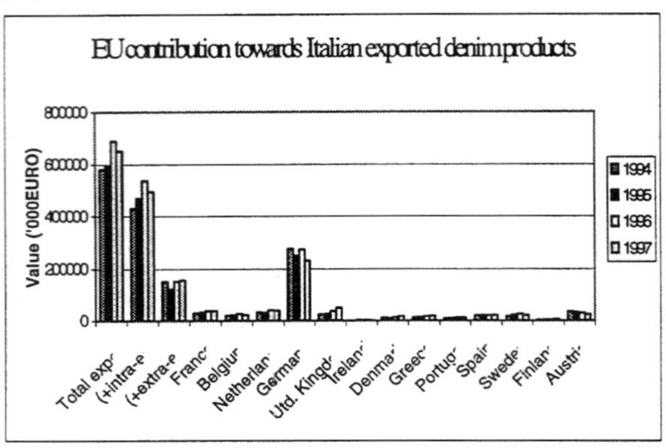

4.81

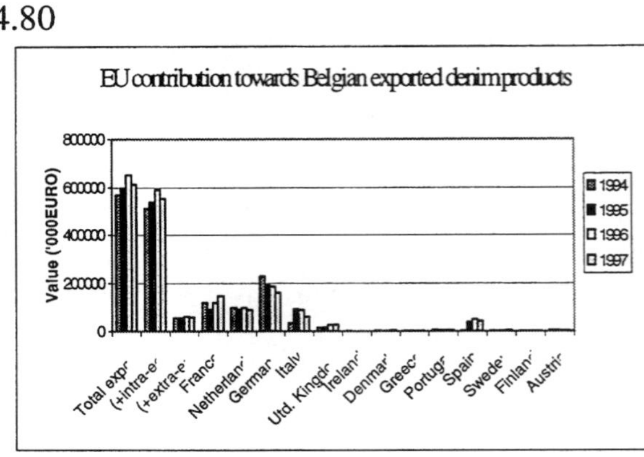

4.82

- Germany was the single key importer, absorbing 6.4 million Euro (67%) worth of Austrian denim jeans in 1997.

<u>Finland (See Fig. 4.80)</u>

- 1.6 million Euro worth of Finnish jeans was exported to the world in 1997, rising sharply from 1.1 million Euro in 1995.
- Finland was the only EU country which exported more to non-EU importers. In 1997, 72.1% of Finnish jeans were exported to non-EU destinations. What's more, the trend was upward after Finland joined the EU in 1995.
- The main EU importer was Sweden, which imported 25% of Finnish denim jeans in 1997. (In terms of EU imports alone, it accounted for 90%.) Other EU

countries' shares were very low--the United Kingdom with 0.9% and Germany with 0.8%.

4.2.2.3 Denim products

Italy (See Fig. 4.81)

- Italy was the most important exporter in the EU, both in terms of denim fabric and jeans exports. The export value grew from 1994 and reached a peak in 1996 of 689 million Euro. In 1997, the value shrank to 650.6 million Euro, accounting for 27.7% of EU members' total exports.
- Denim jeans accounted for 75.8% of the total denim product exports.
- Exports to the EU accounted for 76.0% of Italian total exports, growing at the rate of 3.75% per year.
- Italian denim products had a wide range of export destinations. The most important EU importer was Germany, which imported about 35.4% of Italian denim product exports in 1997. The next three were the United Kingdom (7.5%), Netherlands (6%), and France (6%). The other EU members also imported denim products from Italy, but with a market share less than 3.8%.

Belgium (See Fig. 4.82)

- Belgium was the second most important exporter in the EU denim export market, amounting to 612.7 million Euro in 1997, or 26.2% of EU members' total denim exports.
- Denim jeans exports accounted for 77.3% of all Belgian denim product exports.
- Its export market mainly concentrated on the EU, which captured about 90% of Belgium's denim product exports.
- Germany was again the most important destination, accounting for about 26.3% of the market. France ranked second with 24.3%, followed by Netherlands (14.7%), Italy (10.4%) and Spain (7.2%).

Germany (See Fig. 4.83)

- Though Germany was the No. 3 exporter in the EU market, its export value was only about one third of that of Italy and Belgium. In 1997, the value reached 215.7 million Euro with a growth rate of 3.2% per year between 1994 and 1997.
- Denim jeans constituted 69.4% of the total denim product exports.
- Exports to non-EU importers were larger than those to EU members in 1994. However, with an annual growth rate of 17.3%, the EU importers overtook non-EU importers in the following three years, eventually accounting for 57.0% of German denim product exports.
- Austria was the leading importer, holding 21.5% of the German denim product exports, followed by Netherlands (9.8%), Belgium (5.3%), and France (4.9%) in 1997.

Spain (See Fig. 4.84)

- Denim product exports peaked in 1996 with 213 million Euro and then dropped to 205 million Euro in 1997. The average growth rate was 17.5% every year.
- Denim jeans were responsible for about 52.8% of the total denim product exports.
- The proportion of EU imports from Spain continuously increased from 1994, indicating that the EU market has become more and more important. The

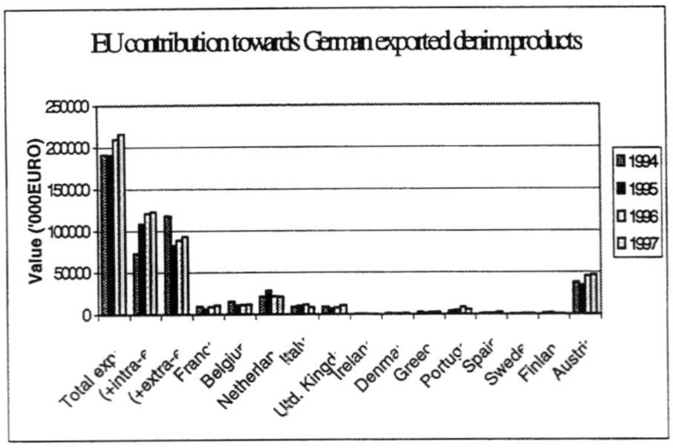

4.83

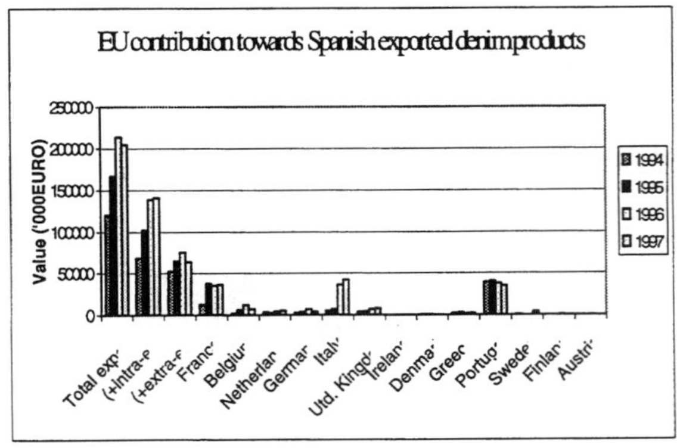

4.84

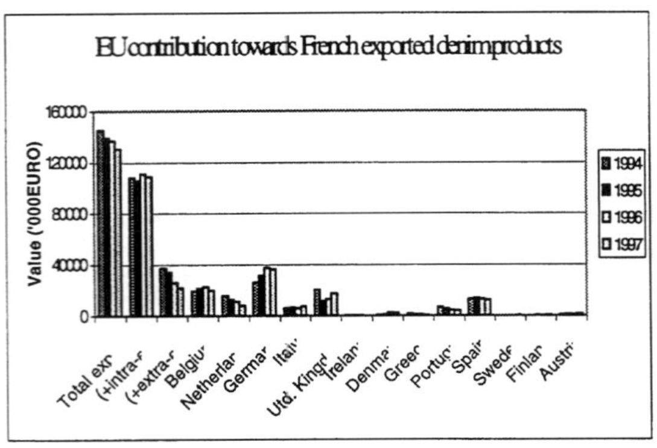

4.85

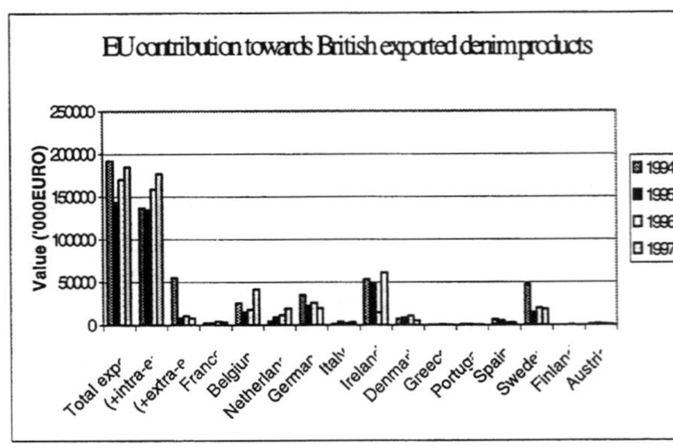

4.86

percentage was 56.4% in 1994, rising to 68.8% at an annual average rate of 26.8%. Exports to non-EU importers also increased, but at a slower speed. Consequently, its market share gave way to that of the EU importers.

- Italy, France, and Portugal were the top three importers in 1997, accounting for 20.4%, 17.7%, and 16.9% of Spanish denim product exports. Exports to Portugal were relatively stable, while those to Italy and France experienced strong growth since 1995.

France (See Fig. 4.85)

- French denim product exports decreased from 145 million Euro in 1994 to 130 million Euro in 1997 with the average reduction rate of 2.55% per year.
- Denim jeans accounted for 80% of the total denim product exports.
- Exports to EU countries played an important role, accounting for 83.5% of French total denim exports. The proportion grew 3.1% per year from 1994 to 1997. On the other hand, exports to non-EU countries diminished faster at the average annual rate of 10.6%.
- The main EU destination was Germany which imported 27.7% of the total denim product exports in 1997, growing at the annual rate of 14.3%. Other EU importers included Belgium (15.1%), the United Kingdom (13.28%), Spain (9.3%), Netherlands (6.1%) and Portugal (3.1%).

The United Kingdom (See Fig. 4.86)

- Denim product exports reached a peak of 191.9 million Euro in 1994 but fell sharply to 143.2 million Euro in 1995. The export performance picked up in 1996 and its 1997 value was close to the 1994 level, amounting to 185.2 million Euro.

- Denim jeans accounted for 97.1% of British denim product exports.
- Exports to the EU took the lion's share, especially from 1995. In 1997, 95.5% of the total exports went to the EU countries. However, exports to non-EU markets slumped in 1995 and then remained at a similar level.
- The most important customers were Ireland (33.9% of British total exports) and Belgium (22.5%). Another three EU members, namely Germany, Sweden and Netherlands, each imported about 10.0% respectively.

<u>Netherlands (See Fig. 4.87)</u>
- The value of denim product exports in 1997 was 129.7 million Euro, an increase of 36.6 million Euro over 1994.
- Denim jeans exports accounted for approximately 93.4% of the total denim product exports.
- More than 90% of Dutch denim products were exported to other EU members.
- The important customers were Belgium, the United Kingdom and Germany, having shares of 35.3%, 24.8%, and 20.4% respectively in 1997.

<u>Ireland (See Fig. 4.88)</u>
- Irish denim product exports fell from 77.4 million Euro in 1994 to 62.9 million Euro in 1997 at the annual rate of 4.7%.
- Denim jeans accounted for 33.4% of the total denim product exports.
- Irish denim product exports to the EU took a downward trend since 1995, decreasing from 90.1% in 1994 to 73.7% in 1997 at the rate of 8.5% per year. On the other hand, exports to non-EU markets had a steady upward trend at the rate of 34.9% per year.
- The United Kingdom was the leading importer with 47.7% of the market, far exceeding Belgium which was the No. 2 importer with 15.5%.

<u>Denmark (See Fig. 4.89)</u>
- About 42 million Euro worth of Danish denim products were exported in 1997, 7 million Euro down from the 1996 peak.
- Denim jeans took the overwhelming share of 99.6% of the denim product exports.
- Denim product exports to EU countries increased significantly in the past few years. Their proportion doubled from 35.6% in 1994 to 78.2% in 1997. The export value grew at the rate of 39.9% per year. In contrast, exports to non-EU markets reduced since 1994 at the rate of 15.0% per year.
- In terms of EU importers, a large proportion (42.7%) of Danish denim products were exported to Germany in 1997, followed by Sweden (14.9 %), United Kingdom (8.1%) and Finland (7.0%).

<u>Sweden (See Fig. 4.90)</u>
- Denim product exports grew at a very striking rate of 290% after Sweden joined the EU in 1995, amounting to 41.8 million Euro in 1997.
- Just as with Denmark, denim jeans accounted for 99.8% of Sweden's total denim product exports.
- Though exports to EU importers dominated the export market in 1996, they reduced quickly from 82.4% in 1996 to 52.5% in 1997. On the other hand, non-EU importers significantly increased their sourcing from Sweden in 1997.

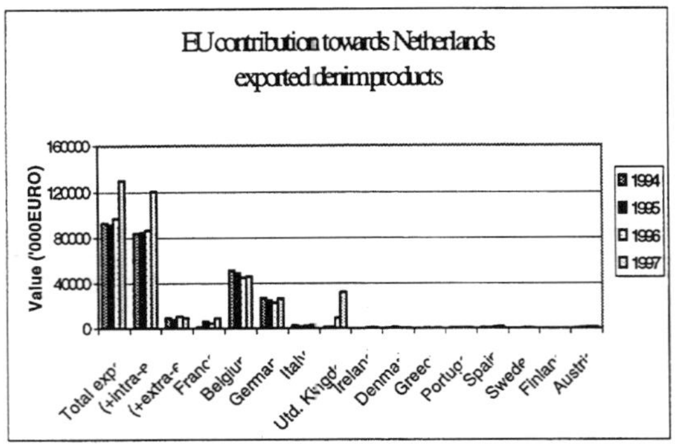

4.87

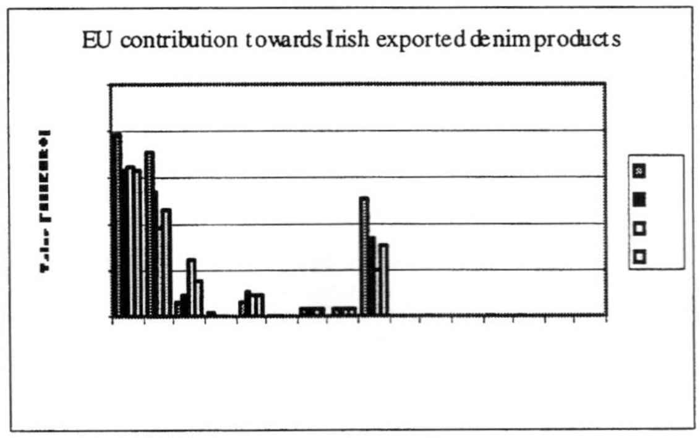

4.88

- Denmark and Finland were the leading importers. In 1997, they imported 10.6 and 9 million Euro worth of Swedish denim products respectively.

Greece (See Fig. 4.91)

- Greece exported 67.7 million Euro worth of denim products in 1994 but in 1997, the export value decreased sharply to 37.2 million Euro.
- Denim jeans accounted for 72% of the total denim product exports.
- The poor export performance was mainly due to the sudden slump of exports to EU markets, which was only 18 million Euro in 1997. Exports to non-EU countries were relatively stable, and eventually surpassed the EU markets as the more important destinations for Greek denim products, capturing about 51.6% of the market share.
- Denim product exports to EU members mainly concentrated on Germany, Belgium, and Italy. Germany was the leading importer with a market share of 27.9% in 1997.

Portugal (See Fig. 4.92)

- 21.5 million Euro worth of Portuguese denim products was exported in 1994, reaching a peak in 1996 with 24.7 million Euro. However, the value fell sharply in 1997 to only 16.5 million Euro.
- Denim jeans accounted for 89.3% of the total denim product exports.
- Exports to the EU accounted for 77.6% of the Portuguese denim product exports in 1994, increasing to 96.7% in 1995 and falling to 88.6% in 1997.
- The main EU importers were Germany, Spain, and France. The former imported about 5.7 million Euro (34.3%) worth of Portuguese denim products in 1997.

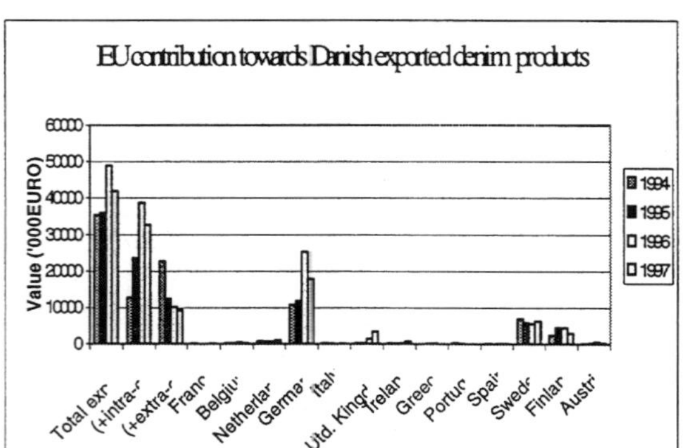

4.89

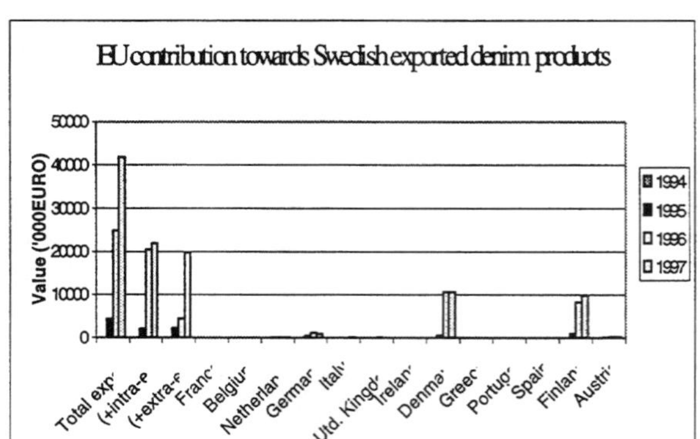

4.90

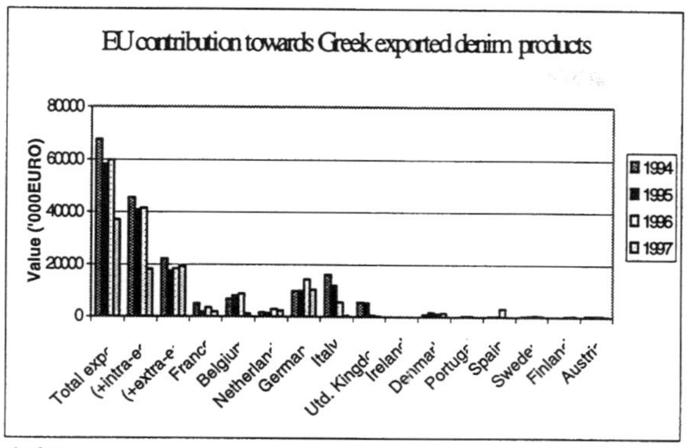

4.91

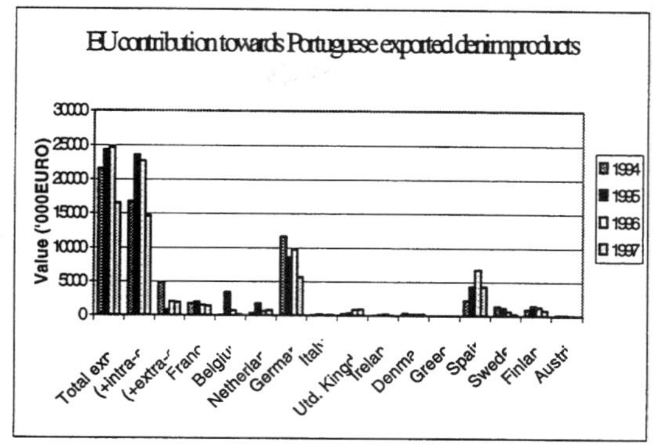

4.92

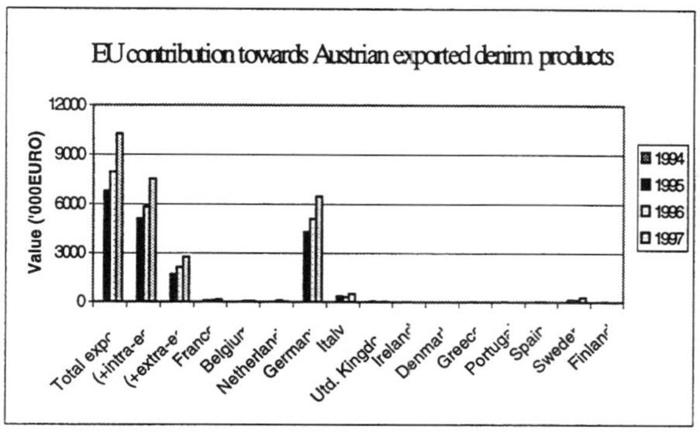

4.93

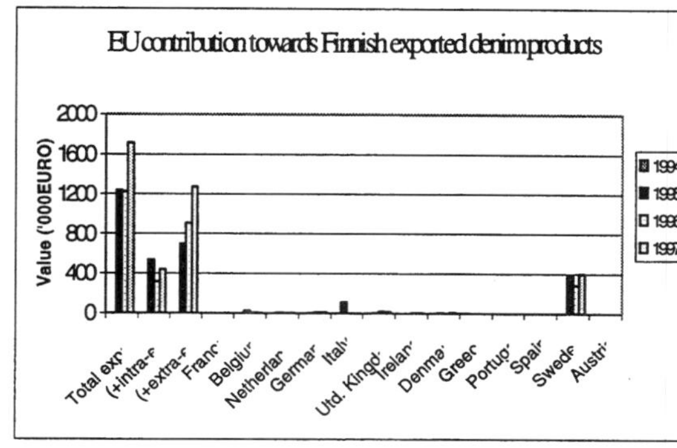

4.94

Austria (See Fig. 4.93)

- Denim product exports were 10 million Euro in 1997. From 1995 to 1997, the annual average growth rate was 17.3%.
- Denim jeans constituted 93.4% of the total denim product exports.
- Exports to EU countries accounted for 73% of the total denim product exports. Both EU and non-EU markets have experienced steady growth since 1995.
- Germany was the single most important destination for Austrian denim products, capturing about 63% of the total denim product exports.

Finland (See Fig. 94)

- Denim product exports were much smaller compared with other EU members, reaching 1.7 million Euro in 1997.
- Denim jeans held about 92.2% of the total denim product exports.
- The export pattern of Finnish denim products was unique in the EU market, with more focus upon non-EU importers. In 1997, exports to non-EU markets accounted for 74.3%.
- As regards EU importers, Sweden was the dominant destination, accounting for more than 95% of the total exports to the EU.

REFERENCES

1. Stogdon, R., Denim and jeans: trends in EU production and Trade. Textile Outlook International, 1996, p.9-30.
2. International, S., The jeans encyclopedia of legend, past and present.1992, Sportswear International.

3. Zimmermann, R., Denim: what might the future hold for the world's hottest textile sector? ITS Textile Leader, 1998, p.34-48.
4. The jeans encyclopedia of legend, past and present, Sportswear International, 1992
5. Europa, Europe Union, 20.Dec. 1998, http://europa.eu.int/

5 Japan Denim Products Trade[1]

OVERVIEW[2]

The Japanese denim industry plays an important role in the world denim arena. Though its development history is much shorter than that of the US, its focus upon product innovation and technological breakthrough make Japan one of the world important denim product manufacturers, featuring superior quality and high added value.

Japan entered domestic denim production in 1971 when Kaihara Textile Mills Ltd. began to supply denim made of rope-dyed yarn [1]. Today, Japanese denim production is concentrated upon three manufactures--Kaihara, Kurabo and Nisshinbo, which account for approximately 90% of Japan's total denim production [2].

The boom of one-washed jeans since 1972 pushed Japan-labeled quality denim into the world market, establishing Japan as one of the most important denim producers and traders in the world. By the end of 1977, denim exports from Japan reached 30 million square yards. It hit the record high of 70.6 million square meters in 1983 due to the very large demand for stretch denim [1].

Figure 5.1 shows a different picture between the 1994-1995 period and 1998-1999 (Jan.-May.) period. Denim exports from Japan to the world amounted to US$ 79.3 million in 1995 with 21.9 million square meters, up by 26.0% over 1994. HK was the most important destination, accounting for 33.6% of Japan's total denim exports in 1995. Korea and Belgium ranked second and third, holding 13.6% and 11.9% respectively. However, the ranking list underwent dramatic changes three years later. Exports to China[3] showed a very strong growth, rising from 1.4 million square meters in 1994 to 2.7 million square meters during the first five months in 1999. China became the No. 1 importer with 34.2% of Japan's total denim exports, followed by the Philippines (14.3%), HK (10.8%), and Belgium (6.9%). From January to May, 1999, Japanese denim exports amounted to US$ 25.5 million with 7.9 million square meters, up by 34.3% over the same period of 1998.

---

1 Detailed and systematic denim-related import and export data about Japan are not available. Therefore, the following analysis is mainly based upon relevant information available in the US, EU and HK trade databases. Since data in these databases are computed with different denominations and approaches, it is difficult to unify them in an integrated picture. Consequently, the import and export analysis for the Japanese denim product trade is carried out individually for the US, EU and HK markets in order to get some in-depth understanding out of the partial picture.

2 Data in this part are from various issues of JTN

3 In this section, "China" refers to the Chinese mainland.

5.1 Japan's denim exports by major markets

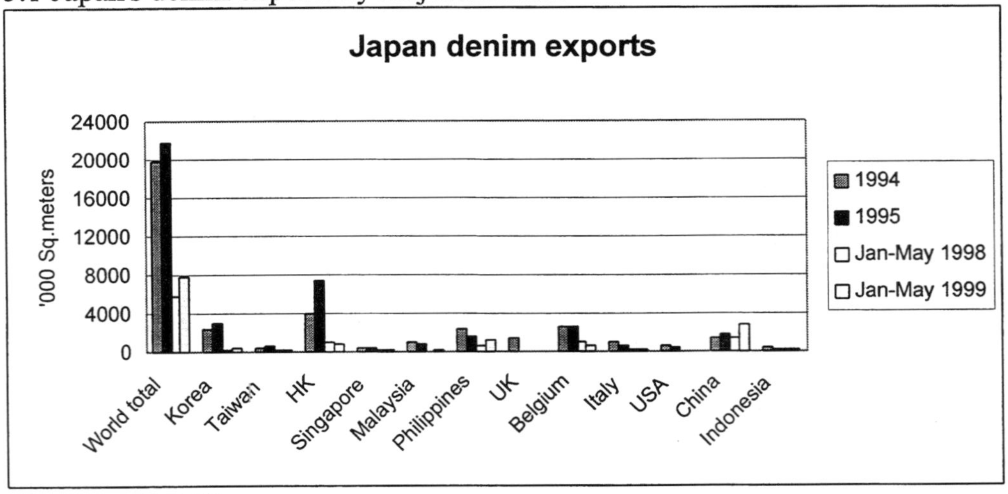

(Source: [3], [4])

The average export price per square meter was around US$ 3.18 to US$ 3.63 during the investigation period. Those destined to the US and Italy were priced higher at US$ 5.68 and US$ 4.06 per square meter while those exported to Malaysia and Indonesia were much lower at US$ 1.60 and US$ 2.40.

Japan started jeans production ten years earlier than its domestic denim production by using imported denim. The foreign entry of Wrangler and Levi Strauss in 1971 further sharpened the market competition [1]. Japanese jeans exports found their way into the world market around 1980 when the stone-washing technique was developed to a very high level by Japanese domestic manufacturers, reaching the peak in 1986. Then it dropped continuously, mainly due to the introduction of chemical washed jeans and the sharp appreciation of the Yen. Today, "Edwin" and "Big John" have become world famous brands, enjoying a high reputation for their quality and fashion style.

Jeans production focused on five enterprises: Edwin, Big John, Bobson, Wrangler Japan and Levi Strauss Japan. Figure 5.2 pictures the production of denim bottom apparel in Japan from 1987 to 1996. It shows that the overall production experienced an upward trend despite the irregular ups and downs during the ten years, increasing from 64 million units in 1987 to 77.8 million units in 1996. Colored jeans showed the strongest growth among the three items, increasing by 61.1% over 1987.

5.2 Japan's denim bottom apparel production from 1987 to 1996

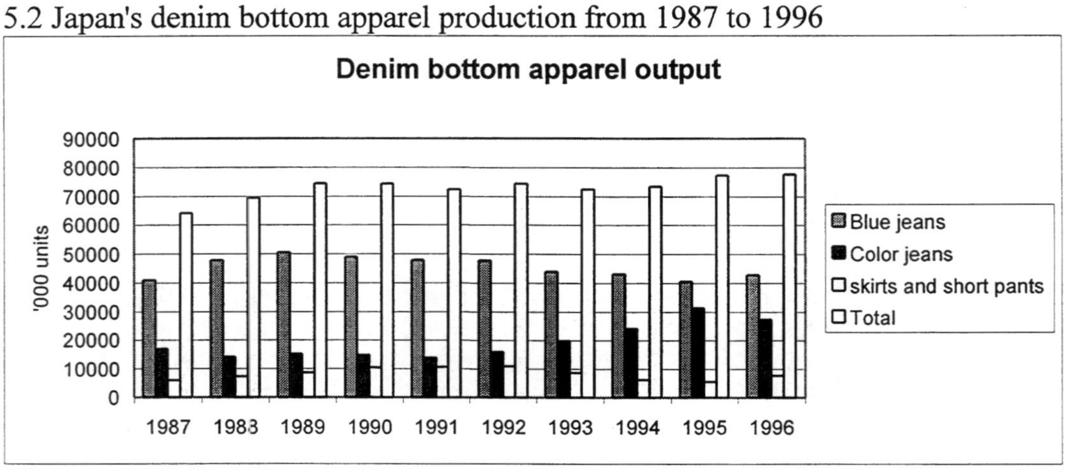

(Source: [5], [3], [6])

5.1 IMPORTS

5.1.1 Denim products imports

5.1.1.1 Imports from the US

Japan imported more denim products from the US than it exported to the US. Therefore, it encountered large trade deficits during the investigation period despite its increasing export efforts. In 1997, the trade deficit was US$ 20.4 million, down from US$ 28.6 million in 1995.

Denim product imports from the US dropped continuously from 1993 to 1997, together with its share in the US denim product export market. In 1993, the total denim product imports from the US amounted to US$ 48.9 million, or 8.2% of US total denim product exports. However, the value slumped to US$ 20.5 million in 1997 with 2.4% of the market share. The average decreasing rate was 14.5% per year.

Among the total denim product imports from the US, denim fabric imports held a very small share. In 1996, it occupied 10.8%, which was the highest proportion during the five years. The ratio dropped to 5.6% in 1997.

5.1.1.2 Imports from the EU[4]

Japan enjoyed a favorable trade balance with the EU. In 1994, the trade balance between the two sides concerning denim products was 57.2 million Euro. Though exports from Japan to the EU dropped quickly during the next three years, the export value was still much bigger than the import value, with the balance reaching 21.7 million Euro in 1997.

According to Table 5.1, Japan was a relatively unimportant export destination for EU denim products, though the value increased by 51.9% from 1993 to 1997. Imports to Japan only accounted for 0.26% of the EU total denim product exports in 1997.

5.3 Denim product imports from the US to Japan between 1993 and 1997

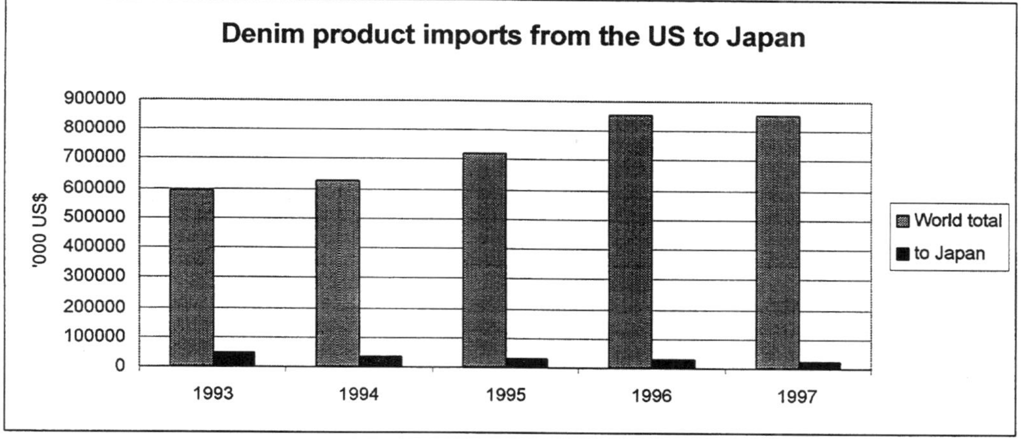

[4] Since denim apparel in the EU database only refers to denim jeans, the denim product trade between Japan and the EU as well as China and the EU will only cover denim fabrics and denim jeans. However in order to unify the whole sector, it will still be titled "denim apparel" instead of "denim jeans".

Table 5.1 Denim product exports from the EU to Japan between 1994 and 1997

Denim product exports from EU to Japan	1994	1995	1996	1997
Total EU exports ('000 Euro)	1744332	1790603	2034903	1971950
To Japan ('000 Euro)	3325	2887	3780	5051
Exports to Japan / Total EU exports %	0.19	0.16	0.19	0.26
Fabric/Total denim product exports to Japan %	5.16	12.02	16.83	4.79

5.4 Denim product exports to Japan from the EU between 1994 and 1997

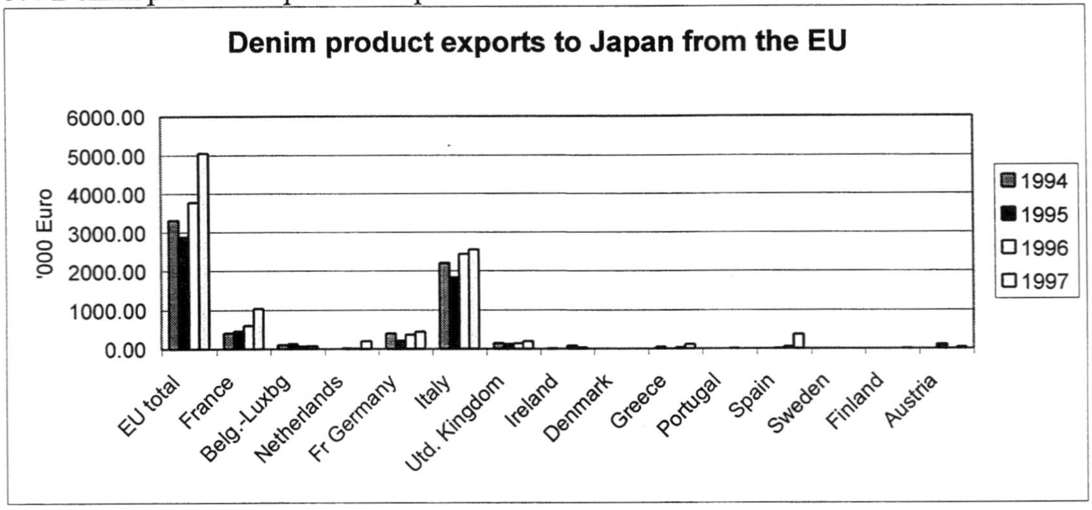

Similar to the imports from the US, denim fabric constituted only a small part of the total denim product imports from the EU. The highest ratio was registered in 1996 with 16.8% and then declined sharply to 4.8% in 1997.

Italy was the leading EU exporter to Japan, capturing 50.6% of total EU denim product exports to Japan. France and Germany ranked second and third with 20.6% and 8.8% respectively. Imports from Spain increased very rapidly in 1997, amounting to another 7.1%.

5.1.1.3 Imports from HK

Owing to HK's unique role in re-export activities, which constitute a large share in HK total trade, denim product trade between Japan and HK is actually composed of four parts: HK domestic exports to Japan, re-exports from third countries via HK to Japan, imports from Japan to the HK market and re-exports from Japan via HK to third countries. The former two constitute the total denim products entering Japan from HK while the latter two make up the total denim products exported from Japan. (See Fig 5.5)

Direct imports from HK were very small in value, too small to be reflected in Figure 5.5. The value underwent a serious drop in 1998, down from HK$ 10.4 million in 1997 to HK$ 3.1 million in 1998. It held only 0.08% of HK total denim product domestic exports. Of the total denim products directly imported from HK, denim fabric imports took a small share, which was 16.7% in 1998.

As regards re-exports into Japan via HK, the value is much larger with HK$ 317 million in 1998, accounting for 4.2% of HK total denim product re-exports. The ratio of denim fabric increased a little from 1.6% in 1992 to 4.4% in 1997.

5.5 Denim product exports to Japan from HK between 1992 and 1998

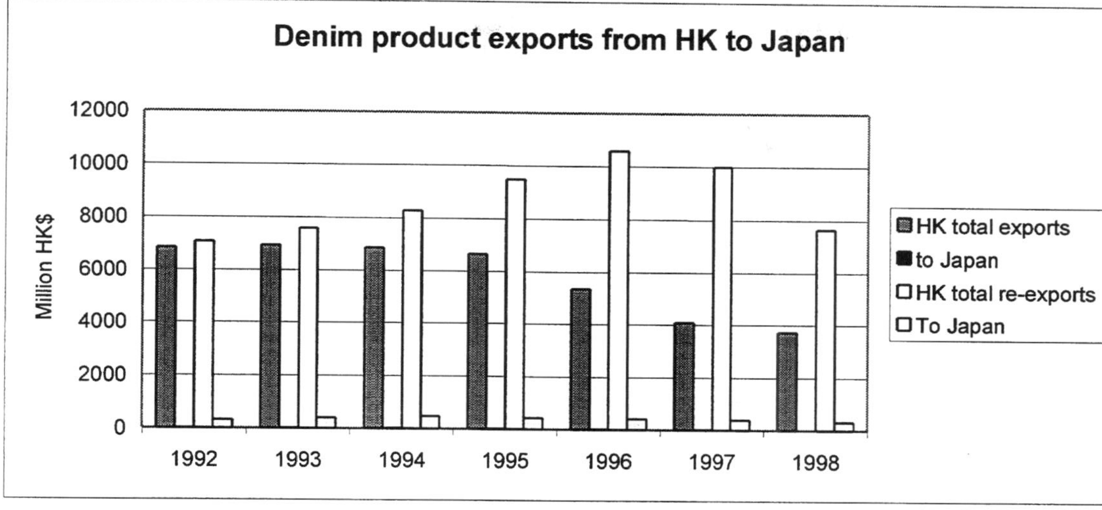

5.6 Denim fabric exports from the US to Japan between 1993 and 1997

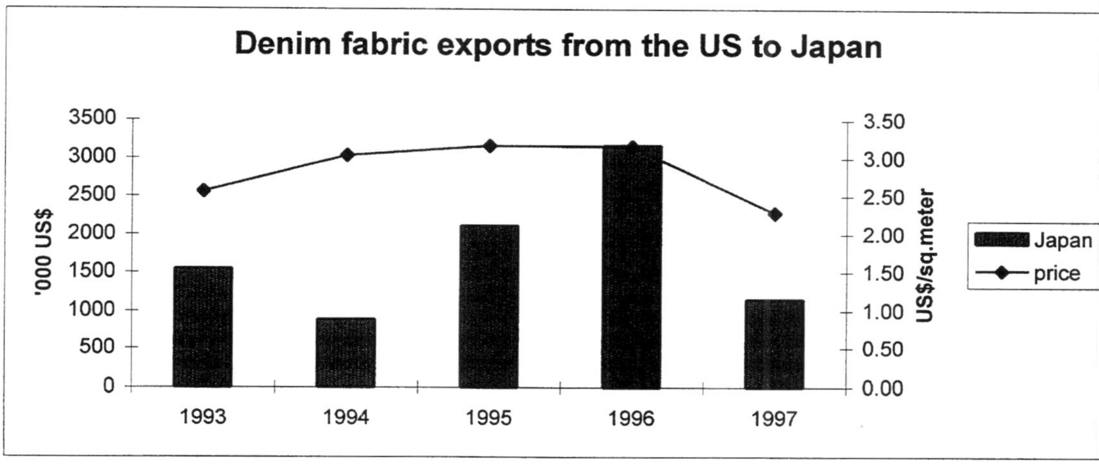

5.1.2 Denim fabric imports

5.1.2.1 Imports from the US

In 1997, imports from the US amounted to US$ 1.15 million, down from US$ 1.55 million in 1993. The import level peaked in 1996 with US$ 3.17 million, or 1.2% of US total denim fabric exports. However, the very sharp drop in 1997 made the market share shrink to 0.4%. (See Fig. 5.6)

The average import price from the US was US$ 2.28 per square meter in 1997, lower than that of all the previous years.

5.1.2.2 Imports from the EU

Denim fabric imports from the EU were 145.8 thousand Euro in 1994, peaking in 1996 with a value of 636.2 thousand Euro and plunging to 242.2 thousand Euro in 1997. In 1997, EU denim fabrics sent to the Japanese market only captured 0.04% of the total EU denim fabric exports and 0.08% of those to non-EU destinations.

Among the individual EU suppliers, Italy took the largest share, which was 73.0% in 1997. Belgium ranked second with a market share of 24.8%. The top two suppliers were responsible for 97.8% of total EU denim fabric exports to Japan. Ireland once

supplied 75.0 thousand Euro worth of denim fabrics in 1996, or 11.8% of the total EU denim fabric imports into Japan. But there were no imports from Ireland in 1997.
The average import price per square meter ranged between 6.84 Euro to 9.52 Euro. Generally speaking, imports from Greece and Portugal were lower than the average level while those from Italy and Germany were of high added value, reaching 12.4 and 13.7 Euro respectively in 1997.

5.1.2.3 Imports from HK

A. HK domestic exports to Japan

Denim fabric imports from HK hit a low in 1993 with a value of HK$ 166 thousand. The year 1998 witnessed another slump from the peak level of HK$ 3.4 million in 1997, decreasing by 85% to HK$ 511 thousand. Japan was not the major destination for HK domestic denim fabric exports, only accounting for 0.15% in 1997 when the export value to Japan reached its peak.

The average import price was around HK$ 20 per square meter in 1998, almost the same level as that in 1992. (See Fig. 5.8)

B. Re-exports through HK to Japan

Figure 5.9 shows that denim fabric re-exports via HK to Japan registered a record high in 1998, amounting to HK$13.8 million. The growth momentum was especially

5.7 Denim fabric exports from the EU to Japan between 1994 and 1997

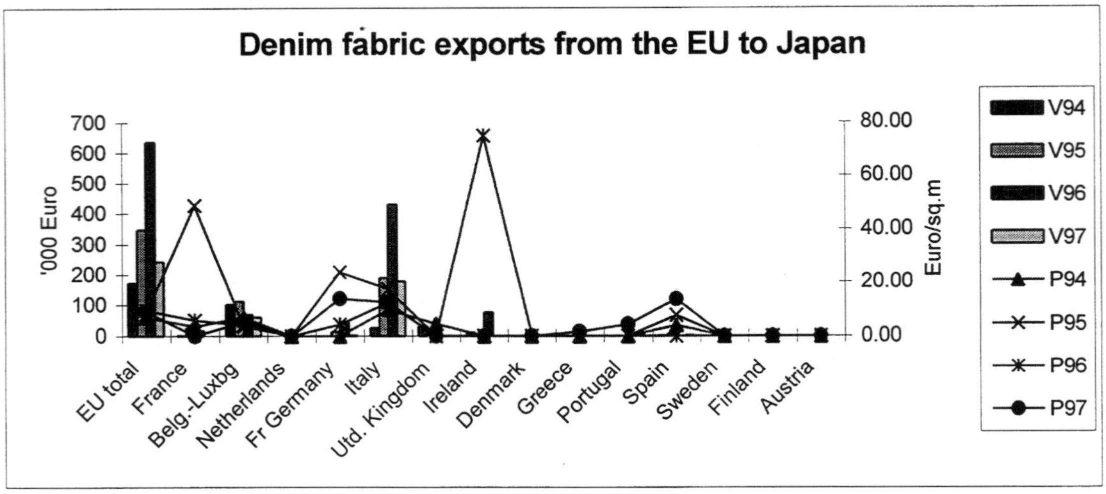

5.8 Denim fabric domestic exports from HK to Japan between 1992 and 1998

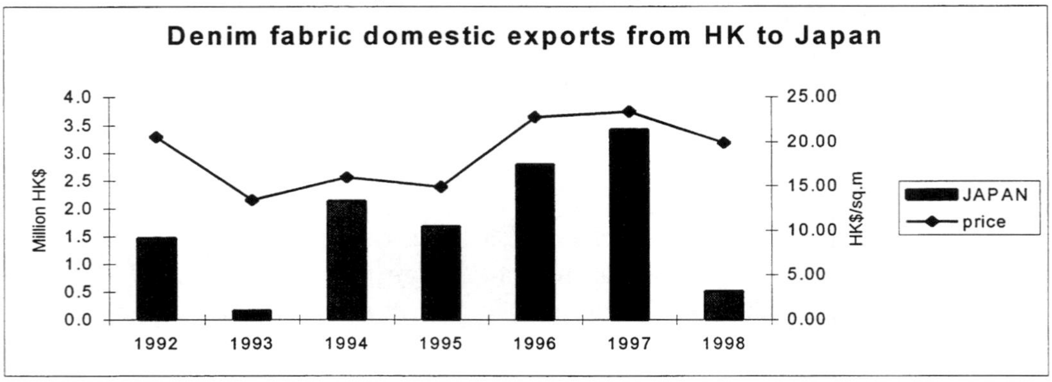

5.9 HK denim fabric re-exports from HK to Japan between 1992 and 1998

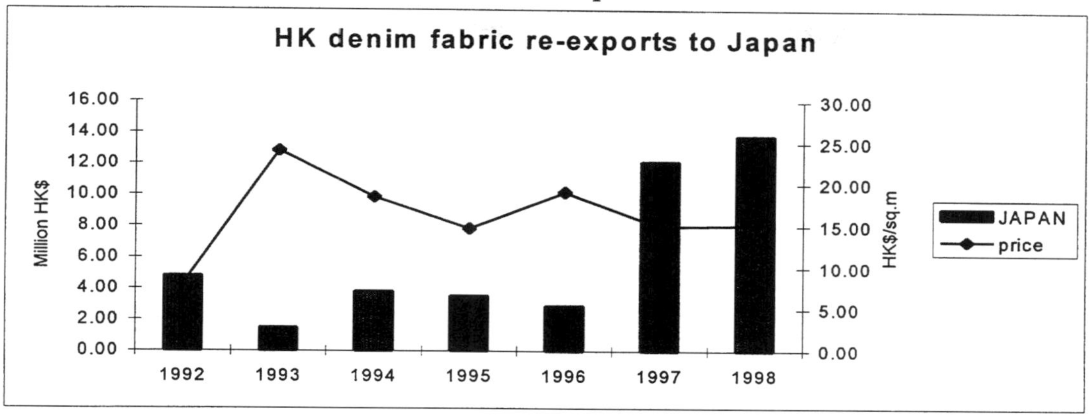

5.10 Denim apparel exports from the US to Japan between 1993 and 1997

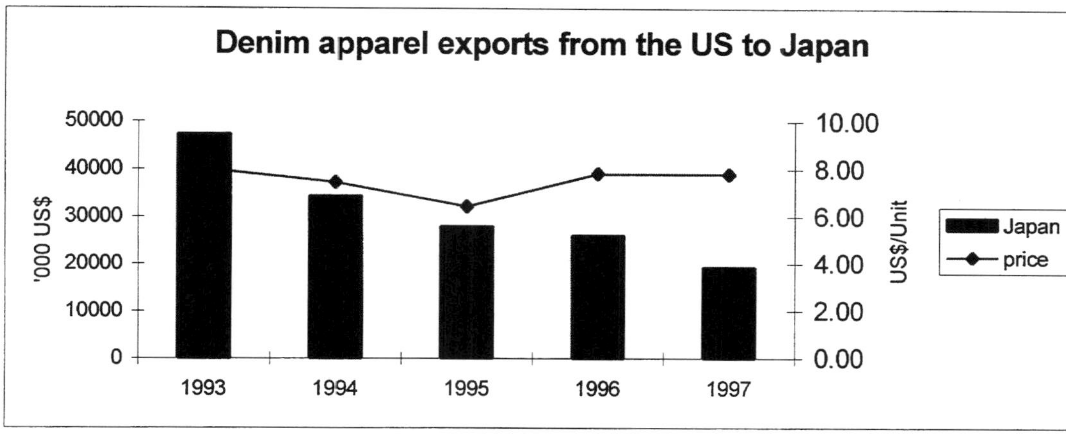

strong between 1996 and 1997 at the rate of 419.5%. Compared with the direct imports figure, re-exports via HK to Japan were much larger, accounting for 96.4% of the total denim fabrics entering Japan from HK in 1997, regardless of the export origin.

The average re-export price for denim fabrics from third countries was HK$ 15.2 per square meter in 1998, lower than the price for HK domestic exports. This may indicate that these denim fabrics were mainly from developing countries rather than developed countries such as Italy and France.

5.1.3 Denim apparel imports

5.1.3.1 Imports from the US

Denim apparel imports from the US have shown a very clear downward trend since 1993. In 1997, only US$ 19.4 million worth of denim apparel was imported into Japan, plunging from US$ 47.3 million in 1993. As a result, its share in the US denim apparel export market shrank considerably from 11.6% to 3.4% in the investigation period. The average import price for US denim apparel was US$ 7.8 per unit or US$ 93.5/dozen in 1997.

5.1.3.2 Imports from the EU

EU jeans exports to Japan amounted to 4.8 million Euro in 1997, increasing by 52.5%

over 1994 at an annual average growth rate of 17.5%. Compared with the denim fabric imports from the EU, jeans imports to Japan took a relatively larger share of the EU export market. They accounted for 0.3% of the total EU jeans exports and 2.6% of those destined to Non-EU countries. (See Fig. 5.11)

Italy has been the leading supplier among the EU members since 1994. However, its importance in the Japanese jeans import market faced great challenges from other EU countries, such as France and Spain. In 1994, Italy captured more than two thirds of the total EU jeans imports into Japan but it slumped to 49.4% in 1997. By sharp contrast, France increased its share from 13.0% in 1994 to 21.6% in 1997 and at the same time, Spain emerged as the No. 4 supplier with a market share of 7.5%. Germany, Netherlands and the United Kingdom held another 9.2%, 4.2% and 4.2% respectively in 1997.

The average import price from the EU was 32.3 Euro/Unit in 1997. As for the denim fabric imports from the EU, jeans from Greece, Portugal and Spain were sold at lower prices, ranging from 14.8 to 22.7 Euro/Unit while those from Germany and Italy were priced higher, reaching 40.5 and 43.7 Euro/Unit respectively.

5.1.3.3 Imports from HK

A. HK domestic exports to Japan

Direct denim apparel imports from HK to Japan reached HK$ 2.6 million in 1998, hitting a record low since 1992 when imports peaked with a value of HK$ 10.6 million. As with denim fabric imports from HK, Japan played a trivial role in HK denim apparel domestic exports. In 1998, those exported to Japan from HK only accounted for 0.14%. The average import price was HK$ 88.89 per unit in 1998, lower than the previous years' level. (See Fig. 5.12)

B. Re-exports through HK to Japan

Japan imported a lot of denim apparel from third countries via HK, though the value was continuously down since 1994. In 1998, the import value through this

5.11 Denim jeans exports from the EU to Japan between 1994 and 1997

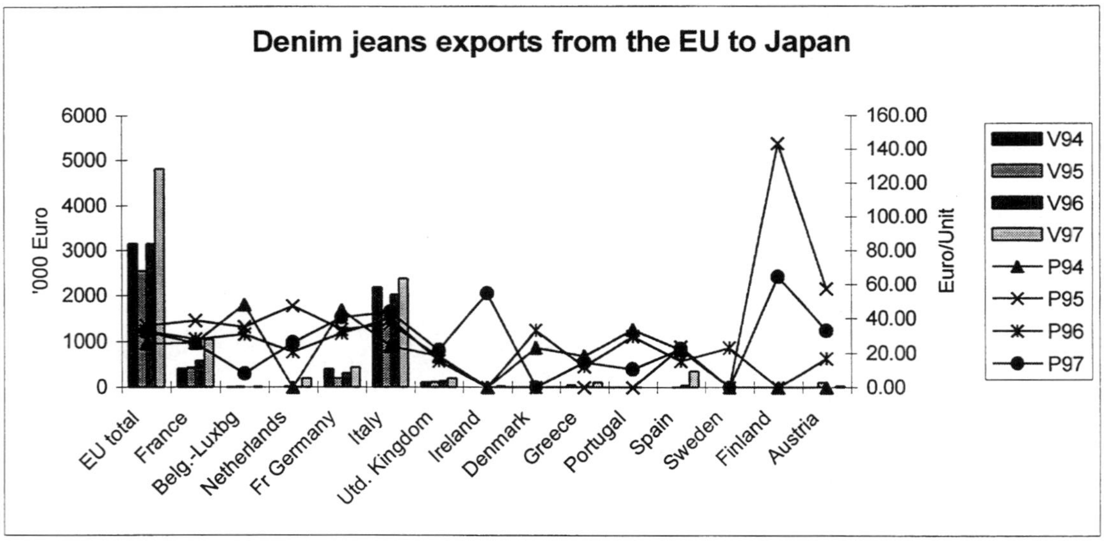

5.12 Denim apparel domestic exports from HK to Japan between 1992 and 1998

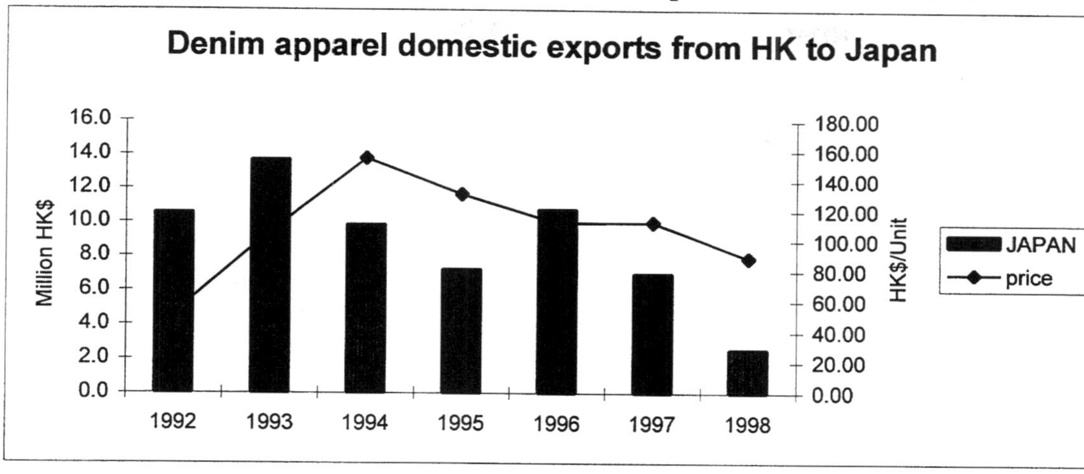

5.13 Denim apparel re-exports from HK to Japan between 1992 and 1998

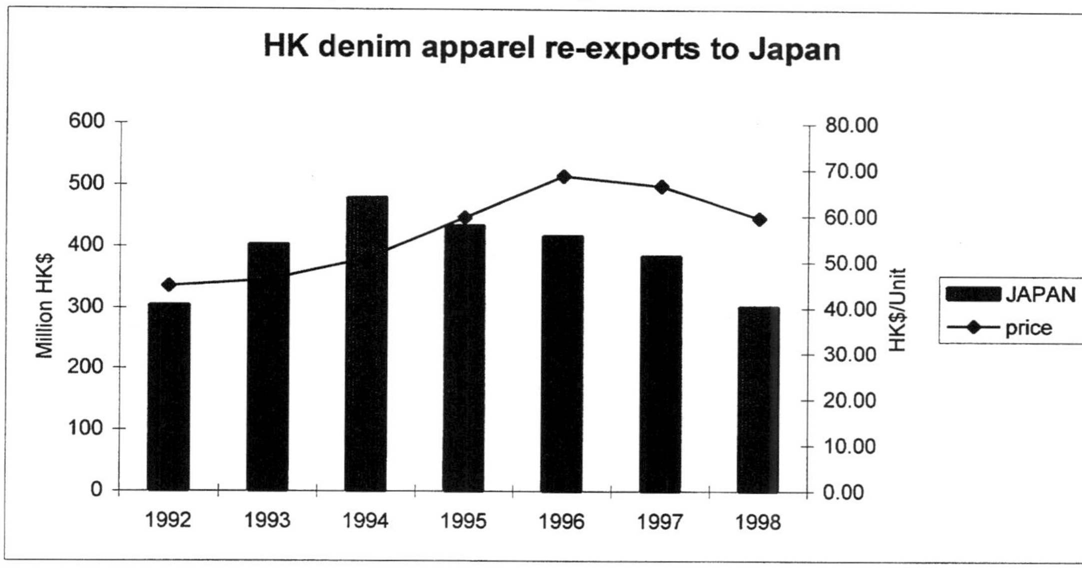

approach was HK$ 303.2 million, accounting for 99.2% of total denim apparel entry into Japan from HK. Such an import pattern, both for denim fabric and denim apparel, clearly indicates that HK acts more as an important intermediate between Japan and other countries than as a direct supplier to the Japanese market. (See Fig. 5.13)
The average import prices for these re-exports were lower than those for direct imports from HK, ranging from HK$ 44.8 to HK$68.6 per unit. In 1998, the figure was HK$ 59.6.

5.2 EXPORTS

5.2.1 Denim products exports

5.2.1.1 Exports to the US

Exports from Japan to the US were insignificant in the US denim product import market, with a market share of less than 0.1%. Despite the increasing trend of overall

US denim product imports, those from Japan experienced a very serious drop. The 1997 level was only 6.1% of the 1995 level.

In 1995, almost all the exports from Japan were composed of denim apparel. Two years later, denim fabric exports increased their share to 2.2%.

5.2.1.2 Exports to the EU

As mentioned earlier, Japan exported more denim products to the EU than it imported from the EU. By contrast to the increasing trend in the EU import market, those from Japan fell tremendously in 1995 and the year 1997 registered a record low of 26.8 million Euro. Consequently, the market share of Japanese denim product exports shrank from 2.5% in 1994 to 1.0% in 1997.

Denim fabric exports remained a relatively small share. It was 15.5% in 1997, a bit lower than that in 1994.

Among the EU importers, Germany was the most important destination for Japan denim products, followed by Italy, Netherlands and Belgium. These four accounted for 98.3% of total Japan imports into the EU market in 1997, among which Germany held 77.5%, Italy 8.9%, Netherlands 8.0% and Belgium 3.9%. One point worth mentioning is that the United Kingdom has sharply reduced its imports from Japan since 1995 and its market share dropped from 8.8% in 1994 to a mere 0.56% in 1997.

Table 5.2 Denim product exports from Japan to the US between 1995 and 1997

Denim product exports from Japan to the US	1995	1996	1997
Total US imports ('000 US$)	1604967	1853674	2003791
From Japan ('000 US$)	1518	921	93
Imports from Japan / Total US imports %	0.095	0.050	0.005
Fabric/Total Japan imports %	0.00	1.30	2.15

Table 5.3 Denim product exports from Japan to the EU between 1994 and 1997

Denim product exports from Japan to EU	1994	1995	1996	1997
Total EU imports ('000 Euro)	2452161	2357517	2771175	2648607
From Japan ('000 Euro)	60503	37447	28019	26750
Imports from Japan / Total EU imports %	2.47	1.59	1.01	1.01
Fabric/Total denim product imports from Japan %	16.96	20.59	16.25	15.51

5.2.1.3 Exports to HK

Exports from Japan to HK, regardless of their final destinations, had some weight in both HK import and re-export markets. Exports to the HK domestic market dropped from HK$ 514.5 million in 1992 to HK$ 377.8 million in 1998, capturing 4.6% of HK total denim product imports. At the same time, those to third markets via HK increased a little from HK$ 232.0 million to HK$ 298.2 million, holding 3.9% of HK total denim product re-exports.

There is one point in common between Japanese exports to HK and to third countries via HK, that is, denim fabrics dominated the export product mix throughout the investigation period. Take the year 1998 for example: Japanese denim fabric exports to the HK domestic market accounted for over 99% of Japan's total denim

5.14 Denim product exports from Japan to the EU by country

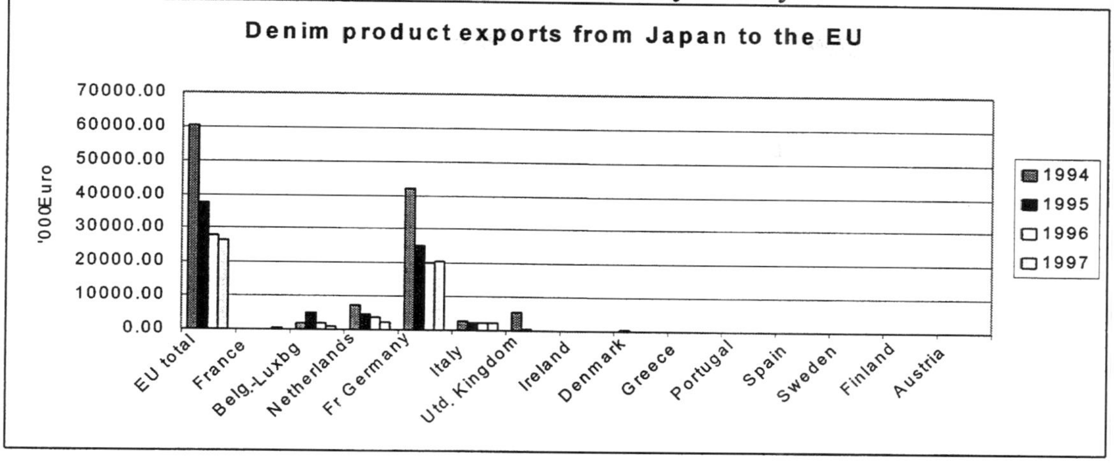

5.15 Denim product exports from Japan to HK between 1992 and 1998

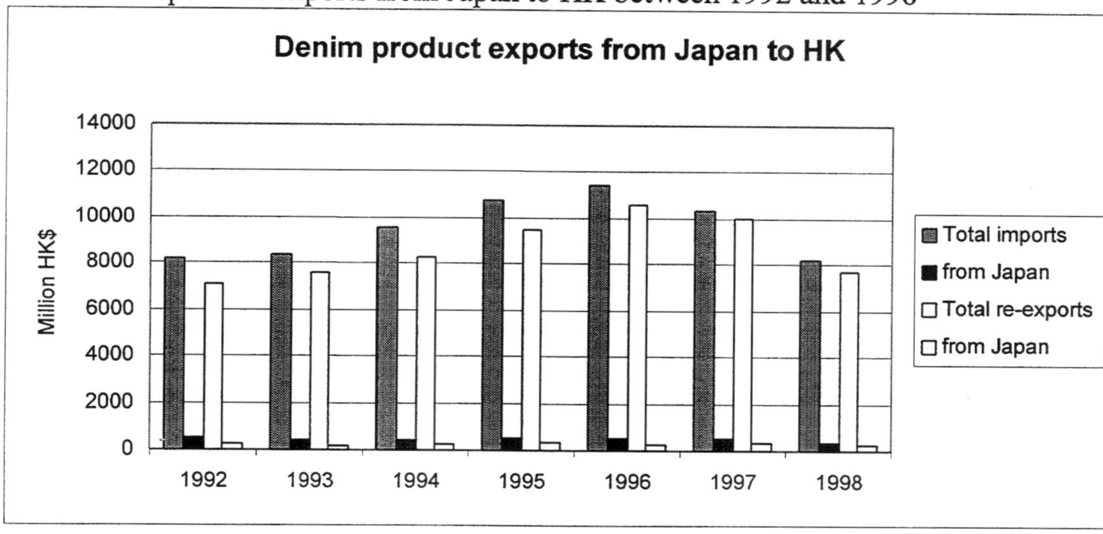

5.16 Denim fabric exports from Japan to the US between 1995 and 1997

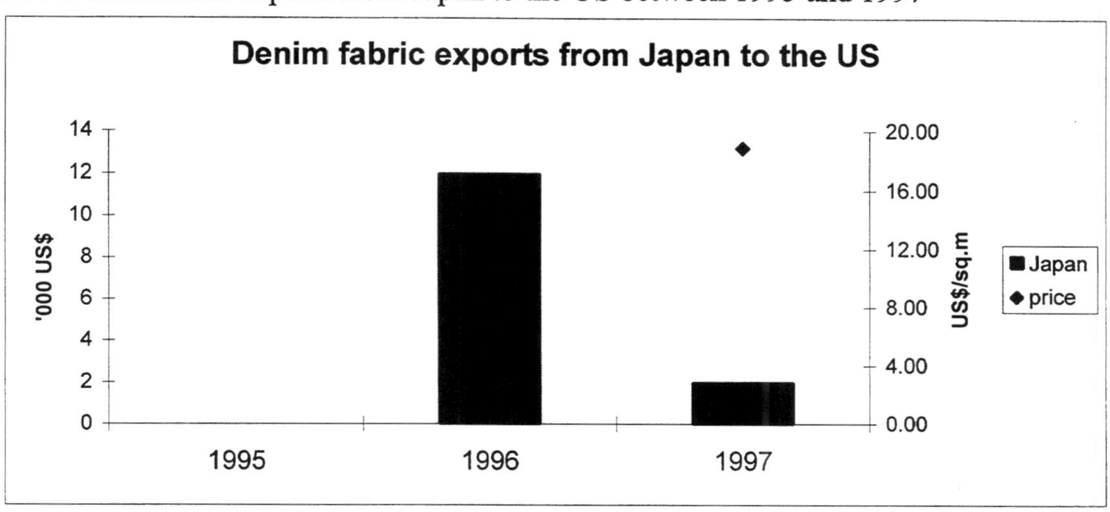

product exports to HK and those to the third countries via HK also held about 98.8% of Japan's total denim product re-exports to HK.

5.2.2 Denim fabric exports

5.2.2.1 Exports to the US

Denim fabric exports to the US were insignificant. During the investigation period from 1995 to 1998, there were only US$ 14 thousand worth of denim fabric exports in total, zero for 1995, US$ 12 thousand for 1996 and US$ 2 thousand for 1997. (See Fig. 5.16)

Despite the small volume, the average export price was high, reaching US$ 18.9 per square meter in 1997. This indicates that Japan domestic production is now focusing more upon high value added items.

5.2.2.2 Exports to the EU

Denim fabric exports to the EU experienced sharp decline during the investigation period, plunging from 10.3 million Euro in 1994 to 4.15 million Euro in 1997. The average decrease per year hit 20%. As a result, Japan held only 0.9% of the EU total denim fabric imports in 1997, down from 2.1% in 1994. As regards the total EU denim fabric imports from non-EU countries, Japan's market share also fell from 4.9% in 1994 to 1.9% in 1997.

5.17 Denim fabric exports from Japan to the EU between 1994 and 1997

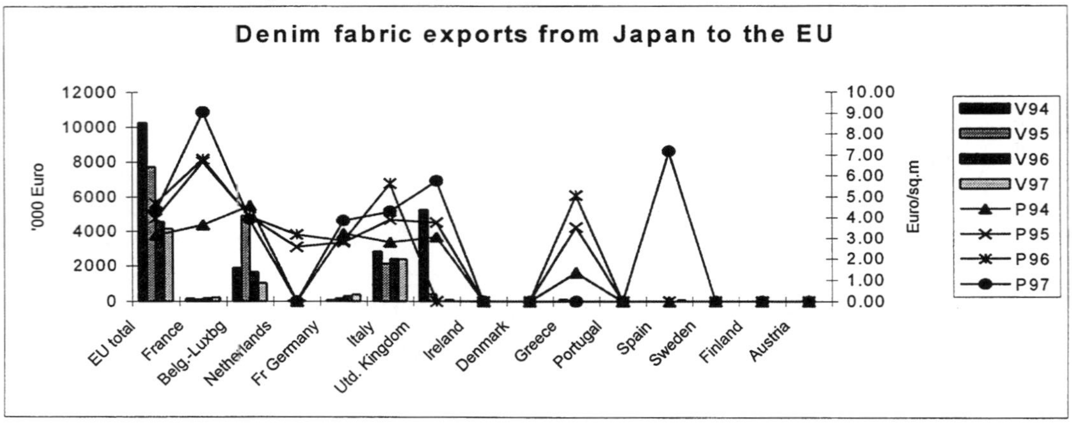

5.18 Denim fabric exports from Japan to HK between 1992 and 1998

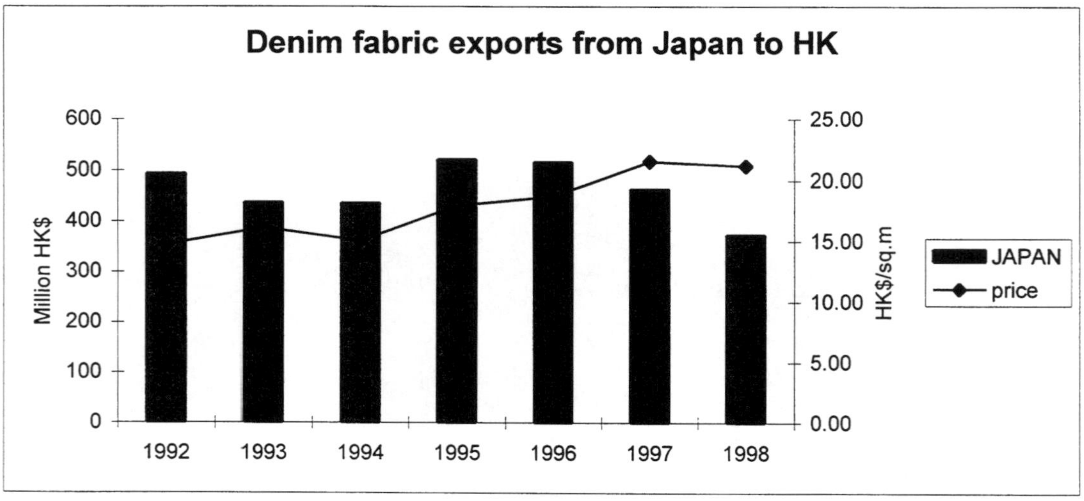

The United Kingdom was once the most important EU importer, in 1994 absorbing over a half of Japan's denim fabric exports into the EU market. However, its import value from Japan slumped to 79.7 thousand Euro in 1997, only accounting for 1.9% of the market share. Belgium replaced the UK as the leading importer in 1995, taking 63.7% of the market share, but it was overtaken by Italy one year later. In 1997, Italy imported 2.4 million Euro worth of Japanese denim fabrics, holding 57.1% of the market share. It was followed by Belgium (24.9%), Germany (9.4%) and France (5.2%).

The average export price was around 3.2 Euro to 4.7 Euro per square meter. Exports to France fetched the highest price, reaching 9.1 Euro per square meter in 1997. A point worth mentioning is that exports to Greece in 1996 and to Spain in 1997 were sold above the average level, topping 5.1 and 7.2 Euro per square meter respectively.

5.2.2.3 Exports to HK

A. Exports to the HK domestic market

Japan supplied HK$ 671.5 million worth of denim fabrics to HK in 1998, of which 44.4% was further directed towards other third countries.

Japanese denim fabric exports to HK were of some importance to the HK import market. In 1998, the export value reached HK$ 373.3 million, i.e. about 6.6% of HK's total denim fabric imports. The export value fluctuated from 1992 to 1998, showing a downward trend since 1995. (See Fig. 5.18)

The average export price was about HK$ 21.2 per square meter in 1998, demonstrating an upward trend from HK$ 14.8 per square meter in 1992.

B. Re-exports to third countries through HK

Re-exports from Japan to other destinations via HK amounted to HK$ 298.2 million in 1998, rising by 29.2% over 1992. It reached the peak in 1995 with a value of HK$ 345.6 million. Its development pattern during the eight years corresponded with that of direct exports to HK. (See Fig. 5.19)

The re-export price was up from HK$ 11.2 per square meter in 1992 to HK$ 17.2 per square meter in 1998.

5.19 Denim fabric re-exports to third countries via HK between 1992 and 1998

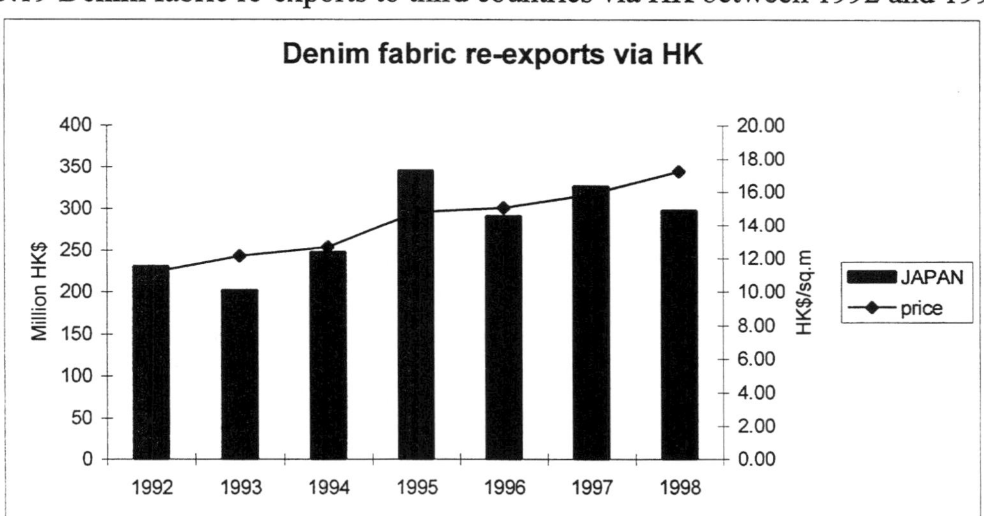

5.20 Denim apparel exports to the US from Japan between 1995 and 1997

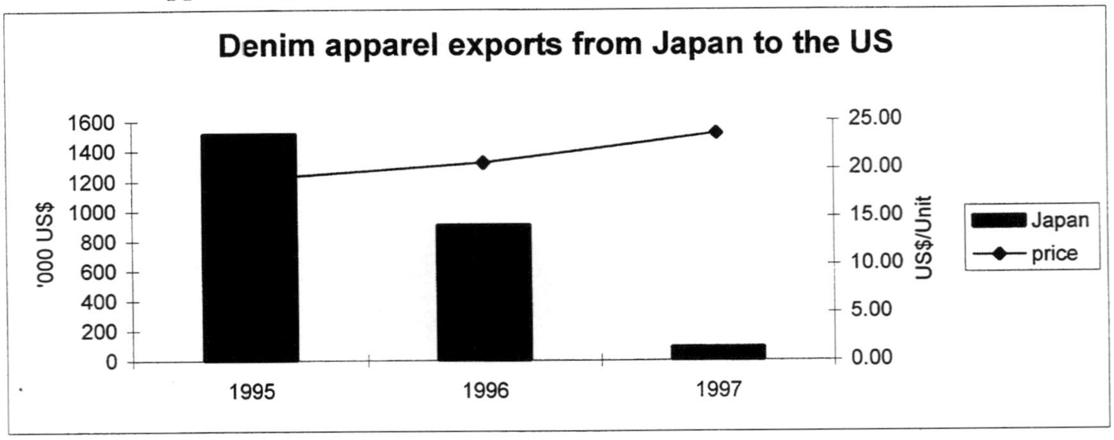

5.21 Denim jeans exports from Japan to the EU between 1994 and 1995

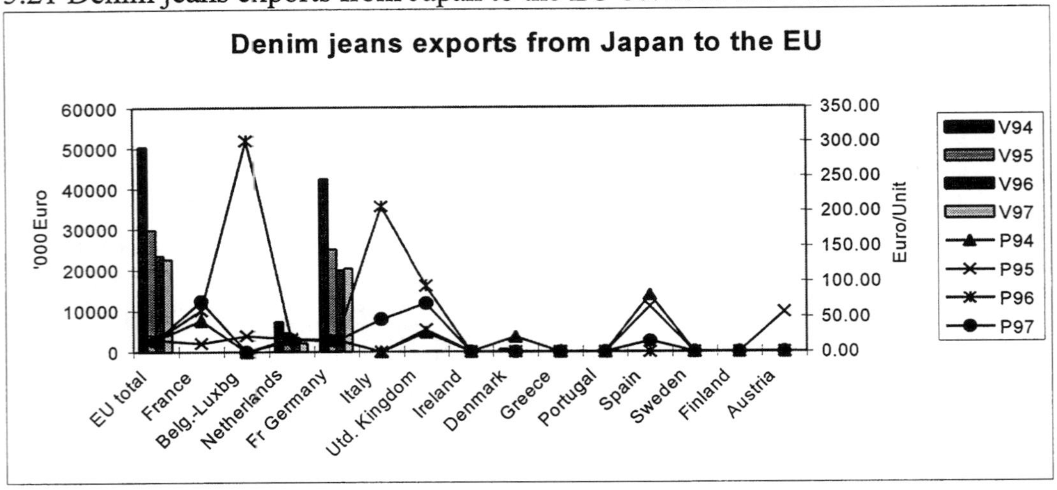

5.2.3 Denim apparel exports

5.2.3.1 Exports to the US

Japanese denim apparel exports to the US had a poor performance in 1997, down from US$ 1.5 million in 1995 to US$ 91 thousand in 1997. The average decrease was 31.3% per year. The average export price topped US$ 23.7 per unit in 1997, up by US$ 4.7 over 1995. (See Fig. 5.20)

5.2.3.2 Exports to the EU

Jeans exports to the EU from Japan underwent the same experiences as Japanese denim fabric exports from 1994 to 1997. The value slumped suddenly from 50.24 million Euro in 1994 to 29.7 million Euro in 1995 and went on decreasing to 22.6 million Euro in 1997. Japan's market share in the total EU jeans imports declined from 2.1% in 1994 to 0.8% in 1997. Of the total EU jeans imported from non-EU suppliers, the share of Japan dropped from 4.5% to 1.7%.

Japanese jeans exports to the EU were highly concentrated. The top two importers together absorbed 99.5% of the total. Germany has been the most important EU importer since 1994, though its import value plunged significantly during the four-year period. It accounted for 90% of the total Japan jeans exports to the EU market. Netherlands ranked second with 9.5%. (See Fig. 5.21)

The average export price was around 16.1 Euro/Unit in 1997. Though France, Italy and the UK imported a very small volume compared with the top two importers, their import prices were much higher than the average level--72 Euro/Unit for France, 46.5 Euro/Unit for Italy and 68.7 Euro/Unit for the United Kingdom. Those to Netherlands and Germany were sold at the average level.

5.2.3.3 Exports to HK

A. Exports to the HK market

Japan supplied HK$ 4.5 million worth of denim apparel to HK in 1998, among which 99.8% was headed directly to the HK domestic market. Compared with the re-exports situation mentioned in previous sectors, direct exports here played an overwhelming role.

Denim apparel exports to HK plunged heavily in 1994 by about 50% and slumped to HK$ 3.6 million in 1996. Though it almost doubled in 1997, it fell to HK$ 4.5 million in the following year. Japan's denim apparel exports occupied a very small share in HK total denim apparel imports, only 0.17% in 1998. (See Fig. 5.22)

The average export price was HK$ 143.7 per unit in 1998, higher than that from HK. (See Fig. 5.22)

5.22 Denim apparel exports from Japan to HK between 1992 and 1998

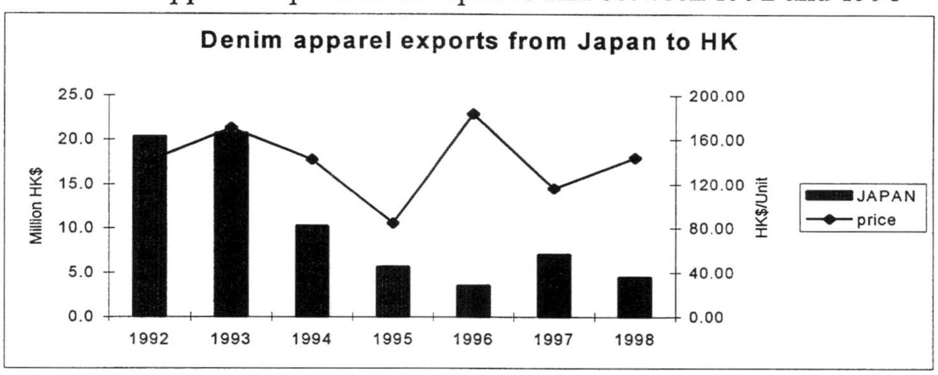

5.23 Japan denim apparel re-exports via HK between 1992 and 1998

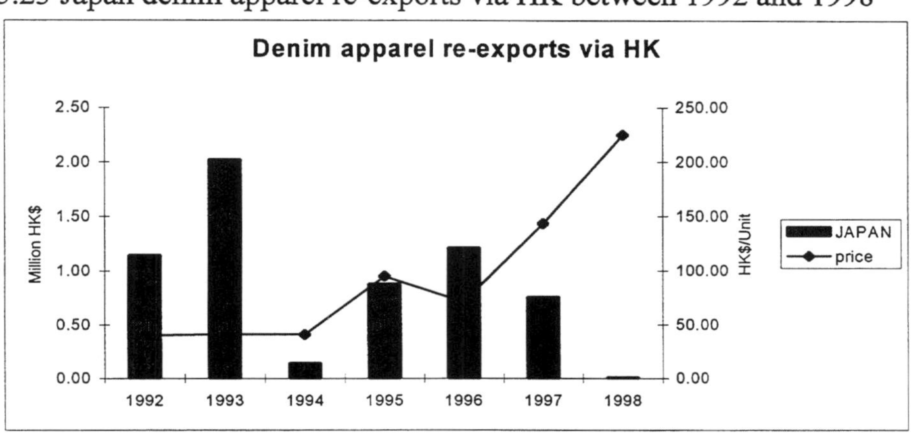

B. Re-exports to third countries through HK

Re-exports to third countries via HK slumped from 1992 to 1998, dropping from HK$ 1.1 million to HK$ 11 thousand. Japan's share of HK total denim apparel re-exports was negligible. However, the re-export price underwent a sharp increase from HK$ 40.3 per unit in 1992 to HK$ 224.9 per unit in 1998. This was much higher than that of those exported to the HK market. (See Fig. 5.23)

REFERENCES

1. Anon, Japanese jeans & Denim: Milestones of Japanese jeans and denim exports; Profile of Japanese denim and jeans makers. JTN Monthly, 1990 ,p.57-59.
2. Anon, Japan's three largest denim makers. JTN Monthly, 1990 ,p.60.
3. Anon, Signs of denim & jeans market revival. JTN Monthly, 1996 ,p.67-70.
4. Anon, Denim operations: intensifying product development for revival of blue denim. JTN Monthly, 1999 ,p.37.
5. Anon, Great changes in Japan's jeans industry. JTN Monthly, 1996 ,p.69-70.
6. Anon, Revival of blue denim in Japan. JTN Monthly, 1997 ,p.47.

6 Hong Kong Denim Products Trade[1]

OVERVIEW

HK began its denim production in the 1950s when it imported automatic shuttle looms from the Chinese Mainland and Japan. With the introduction of shuttleless looms in the late 1970s, denim produced in HK has undergone great quality improvement and denim trade with other countries shows a health growth [1].

Hong Kong imported HK$ 8155 million worth of denim products in 1992 and the amount expanded by 10.2% per year on the average. The import value peaked in 1996 at HK$ 11476 million. However, the economic recession in HK and lackluster consumer spending cast a shadow upon the import markets. Both Hong Kong denim fabric and apparel imports declined between 1996 and 1998. After a 9.70% decline in 1997, Hong Kong's total denim product imports dropped by 20.6% in 1998. Denim apparel held the lion's share, capturing around 70% of total denim product imports. (See Fig. 6.1)

Between 1992 and 1995, Hong Kong domestic exports remained steady at around HK$ 6780 million. However, they declined continuously in the next three years--by 19.4%, 23.5% and 8.7% in 1996, 1997 and 1998 respectively. In 1996 and 1997, Hong Kong domestic exports shrank in the US, the United Kingdom, and Germany, which were once the most important overseas markets for Hong Kong denim products. In 1998, the export value continued to drop, mainly due to the sharp decline by 15.6% in the Chinese mainland market. By contrast, exports to the US and the United Kingdom rebounded. Hong Kong denim fabric exports increased their share from 41.2% of Hong Kong domestic exports in 1992 to 49.8% in 1998.

Due to its unique geographic location, remarkable strength in the service sector and economic restructuring in the 1980s, re-exports play a key role in HK's foreign trading regime. In 1992, HK$ 7071 million worth of denim products was re-exported through Hong Kong. The value escalated at an average annual growth rate of 12.4%, reaching its peak in 1996 with HK$ 10578 million. However, negatively influenced by the Asian financial crisis, Hong Kong re-exports suffered a sharp drop, by 23.7% in 1998.

The most important trading partner was the Chinese Mainland, which was not only the most significant supplier but also the most crucial consumer for HK denim products. The unique relationship between the two sides has been explained in detail in chapter 2.

---

[1] The data in this section was from HK Census and Statistics Department

6.1 Hong Kong denim product trade

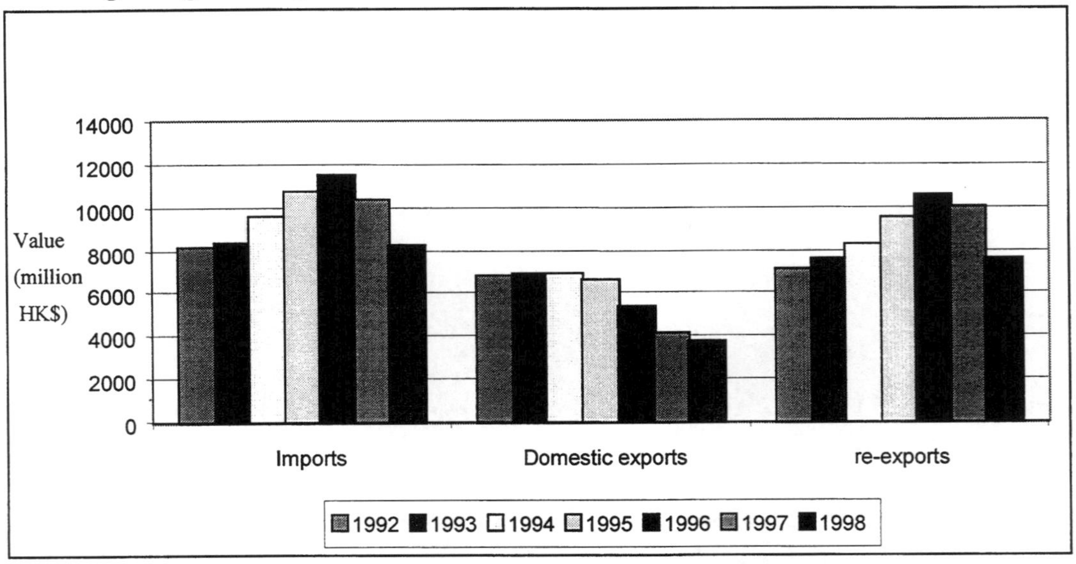

6.2 Hong Kong denim product imports by major markets

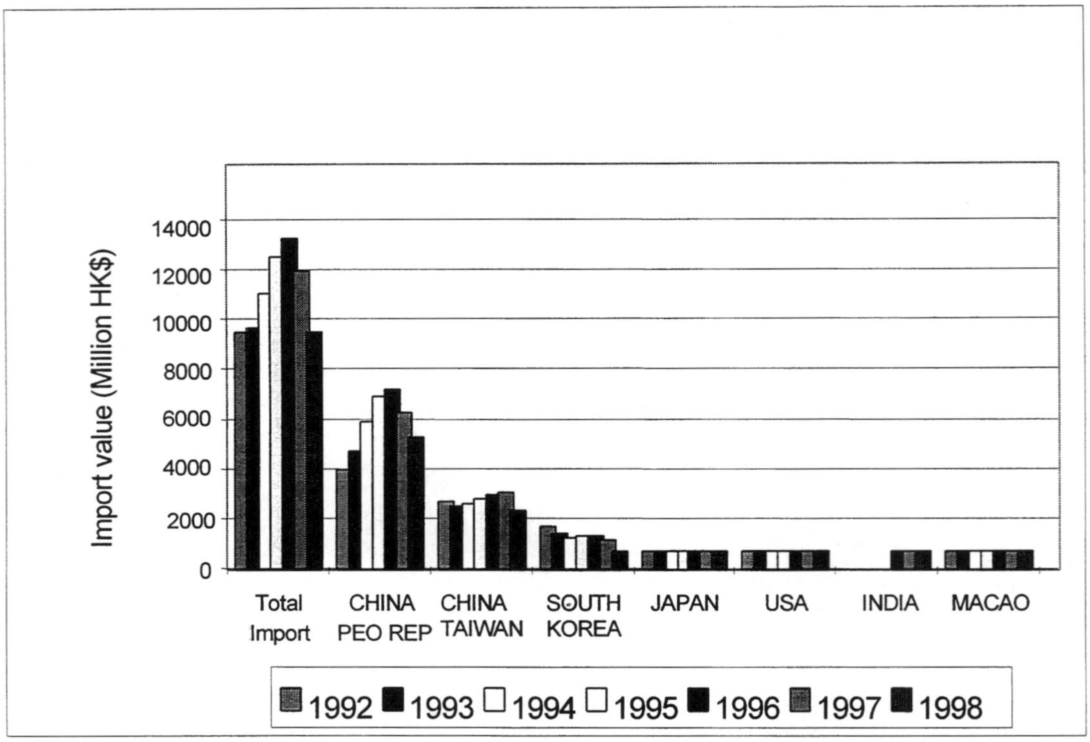

6.1 IMPORTS

Hong Kong's denim product imports reached HK$ 8155 million in 1992, peaking in 1996 with HK$ 11476 million. The total import value declined sharply from 1997 to 1998, falling to HK$ 8225 million in 1998, down by 20.6% over 1997. The major suppliers were the Chinese Mainland, China Taiwan, South Korea, Japan, USA, India and Macao.

6.3 Hong Kong denim product imports

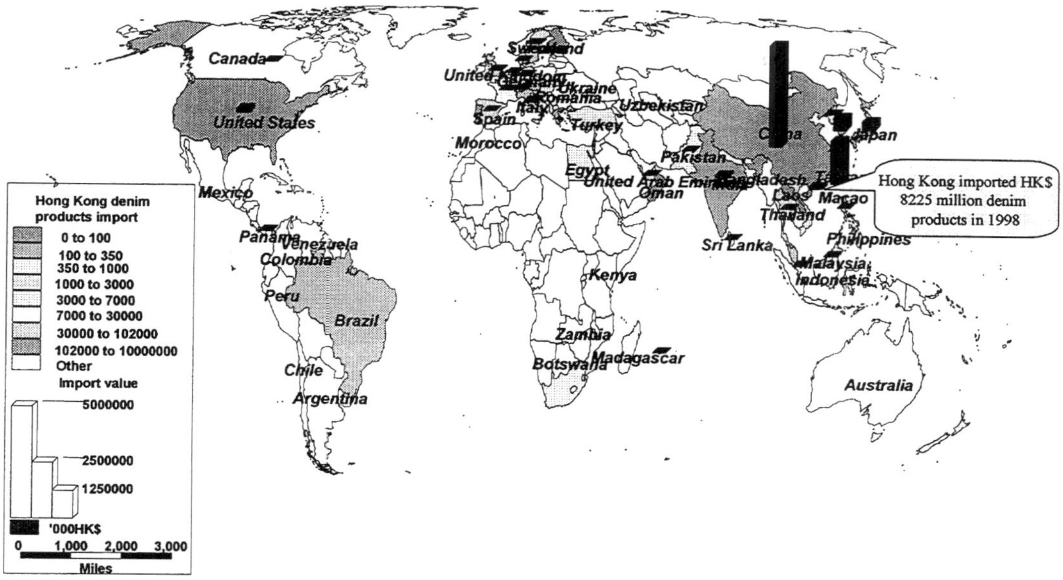

The Chinese Mainland was the most important supplier. HK$ 3392 million worth of denim products were imported from the Chinese Mainland in 1992. The value peaked in 1996 at HK$ 6241 million and then slumped to HK$ 4516 million in 1998. Despite this relatively poor export performance of the Chinese mainland, it still captured 54.9% of Hong Kong total imports in 1998.

China Taiwan, the second most important supplier, held a 23.8% share in 1998. With a growth rate of 2.7% on the average, imports from China Taiwan increased from HK$ 2315 million in 1992 to HK$ 2621 million in 1997. However, the value reduced to HK$ 1955 million in 1998 due to the overall slack performance of the East Asian countries in 1998.

South Korea was the No. 3 supplier, holding about 7.0% of Hong Kong's total imports. From 1992 to 1996, the South Korean export value remained over HK$ 1000 million. However, it dropped rapidly in the next two years, down to HK$ 575 million in 1998.

Other suppliers included Japan (4.6%), USA (2.0%), India (1.2%) and Macao (1.2%). These seven suppliers accounted for about 95% of Hong Kong total imports between 1992 and 1998.

6.1.1 Denim fabric imports

Denim fabric imports comprised 28%~34% of Hong Kong total denim product imports between 1992 and 1998. The value peaked in 1996 with HK$ 7848 million, falling to HK$ 7304 million and HK$ 5635 million in 1997 and 1998.

The important sources of Hong Kong import denim fabric were the Chinese Mainland, China Taiwan, South Korea, Japan, USA, India and Macao. They supplied more than 94% of Hong Kong denim fabric imports during the investigation period.

The Chinese Mainland, replacing China Taiwan as the leading supplier since 1995, accounted for 36.8% of Hong Kong total imports of denim fabric with a value of HK$ 2072 million in 1998.

6.4 Hong Kong denim fabric imports

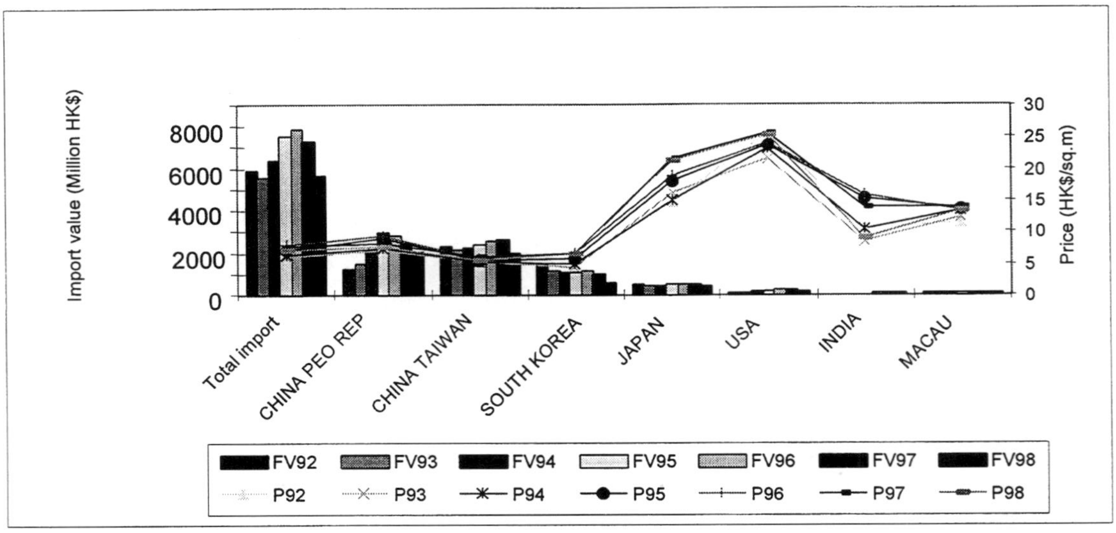

China Taiwan was the second most important supplier in 1998, and exported HK$ 1951 million worth of denim fabric to HK, accounting for 34.6% of Hong Kong total denim fabric imports.

South Korea was the third most important supplier, holding another 10.1%. It was followed by Japan (6.6%), USA (2.3%), India (1.8%) and Macao (1.8%).

The average import price ranged from HK$ 6.4 to HK$ 7.7 per square meter. Imports from the Chinese Mainland were around HK$ 7.2 to HK$ 9.2 per square meter, a little higher than the average import price. Those from China Taiwan were at HK$ 5.0 to HK$ 6.0 per square meter. Denim fabrics from the US were imported at the highest price, which was about HK$ 25 per square meter. Prices from Japan, Macao and India were HK$ 21, HK$ 13.3, and HK$ 9.0 per square meter respectively in 1998.

6.1.2 Denim apparel imports

Hong Kong imported HK$ 2301 million worth of denim apparel in 1992. The import value peaked in 1996 with HK$ 3626 million, growing by 14.4% per year from 1992 to 1996 on the average. However, it fell sharply in the next two years, dropping to HK$ 3058 million and HK$ 2588 million in 1997 and 1998.

The Chinese Mainland played a dominant role in the Hong Kong denim apparel import market. About 94% of Hong Kong denim apparel imports were from the Chinese Mainland between 1992 and 1998.

Compared with the overwhelming share of the Chinese mainland, other suppliers were insignificant, such as Italy (2.0%), USA (1.4%), and Philippines (0.6%).

The average import price of denim apparel was at HK$ 31 to 37 HK$ per unit during the investigation period. Denim apparel with the lowest import price was from the Chinese Mainland, which was at HK$ 30-odd per unit. Imports from Panama, South Korea and Singapore were priced at HK$ 40 to HK$ 80 per unit, while those from Italy had the highest price, which was over HK$ 300 per unit.

6.5 Hong Kong denim apparel imports

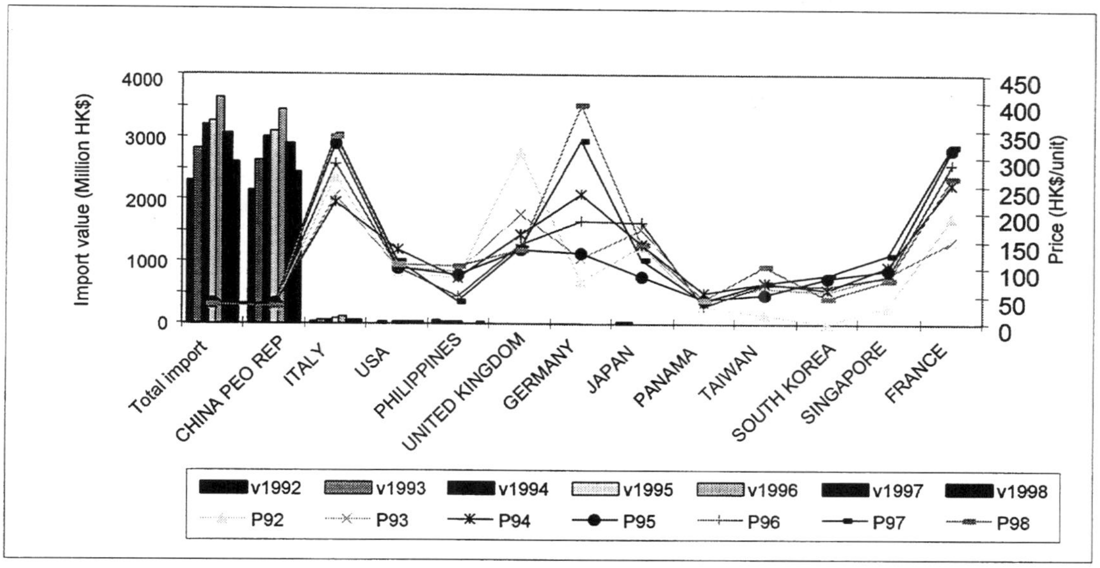

6.2 EXPORTS

Hong Kong, as the second most important supplier in the US market and the fourth largest non-EU supplier in the EU market, played an important role in the international denim trade arena.

Hong Kong's total denim product exports (domestic exports and re-exports) amounted to HK$ 13.9 billion in 1992 but fell to HK$ 11.4 billion in 1998. The ratio of domestic exports was down from 49.2% in 1992 to 32.9% in 1998; while that of re-exports increased sharply from 50.8% in 1992 to 67.1% in 1998 with an annual average growth rate of 12.4%, accounting for 67.1% of Hong Kong's total exports.

6.2.1 Hong Kong denim products domestic exports

Hong Kong domestic exports went down rapidly from HK$ 6844 million in 1992 to HK$ 3733 million in 1998. The following 11 countries received 89.1% of Hong Kong domestic exports in 1998. (See Fig. 6.6)

The Chinese Mainland was again the most important consumer, importing 32.8% of Hong Kong total domestic exports. With an annual growth rate of 11.8%, Hong Kong exports to the Chinese Mainland increased from HK$ 944 million in 1992 to HK$ 1450 million in 1997. However, they slumped to HK$ 1225 million in 1998.

The US captured 28.1% of Hong Kong domestic exports in 1998. It was once the most important market for Hong Kong before 1996. However, with the quick emergence of the Chinese mainland and its own decline trend, it reduced to the No. 2 place. In 1998, it imported HK$ 1048 million worth of denim products from Hong Kong, much lower than the 1992 level, which was HK$ 2009 million.

In 1992 and 1993, the United Kingdom was the second most important consumer after the US. It then experienced a continuous drop in the following years, by 8.7% in 1995, 41.8% in 1996 and 60.3% in 1997. Though exports to the United Kingdom rebounded by 57.5% in 1998, amounting to HK$ 294 million, its market share was

6.6 Hong Kong denim product exports

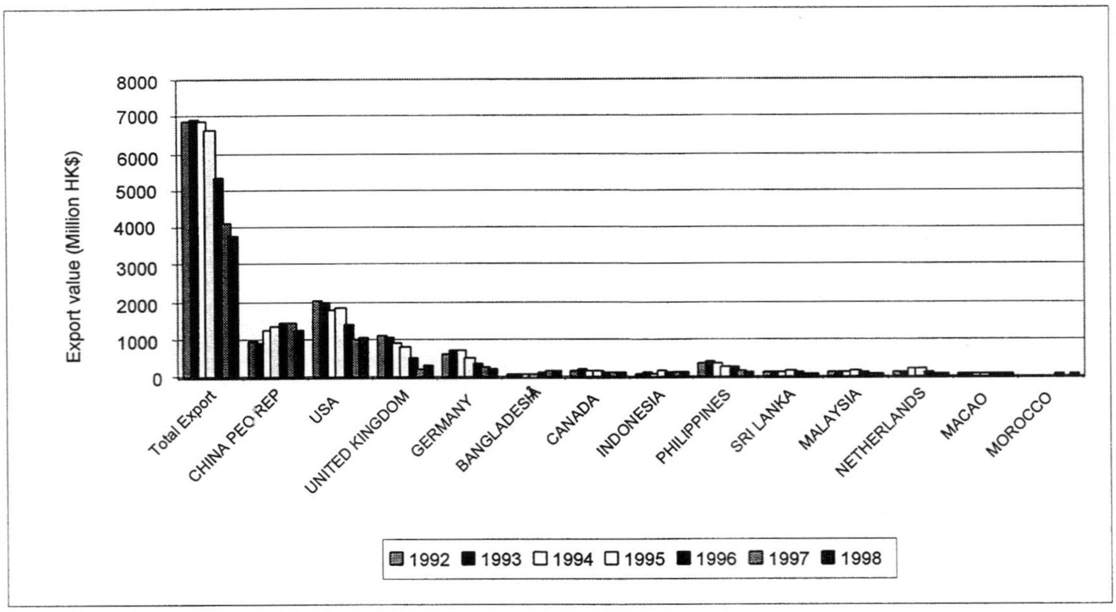

6.7 Hong Kong denim product exports

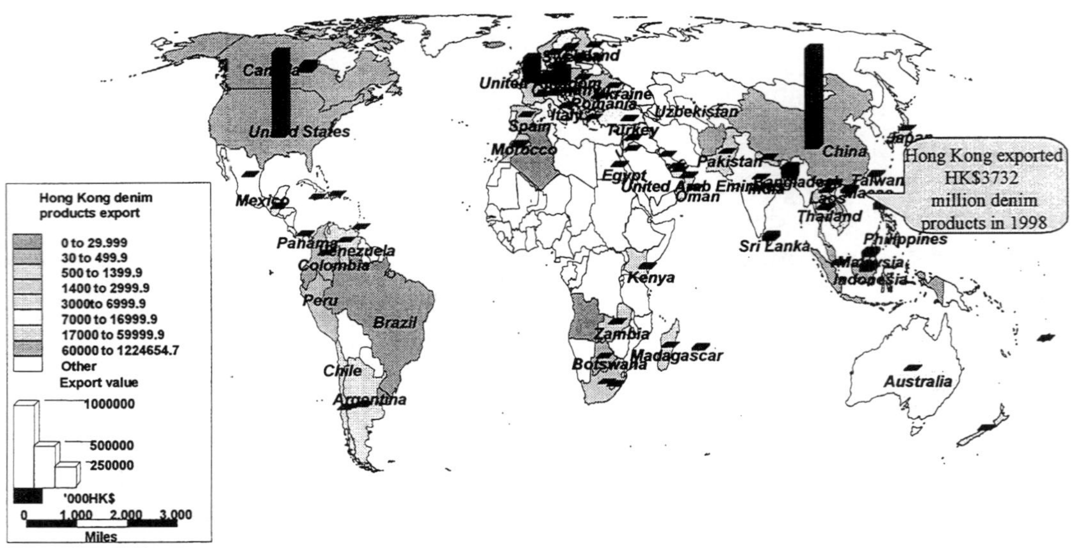

much smaller comparing with the top two suppliers, and it only accounted for 7.9% of Hong Kong domestic exports.

The other major consumers of Hong Kong denim products in 1998 were Germany (5.1%), Bangladesh (3.7%), Canada (2.4%), Indonesia (2.1%), Philippines (2.0%), Sri Lanka (1.8%), Malaysia (1.7%) and Netherlands (1.4%).

6.2.1.1 Denim fabric domestic exports

Though the export value was down from HK$ 2820 million in 1992 to HK$ 1859 million in 1998, its proportion of the total domestic exports was up from 41.2% to 49.8%.

6.8 Hong Kong denim fabric domestic exports

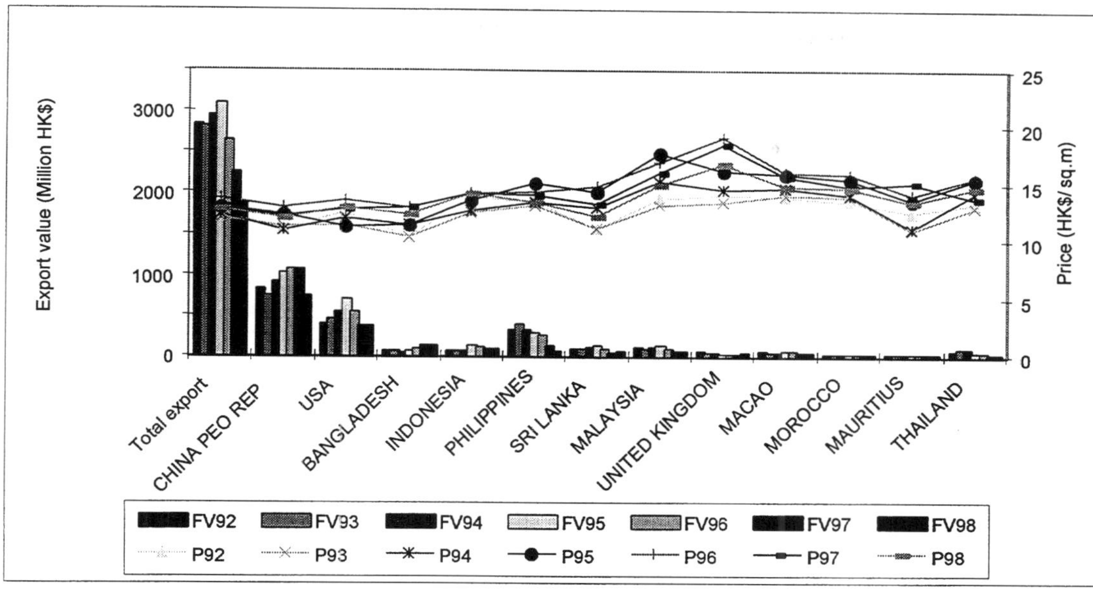

The Chinese Mainland was the most important market for Hong Kong denim fabric, accounting for 39.9% in 1998. Exports to the Chinese mainland grew by 4.9% every year on average between 1992 and 1997. The year 1998 saw the slump by 30.4% over 1997, decreasing from HK$ 1.1 billion to HK$ 741 million.

The US was the second most important market. Dropping from the peak value of HK$ 687 million in 1995, its value was only HK$ 362 million in 1998, accounting for 19.5% of Hong Kong denim fabric exports.

Bangladesh, which imported 7.3% in 1998, was the third most important denim fabric destination for Hong Kong. Its imports from HK grew from HK$ 57 million in 1992 to HK$ 136 million 1998

The average export price of Hong Kong denim fabric was HK$ 12.0 to 13.8 HK$ per square meter during the investigation period.

6.2.1.2 Denim apparel domestic exports

Hong Kong's domestic exports of denim apparel slumped from HK$ 4024 million in 1992 to HK$ 1873 million in 1998, a drop of 46.5%. Coinciding with this downtrend, its proportion of the total domestic exports also went down from 58.8% to 50.2%. (See Fig. 6.9)

The most important market for Hong Kong denim apparel was the US, which absorbed 36.6% of Hong Kong domestic exports in 1998. Its imports from HK reduced sharply from HK$ 1621 million in 1992 to HK$ 686 million in 1998.

Another important market was the Chinese Mainland, growing from HK$ 120 million in 1992 to HK$ 483 million in 1998 with an annual average growth rate of 50.6%. It captured 25.8% of Hong Kong denim apparel exports in 1998. Compared with the 3.0% market share in 1992, this shows that the Chinese mainland has become crucial to HK denim apparel exports in the recent years.

Though the United Kingdom and Germany ranked third and fourth, exports to these two markets fell rapidly between 1992 and 1998. The former imported 13.5% of Hong Kong denim apparel exports in 1998, dropping from HK$ 999 million in 1992 to HK$ 252 million in 1998. The latter held another 10.3% in 1998, also decreasing sharply from HK$ 567 million in 1992 to HK$ 192 million in 1998.

6.9 Hong Kong denim apparel domestic exports

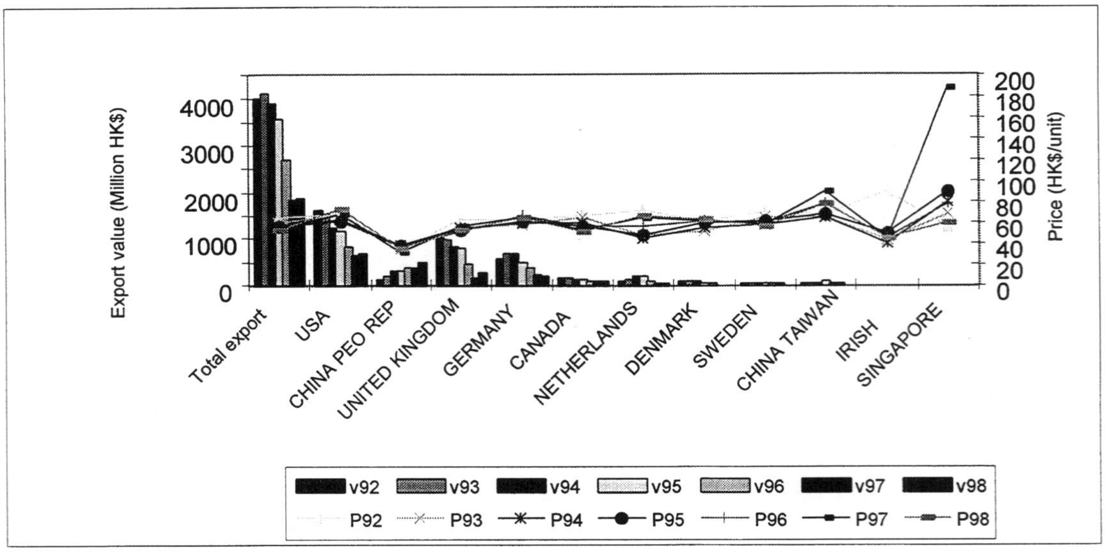

The other major importers in 1998 included Canada (3.7%), Netherlands (2.8%), Denmark (0.9%), Sweden (0.8%), China Taiwan (0.7%), Ireland (0.7%) and Singapore (0.6%).

Hong Kong-made denim apparel is competitive in view of its excellent workmanship, flexibility, responsiveness and sensitivity to fashion trends. The average export price was at HK$ 65.1 per unit in 1992. Between 1992 and 1998, the price decreased year by year and became HK$ 53.9 per unit in 1998. The sharp rise of lower-priced denim apparel exports to the Chinese mainland mainly explains the continuous drop of the average export price.

6.2.2 Re-exports[4]

The peak of Hong Kong denim product re-exports appeared in 1996 with a value of HK$ 10.58 billion with an average growth rate of 9.9% per year between 1992 and 1996. However, the value reduced in 1997 and 1998, amounting to HK$ 7620 million in 1998.

As mentioned earlier, re-exports have played a more and more important role in HK denim product trade. They accounted for 67.1% of Hong Kong's total exports in 1998.

6.2.2.1 Hong Kong denim products re-export by market

The major markets for Hong Kong denim product re-exports were the Chinese Mainland, Panama, Japan, Bangladesh, Philippines, Egypt, the US, Australia, Indonesia and Sri Lanka. (See Fig. 6.10 & 6.11)

The Chinese Mainland was the most important market for Hong Kong denim product re-exports. It imported HK$ 4493 million worth of denim products in 1998, accounting for 59.0% of Hong Kong's total re-exports. Hong Kong re-exports to the

[4] The definition of a Hong Kong Re-export is that it is a product which has previously been imported into Hong Kong and which is re-exported without having undergone in Hong Kong a manufacturing process which has changed permanently the shape, nature, form or utility of the product (TDC).

Chinese Mainland peaked in 1997 with HK$ 6032 million or 60.4% of Hong Kong total re-exports. Between 1992 and 1997, the denim products exported to the Chinese Mainland from other countries via Hong Kong increased by 8.4% per year on the average. However, the value fell by 25.5% in 1998. The second most important Hong Kong re-export market was Panama, which held 8.8% of Hong Kong re-exports in 1998. Growing from HK$ 507 million in 1992, re-exports to Panama reached HK$ 855 million in 1997 and then declined to HK$ 670 million in 1998.

Japan ranked the third with HK$ 317 million or 4.2% of Hong Kong total denim product re-exports in 1998. Re-exports to Japan reached their highest level in 1994 with HK$ 483 million.

Bangladesh, the Philippines and Egypt were the other three important markets, accounting for 2.3%, 2.3% and 2.1% respectively in 1998. These three re-export markets grew rapidly between 1992 and 1998. Re-exports to Bangladesh rose from

6.10 Hong Kong denim product re-exports by major markets

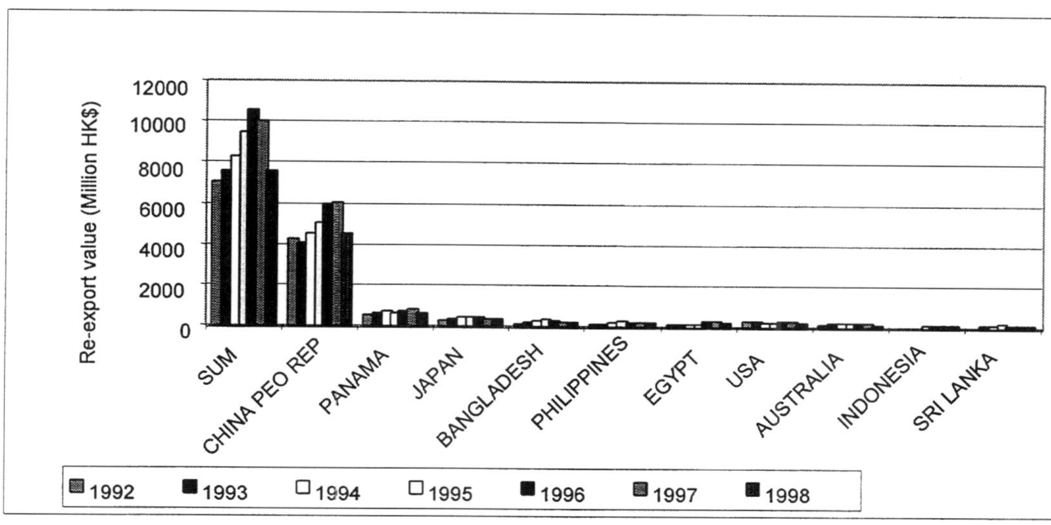

6.11 Hong Kong denim product re-exports by market

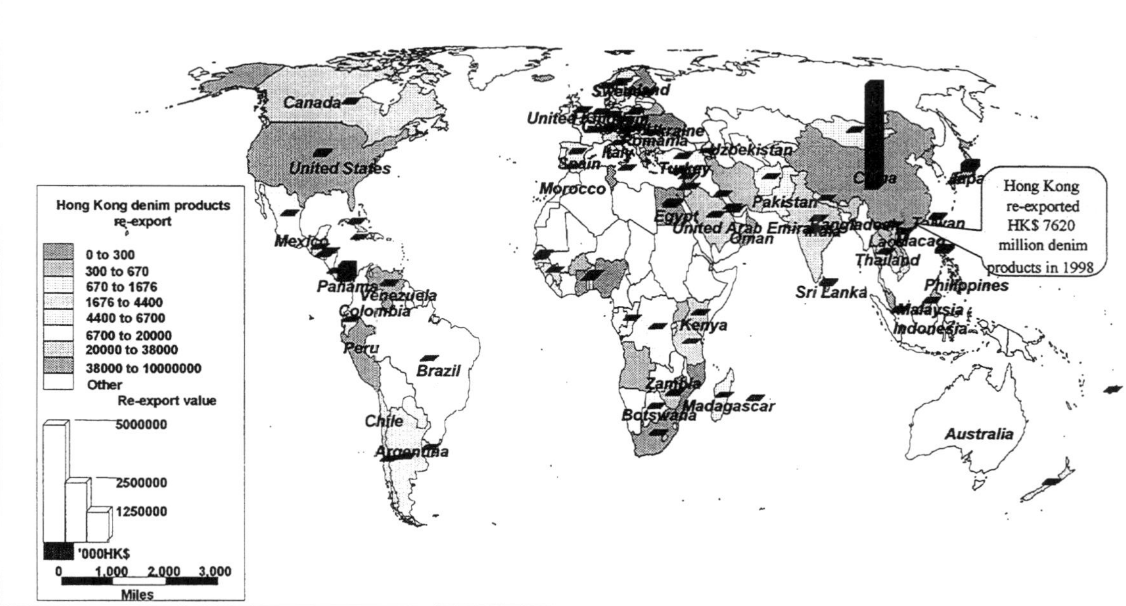

HK$ 94 million in 1992 to HK$ 178 million in 1998 at an annual average growth rate of 14.8%. Those to the Philippines and Egypt enjoyed a faster growth rate of 22.0% and 21.3% per year from 1992 to 1998, reaching HK$ 176 million and HK$ 156 million respectively in 1998.

Other re-export markets included the US (1.8%), Australia (1.6%), Indonesia (1.4%) and Sri Lanka (1.1%). With an average reducing rate of 7.3% every year, re-exports to the US fell from HK$ 280 million in 1992 to HK$ 137 million in 1998. By sharp contrast, Indonesia experienced the strongest growth between 1992 to 1998, increasing from HK$ 33 million in 1992 to HK$ 106 million in 1998 with an annual average growth rate of 37.1%.

The ten above countries captured 79%~85% of Hong Kong total re-exports between 1992 to 1998.

A. Denim Fabric Re-export by Market

Hong Kong denim fabric re-exports peaked in 1996 with a value of HK$ 7227 million. In 1997, the value fell to HK$ 7115 million. It further slumped by 22.2% to HK$ 5539 million in 1998. Unlike the domestic export composition, denim fabric took a dominant share, which was 72.7% of the total re-exports in 1998.

The Chinese Mainland was the most important market, absorbing 80.6% of Hong Kong denim fabric re-exports in 1998. Re-exports to the Chinese Mainland grew from HK$ 4198 million in 1992 to HK$ 6000 million in 1997. However, the re-export value shrunk to HK$ 4463 million in 1998.

Bangladesh and the Philippines were the other two major markets for Hong Kong denim fabric re-exports, accounting for 3.2% and 3.0% respectively in 1998.
Indonesia and Sri Lanka held another 1.9% and 1.5% in 1998. Other markets included Morocco (0.8%), Mauritius (0.7%), Macao (0.7%), Thailand (0.7%), China Taiwan (0.6%) and Panama (0.6%). Exports to most markets shrank between 1997 and 1998, causing the overall denim fabric re-exports to decline sharply.

6.12 Hong Kong denim fabric re-exports by market

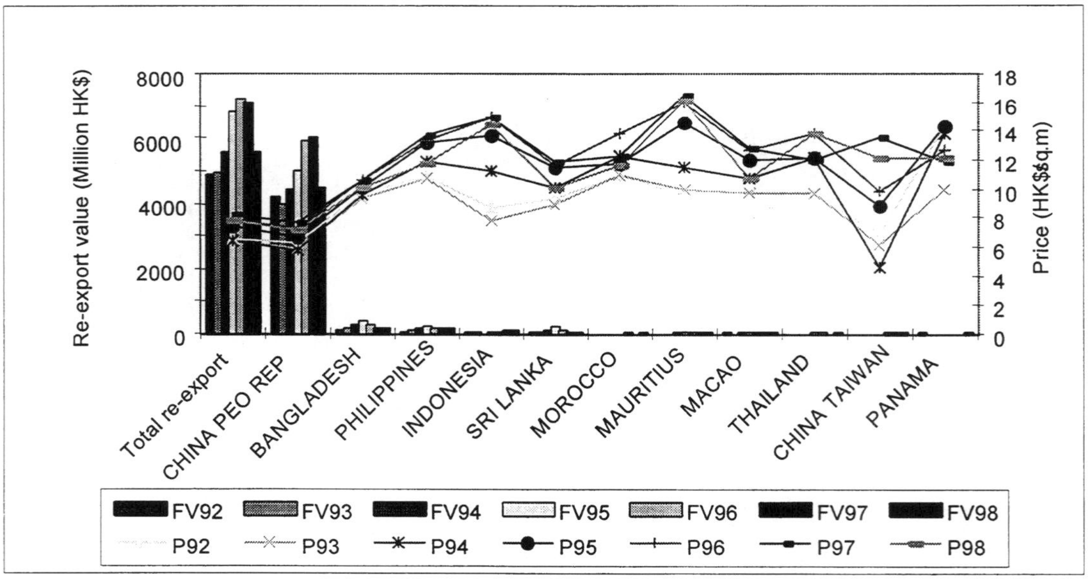

The average re-export price was HK$ 7.7 per square meter in 1998. Those to the Chinese Mainland were around HK$ 7.2 per square meter, which was the lowest price of all Hong Kong denim fabric re-exports. The denim fabrics re-exported to Bangladesh, Philippines, Sri Lanka and Macao were about HK$ 10 per square meter; those to Philippines and Morocco were more than HK$ 11.6 per square meter, and those to Indonesia, Mauritius, Thailand, Taiwan and Panama were over HK$ 12 per square meter in 1998.

B. Denim Apparel Re-Export by Market

The major markets for Hong Kong denim apparel re-exports were Panama, Japan, Egypt, USA and Australia. (See Fig. 6.13)

Panama, holding 30.6% of Hong Kong denim apparel re-exports, was the leading importer in 1998, and re-exports to Panama reached a peak of HK$ 847 million in 1997.

Japan was the second most important market for Hong Kong denim apparel re-exports, accounting for a 14.6% market share. Re-exports peaked in 1994 with 685 million and then dropped to HK$ 303 million in 1998.

Egypt received 7.4% of Hong Kong denim apparel re-exports in 1998. Such re-exports to Egypt grew from HK$ 64 million in 1992 to HK$ 227 million in 1997. In 1998, the value slumped to HK$ 153 million.

6.3% of Hong Kong re-exported denim apparel were to the US in 1998. The re-export value was not steady between 1992 and 1998. It hit a low in 1998 when only HK$ 130 million worth of denim apparel was re-exported to the US. The peak appeared in 1996 with HK$ 276 million.

HK$ 113 million worth of denim apparel was re-exported to Australia in 1998, accounting for 5.4% of Hong Kong denim apparel re-exports. Re-exports to Australia peaked in 1996 with HK$ 187 million.

Other re-export destinations were South Africa, Venezuela and Singapore, capturing 2.1% to 2.5% of Hong Kong denim apparel re-exports in 1998 each. The Chinese mainland played an insignificant role here, only accounting for 1.4% of Hong Kong denim apparel re-exports in 1998.

6.13 Hong Kong denim apparel re-exports by market

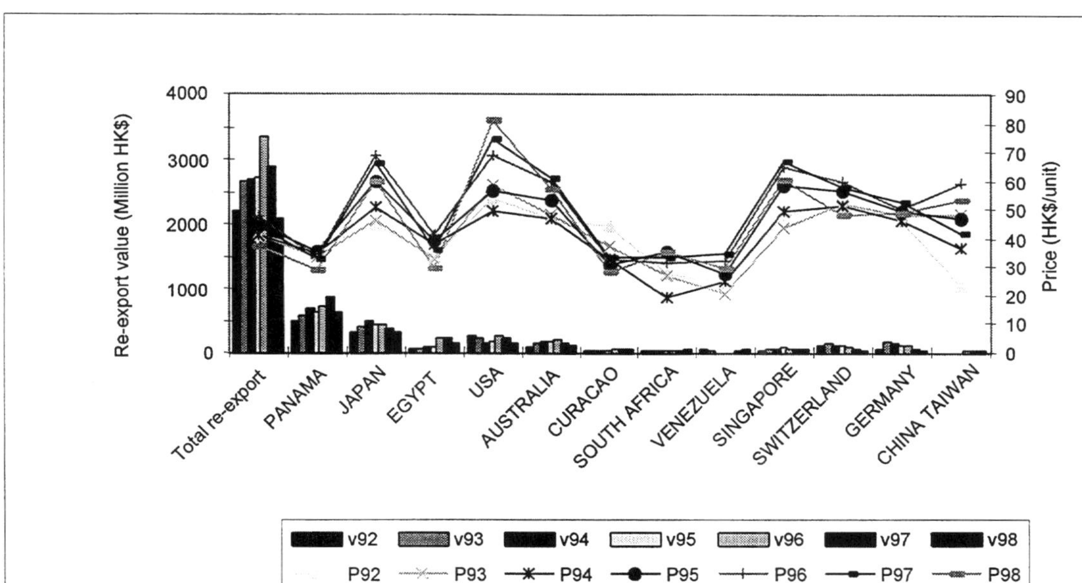

The average price of Hong Kong denim apparel re-exports was about HK$ 37 to HK$ 44 per unit between 1992 and 1998. Denim apparel re-exports to Panama, Egypt and Venezuela were around HK$ 30 per unit, to Japan and Singapore were at HK$ 60-odd per unit, and to the US over HK$ 80 per unit.

6.2.2.2 Hong Kong denim product re-exports by origin

The important origins of Hong Kong denim product re-exports were the Chinese Mainland, China Taiwan, South Korea, Japan, the US and Macao. From 1992 to 1998, these countries (areas) supplied around 95% of Hong Kong total denim product re-exports. (See Fig. 6.14)

Among these suppliers, the Chinese Mainland was the most important origin, supplying 52.5% of Hong Kong denim product re-exports in 1998. Re-exports from the Chinese Mainland increased from HK$ 3176 million in 1992 to HK$ 5692 million in 1996, with an average growing rate of 15.8% per year. However, the value reduced by 16.3% per year after 1996 and became HK$ 4011million in 1998.

The second most important supplier, China Taiwan, held a 28.3% share in 1998. It supplied HK$ 2310 million worth of denim products to Hong Kong for re-exportation in 1992, peaking in 1996 with HK$ 3026 million. In 1998, the value declined to HK$ 2163 million.

South Korea was the third most important source. Re-exports from South Korea reached HK$ 536 million in 1998 with a market share of 7.0%.

Other major sources included Japan (3.9%), USA (2.0%) and Macao (0.9%).

A. Denim Fabric Re-exports by Origin

China Taiwan was the most important source, accounting for 38.9% of Hong Kong denim fabric re-exports in 1998, with a value of HK$ 2163 million. The re-export value reached a peak in 1996 with HK$ 3026 million. (See Fig. 6.16)

The Chinese mainland ranked second with a market share of 35.8% in 1998. The peak appeared in 1995 with HK$ 2419 million, dropping to HK$ 1990 million in 1998.

6.14 Hong Kong denim product re-exports by major origins

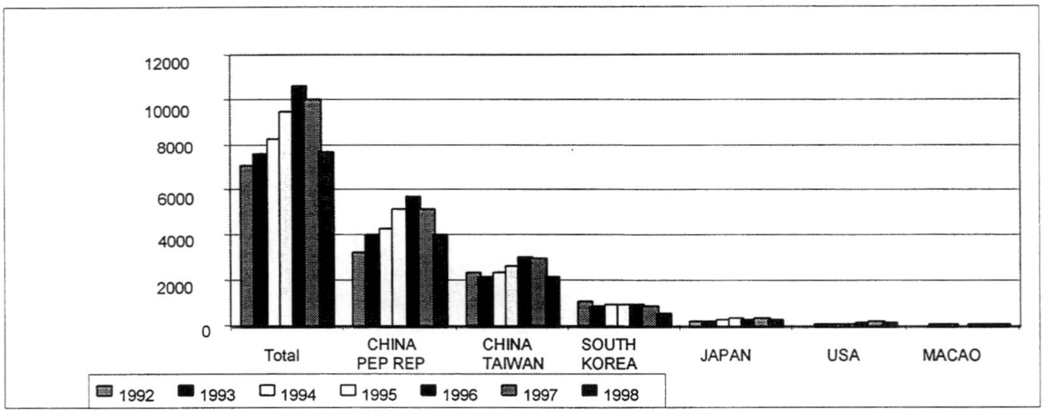

6.15 Hong Kong denim product re-exports by origin

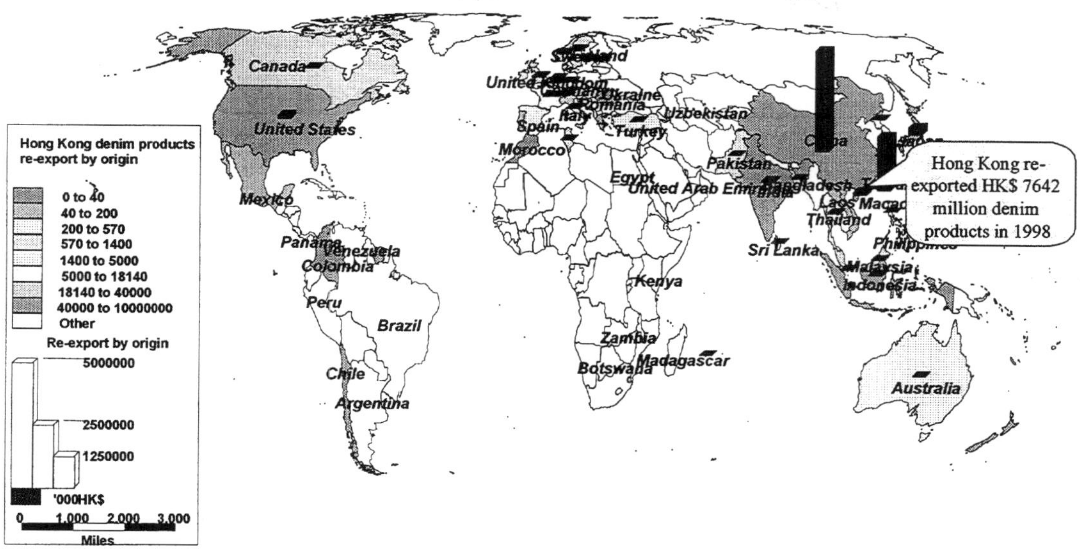

6.16 Hong Kong denim fabric re-exports by origin

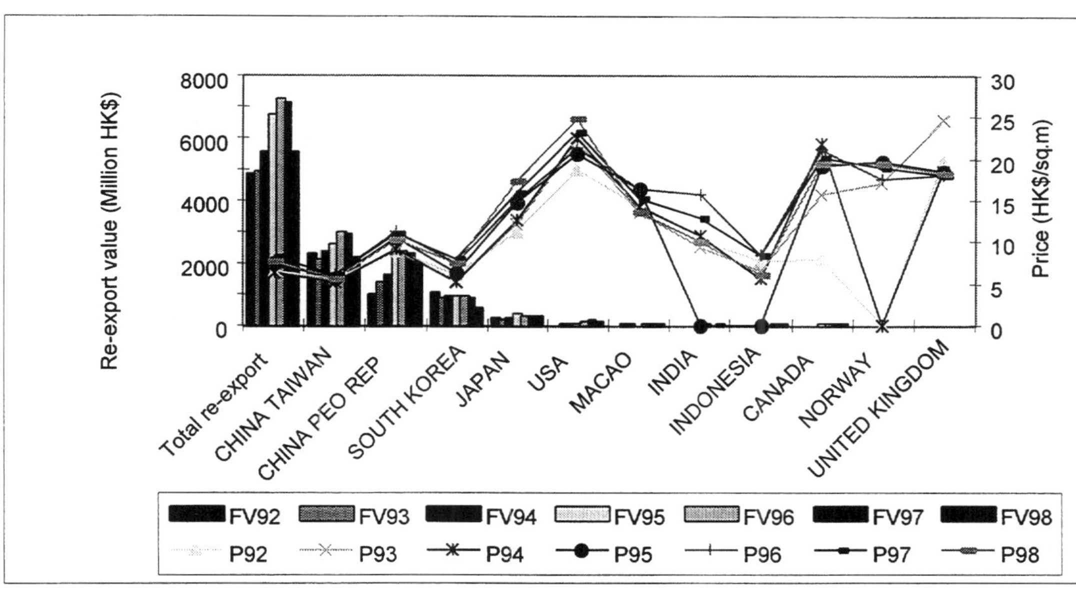

Hong Kong denim fabric re-exports from South Korea experienced a continuous downturn in the past few years. With an average decrease of 8.1% per year, the denim fabric re-exports from South Korea fell from HK$ 1036 million in 1992 to HK$ 535 million in 1998. Therefore, though it was the third most important source for HK denim fabric re-exports, its market share was much smaller, which was about 9.6% in 1998.

Japan held another 5.4% of Hong Kong denim fabric re-exports in 1998. The US accounted for 1.9% and Macao 1.2%.

6.17 Hong Kong denim apparel re-exports by origin

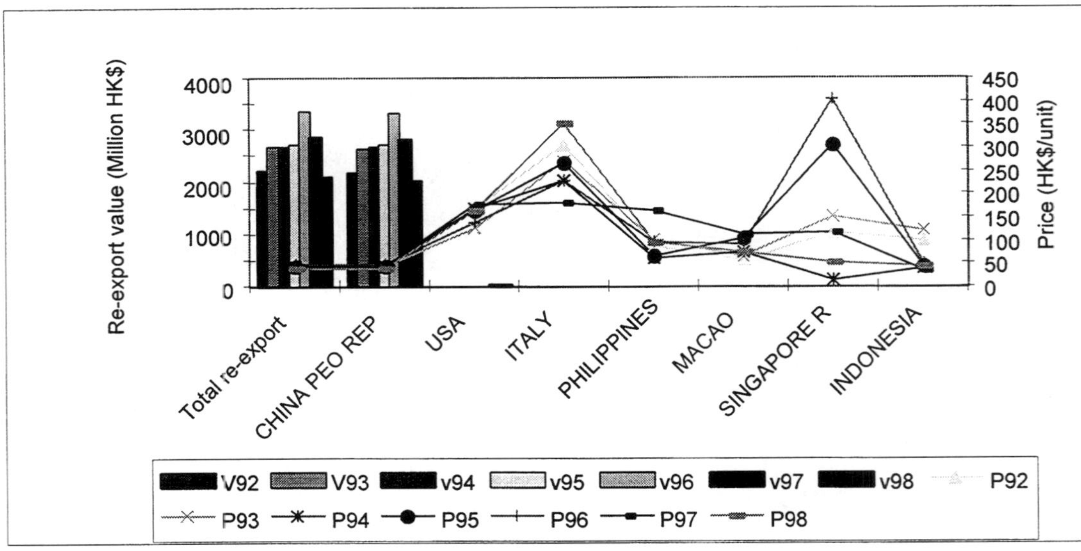

The average price of Hong Kong denim fabric re-exports was around HK$ 6.3 to HK$ 8.1 per square meter between 1992 and 1998. The denim fabrics from China Taiwan were HK$ 5.2 to HK$ 6.1 per square meter, from the Chinese Mainland were HK$ 9.0 to HK$ 11.1 per square meter and from South Korea, India and Indonesia were less than HK$ 10 per square meter. The denim fabrics from Japan, USA, Canada, Norway and the United Kingdom were at a higher price, over HK$ 17 per square meter.

B. Denim Apparel Re-exports by Origin

The Chinese Mainland took the lion's share as the single most important source for Hong Kong denim apparel re-exports. Though re-exports from the Chinese mainland decreased sharply from HK$ 3303 million in 1996 to HK$ 2021 million in 1998, it still accounted for 97.0% of Hong Kong denim apparel re-exports. (See Fig. 6.17)

Another supplier of relative importance was the US, which supplied HK$ 45 million worth of denim apparel in 1998, accounting for 2.1%.

The average price of denim apparel from the Chinese Mainland was around HK$ 36 to 47 per unit, much lower than those from the US and Italy which were HK$ 160 and HK$ 350 per unit respectively. The denim apparel from the Philippines, Macao, Singapore and Indonesia were sold at HK$ 95, 71, 50 and 43 per unit respectively in 1998

REFERENCE

1. Fong, W.S., Denim developments. Textile Asia, 1987, p.186-189.

7 China Denim Products Trade

OVERVIEW

Denim products manufacturing in China has a relatively short history. Jeans have become popular in the Chinese market only since the mid-1980s. However, this sector has undergone rapid development during the past two decades due to China's comparative advantages of an abundant low-cost labor supply and easy access to raw materials such as cotton.

According to a survey on the nationwide denim production from 1995 to 1997 [1, 4], its annual cotton consumption captured 23.3% of the national spun-cotton supplies, or 5% of total cotton consumption. About 96% of denim yarn were pure cotton, 3% blends with linen and 1% blends with polyester or silk. There were about 400 denim manufacturers in 1997 with a production capacity of about 1.2 to 1.3 billion meters, accounting for about 5% of China's annual woven fabric production. In 1997, the sector produced 807 million running meters of woven denim fabric, 108 million less than that in 1995. Domestic consumption experienced strong growth from 158 million meters in 1995 to 388 million meters in 1997.

The denim apparel sector in China was also composed of over 400 manufacturers with an annual average output of 111 million units from 1995 to 1997. With the increasing popularity among Chinese youngsters in the past decade of denim apparel, sales of jeans have been growing at an annual average rate of 10% in the past ten years. According to a 1998 survey in 238 department stores in China, the jeans sales amounted to 26.4 million units [2]. The entry of world-famous brands such as Levi Strauss and Lee greatly sharpened the market competition. Today, foreign labels have occupied almost half of the Chinese market.

By contrast to the boom in the domestic denim market, the export performance of China denim products in 1997 was lackluster. As regards denim fabric exports, it was 420 million meters in 1997, plunging from 757 million meters in 1995. Denim apparel exports suffered similar setbacks, dropping from 65.3 million units to 35.2 million in 1997. The import quotas upon Chinese textile and apparel exports imposed by the US and the EU have been particularly tightened in recent years, causing additional troubles for domestic denim manufacturers which are export-oriented due to the rising quota prices. According to one HK apparel manufacturer's estimation, if costs in China inclusive of quota prices are calculated as 100, then Korea is 80, Sri Lanka 79, Indonesia 77, and Mexico 85 [3]. Therefore, though the manufacturing costs in China are cheaper than in many other low-cost competitors such as Mexico and Indonesia, the overall price level is higher if quota prices are taken into consideration.

7.1 IMPORTS

The analysis of China's denim product trade adopts the same approach as that of the

Japan denim trade due to the lack of sufficient supporting data. Trade with HK is explained in the corresponding section, so this part will focus upon the US and EU markets.

7.1.1 Denim products imports

7.1.1.1 Imports from the US

China enjoyed a very large trade surplus with the US concerning denim products. The gap was US$ 36.3 million in 1995, widening to US$ 43.7 million in 1996 and narrowing back to US$ 36.0 million in 1997. (Table 7.1)

China was not the major export destination for US denim products, only accounting for 0.07% of US total denim product exports in 1997. The import value from the US peaked in 1995 with US$ 1.6 million and then slumped to US$ 607 thousand.

Among the denim product imports from the US to China, denim fabrics occupied a dominant share, increasing from 71.8% in 1993 to 95.6% in 1997.

7.1.1.2 Imports from the EU

Denim product trade between China and the EU was not balanced, with China having a large trade surplus against the EU. The gap was 26.3 million Euro in 1994 and reduced to 17.9 million Euro in 1997 because of sharp fall of both import and export values. (See Table 7.1)

Denim product imports peaked in 1996 with a value of 527.7 thousand Euro, plunging by 86.6% in 1997 to 70.5 thousand Euro. Its share in the EU denim product export market was less than 0.01% in 1995 and 1997.

Compared with the product mix of the US exports to China, the EU denim product exports were quite balanced with the denim fabrics taking a relatively larger share. In 1997, the proportion of denim fabrics was 60.9% (See Tables 7.1 and 7.2)

Figure 7.1 shows that among all the EU suppliers, Italy was the most important with the share of 48.2% in 1997. In 1994, only three countries exported to China. They were Italy (77.0%), the United Kingdom (13.8%) and Germany (9.3%). In 1997, the top three suppliers remained unchanged, with the UK still the No. 2 exporter to

Table 7.1 Denim product exports from the US to China between 1993 and 1997

Denim product exports from the US to China					
	1993	1994	1995	1996	1997
Total US exports ('000 US$)	591675	625678	719365	853902	855767
to China ('000 US$)	1391	265	1607	431	607
Exports to China / Total US exports	0.24	0.04	0.22	0.05	0.07
Fabric / Total China imports from the US	71.75	100.00	99.00	100.00	95.55

Table 7.2 Denim product exports from the EU to China between 1994 and 1997

Denim product exports from EU to China				
	1994	1995	1996	1997
Total EU exports ('000 Euro)	1744331.6	1790602.7	2034903	1971950.3
to China ('000 Euro)	185.32	69.68	527.66	70.52
Exports to China / Total EU exports %	0.011	0.004	0.026	0.004
Fabric/Total denim product exports to China %	56.82	44.98	48.95	60.85

the China market with a 23.6% market share. Germany and France held another 20.5% and 6.1%.

7.1 Denim product exports from the EU to China between 1994 and 1997

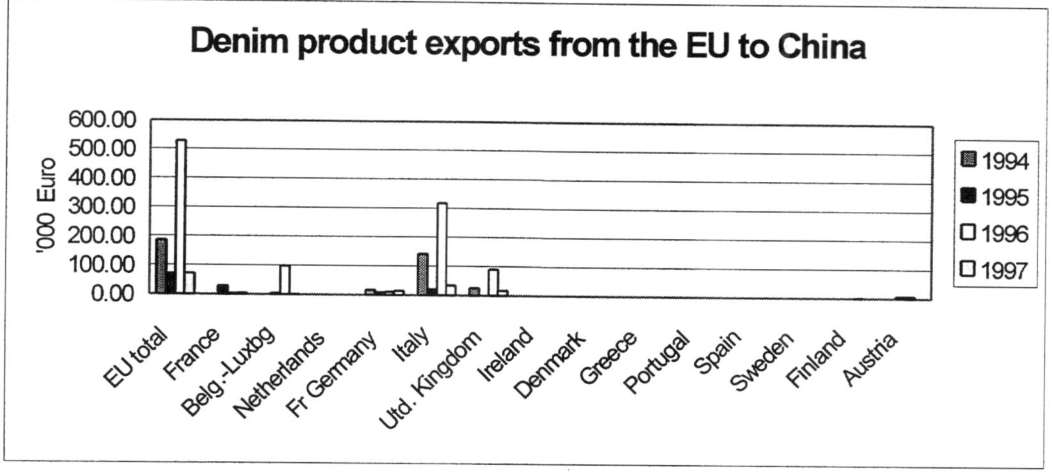

7.2 Denim fabric exports from the US to China between 1993 and 1997

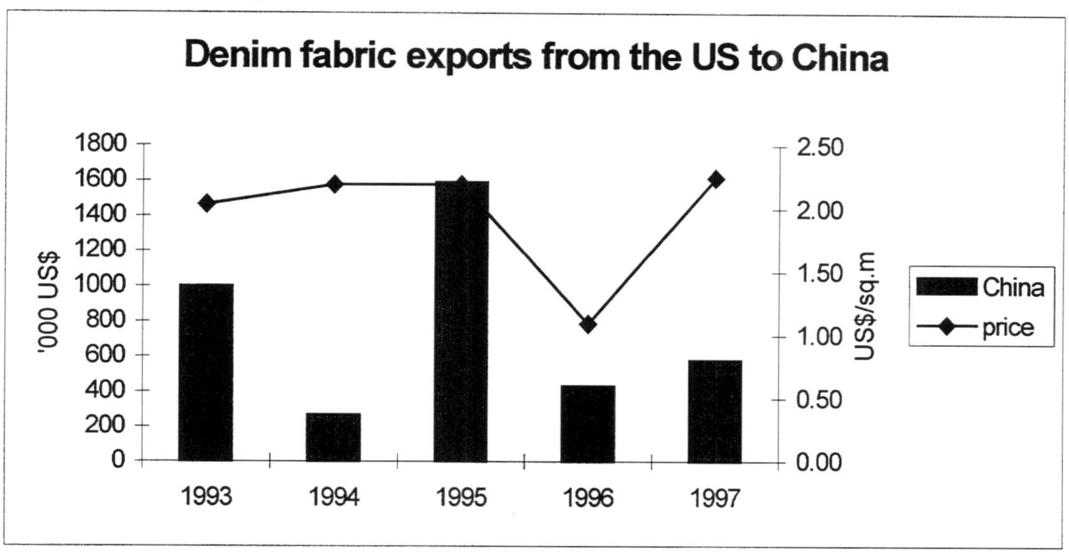

7.3 Denim fabric exports from the EU to China between 1994 and 1997

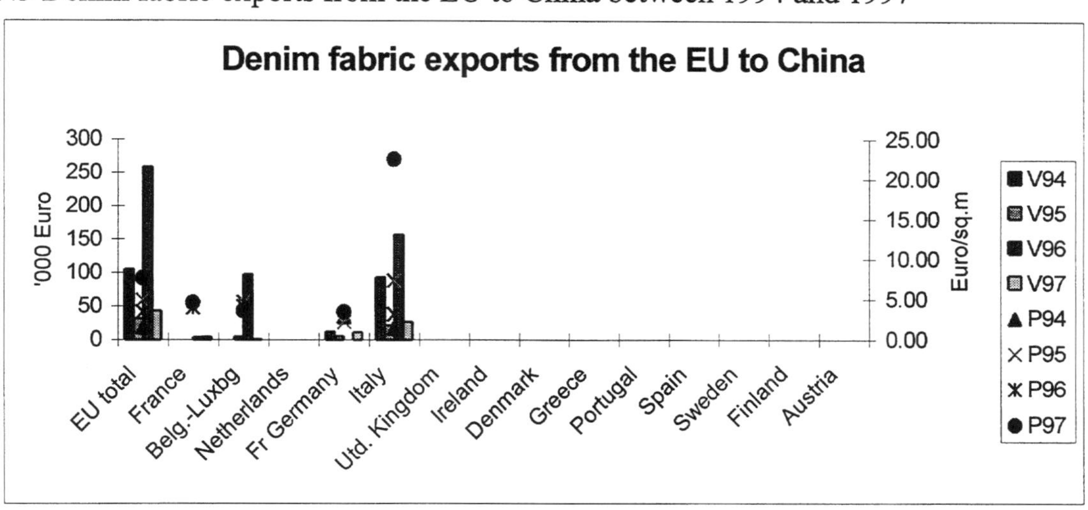

7.1.2 Denim fabric imports

7.1.2.1 Imports from the US

Denim fabric imports from the US were US$ 580 thousand in 1997, only about 36.5% of the peak level in 1995. The share in the US denim fabric export market was continuously down from 0.55% in 1993 to 0.05% in 1997.

The average import price in 1997 was US$ 2.3 per square meter, rising from US$ 2.0 in 1993. This was actually the highest price in the four-year period.

7.1.2.2 Imports from the EU

Denim fabric imports from the EU to China fluctuated violently between 1994 and 1997. The value hit bottom in 1995 with 31.3 thousand Euro and rose sharply and suddenly to 258.3 thousand Euro in 1996 and then slumped to 42.9 thousand Euro in 1997. (See Fig. 7.3)

Italy was the leading EU supplier, accounting for 62.2% of total EU denim fabric exports to China in 1997. Its role was even more dominant in 1994 when it occupied 88.6%. Germany held another 25.7% in 1997. France began exporting to China in 1996 and accounted for 10.0% in 1997. (See Fig. 7.3)

The average import price from the EU was 7.7 Euro per square meter in 1997, rising sharply from 1.6 Euro in 1994. Italian denim fabrics were imported at the highest price which was 22.5 Euro per square meter in 1997. Those from Germany, Belgium and France were around 3.4, 3.6 and 4.6 Euro respectively.

7.1.3 Denim apparel imports

7.1.3.1 Imports from the US

Denim apparel imports from the US to China were negligible--US$ 393 thousand for 1993, US$ 16 thousand for 1995 and US$ 27 thousand for 1997. The small import volume can be mainly attributed to two reasons: one is that China itself is a big apparel producer and exporter; the other is that many foreign labels have successfully penetrated the Chinese market and engaged in local production there.

Accompanying the slump in import value was the sharp rise in import prices. In 1993, the average import price was about US$ 3.27 per unit and, four years later, it jumped to US$ 17.31 per unit. (See Fig. 7.4)

7.1.3.2 Imports from the EU

Denim apparel imports from the EU to China underwent the same development pattern as the denim fabric imports. The value peaked in 1996, reaching 269.4 thousand Euro. It then suddenly plunged to 27.6 thousand Euro in 1997.

Italy, Germany and the United Kingdom were the traditional suppliers among the EU countries, except for the year 1995 when there were no exports from Italy and the UK. Italy was the leading supplier in 1994 and 1996, accounting for 61.6% and 59.6% respectively. The UK replaced Italy in 1997 as the No. 1 supplier, capturing 60.2% of EU total exports to China. Though Germany exported denim apparel to China each year during the four-year period, its value was not large, holding a share of 12.5% in 1997. France was the most important EU supplier in 1995, holding a 74.4% market

7.4 Denim apparel exports from the US to China between 1995 and 1997

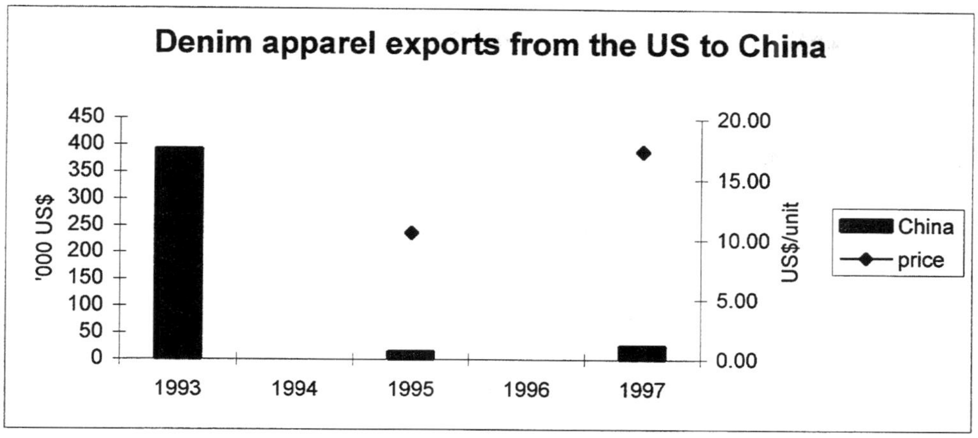

share, but there were no exports from France to China during the other three years. (See Fig. 7.5)

The average import price was not stable. It could be as high as 32 Euro per unit in 1995 and 1997 and as low as 6.0 Euro in 1994. The Italian and British exports in 1997 registered higher unit prices which were 36.3 and 37.1 respectively. Those from Germany and Austria were imported at relatively lower prices--19.2 Euro and 23.3 Euro respectively.

7.2 EXPORTS

7.2.1 Denim products exports

7.2.1.1 Exports to the US

Denim product exports from China to the US were of relatively small consequence in the US denim product import market, accounting for 2.4% in 1995. The export value reached a peak in 1996 with US$ 44.1 million. It dropped by 17.0% in 1997, then holding a 1.8% market share.

Totally different from its import product mix, China denim product exports to the US were almost exclusively composed of denim apparel. There were only US$ 24 thousand worth of denim fabric exports during the three-year period.

7.2.1.2 Exports to the EU

Denim product exports from China to the EU were much larger than its imports from the EU. The export value reached 31.2 million Euro in 1996 and then fell to 18.0 million Euro in 1997. Its share in the EU denim product import market shrank from 1.1% in 1994 to 0.7% in 1997. (See table 7.3)

Denim fabrics were of little consequence in the total denim product exports from China to the EU, though its overall share increased a little to 1.2% in 1997.

Almost all the EU members imported denim apparel from China, except Greece and Portugal, who stopped imports in certain years. Germany held 31.3% of the EU total denim product imports from China in 1997, followed by Denmark (13.1%), Netherlands (12.9%), the UK (11.4%) and Italy (10.6%). France imported less denim products from China since 1995 and its market share shrank to 4.8% in 1997. By

7.5 Denim apparel exports from the EU to China between 1994 and 1997

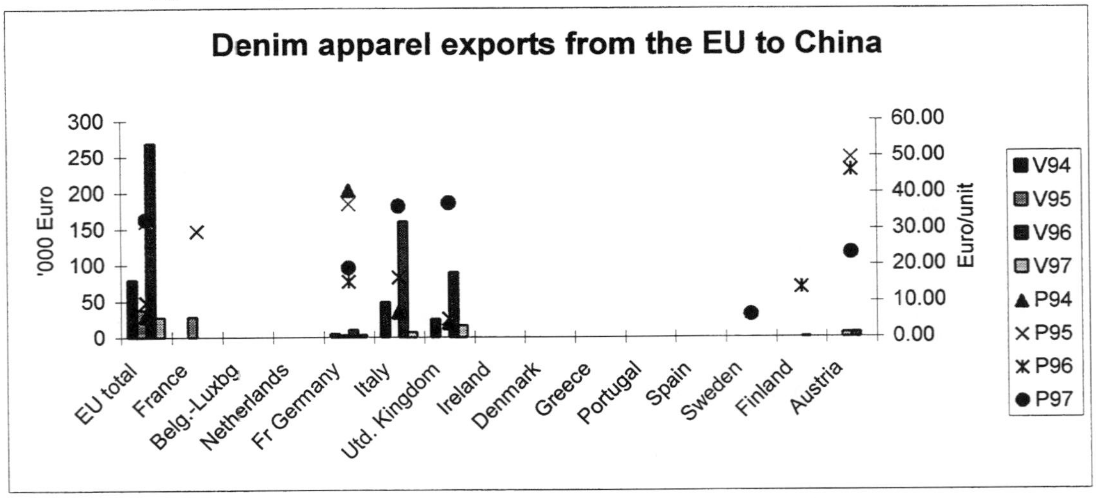

7.6 Denim product exports from China to the US between 1995 and 1997

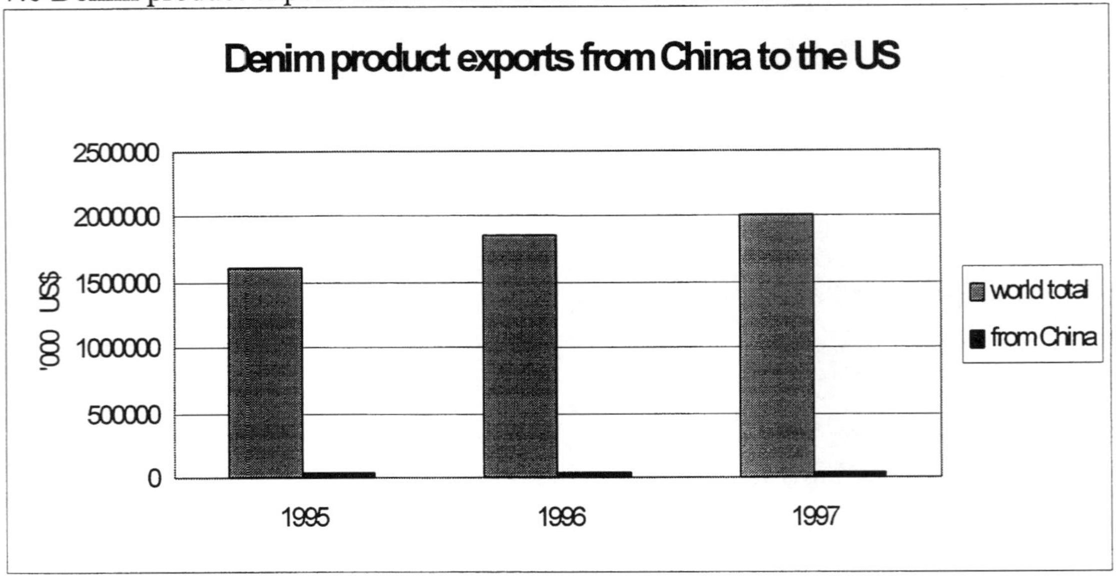

sharp contrast, those destined to Denmark showed a clear uptrend and peaked in 1997 with 2.4 million Euro.

7.2.2 Denim fabric exports

7.2.2.1 Exports to the US

As mentioned earlier, denim fabric exports to the US were very small. There were no exports in 1995 and 1997. In 1996, the value was US$ 24 thousand, accounting for 0.003% of US total denim fabric imports.

The average export price per square meter was US$ 1.2, indicating the low value-added feature of China-label denim fabrics.

7.2.2.2 Exports to the EU

Denim fabric exports from China to the EU were small in value. It dropped from 205.5 thousand Euro in 1994 to 128.6 thousand Euro in 1996 and then picked up to

Table 7.3 Denim product exports from China to the EU between 1994 and 1997

Denim product exports from China to the EU	1994	1995	1996	1997
Total EU imports ('000 Euro)	2452160.7	2357516.9	2771175.2	2648607.1
From China ('000 Euro)	26481.48	17837.44	31248.1	17989.07
Imports from China / Total EU imports %	1.08	0.76	1.13	0.68
Fabric/Total denim product imports from China %	0.78	0.86	0.41	1.15

7.7 Denim product exports from China to the EU between 1994 and 1997

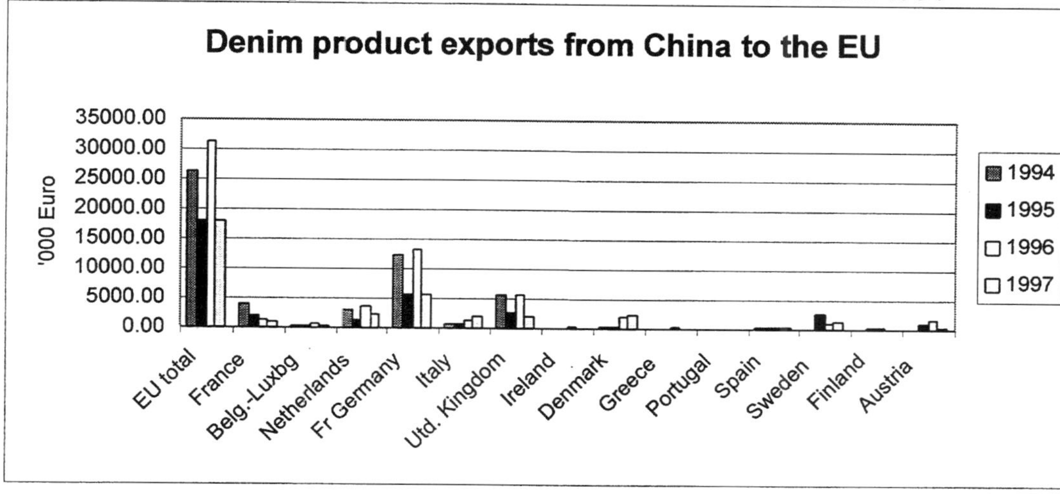

206.5 thousand Euro in 1997. The share of the EU denim fabric import market was insignificant, less than 0.1% throughout the investigation period. (See Fig. 7.9) The United Kingdom was the leading EU importer except for the year 1995 when Italy took the No. 1 place with 91.1% of the market share. The UK's share increased from 57.0% in 1994 to 72.7% in 1997, while that of Italy dropped to 24.8% in 1997. The two were responsible for almost all the Chinese denim fabric exports. Germany also imported from China since 1995, but its import value was really negligible.

The average export price was relatively stable during the four years, from 1.5 Euro per square meter in 1994 to 1.3 Euro in 1997. Denim fabrics exported to Germany registered the highest unit export price, which was 4.4 Euro in 1997. Those to Italy were above the average level as well, reaching 1.9 Euro in 1997 but those to the UK were under the average level, only 1.2 Euro in the same year.

7.2.3 Denim apparel imports

7.2.3.1 Exports to the US

China's strength in apparel production is well illustrated by its export performance. Its denim apparel exports to the US reached US$ 36.6 million in 1997, capturing 2.2% of the US total denim apparel imports. The export value peaked in 1996 at US$ 44.1 million.

The average export price showed an upward trend, from US$ 7.0 per unit in 1995 to US$ 10.4 in 1997. (See Fig. 7.10)

7.2.3.2 Exports to the EU

Chinese denim apparel achieved a much more successful penetration into the EU market than its denim fabrics did, though its market share of EU total denim apparel

7.8 Denim fabric exports from China to the US between 1995 and 1997

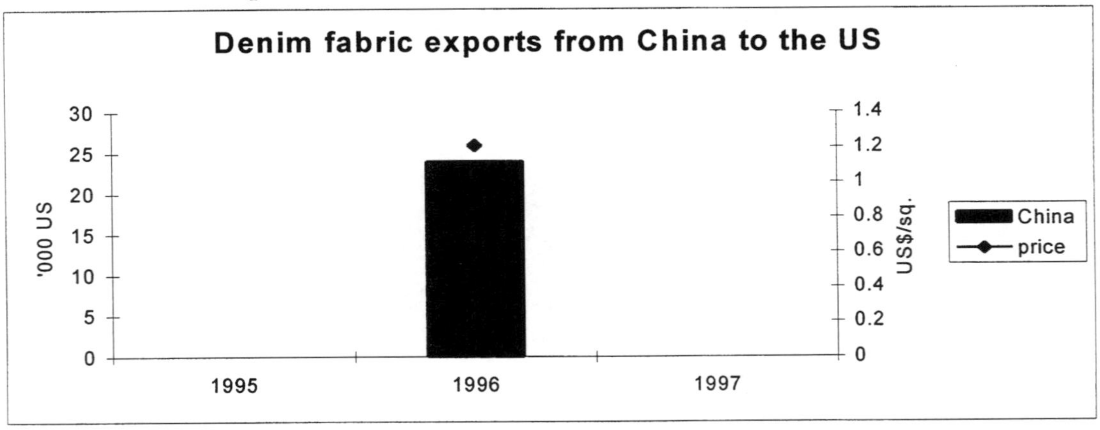

7.9 Denim fabric exports from China to the EU between 1994 and 1997

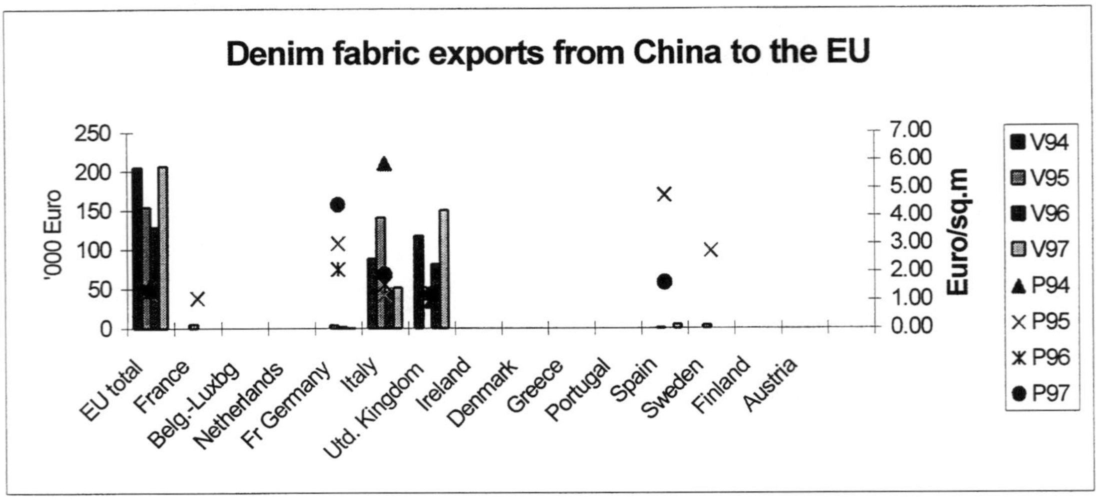

7.10 Denim apparel exports from China to the US between 1995 and 1997

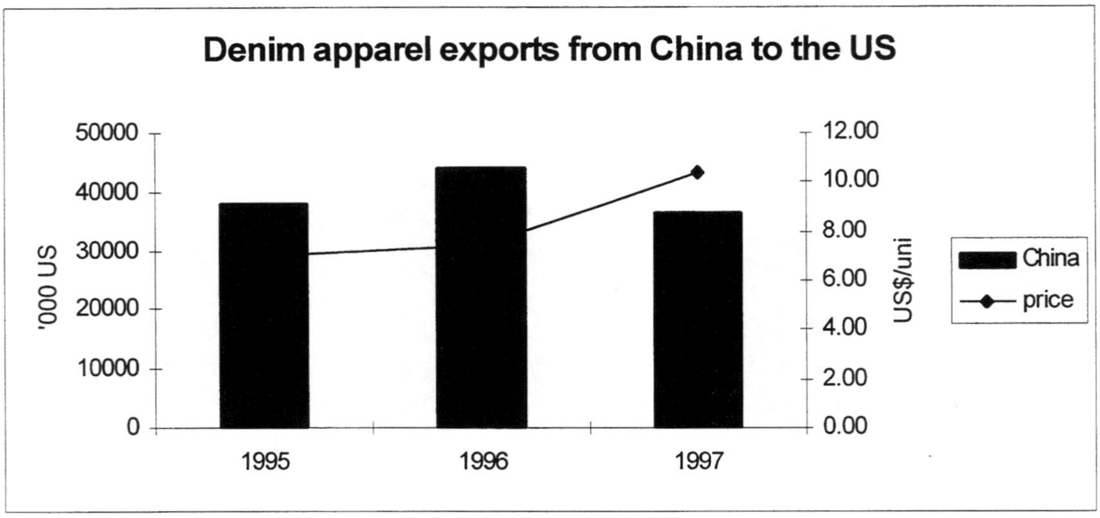

7.11 Denim apparel exports from China to the EU between 1994 and 1997

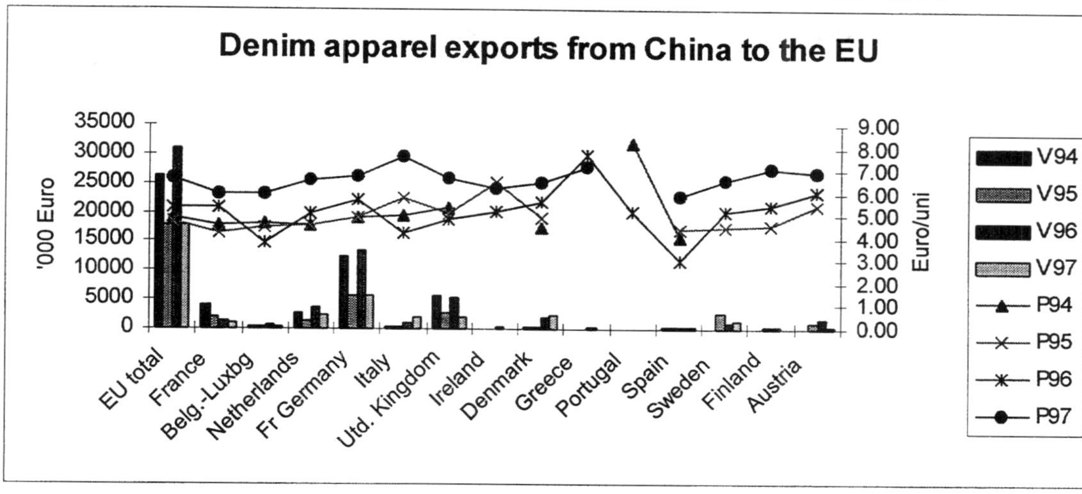

imports shrank from 1.1% in 1994 to 0.6% in 1997. As mentioned earlier, denim apparel was the major item in China's denim product exports to the EU. The export value reached a peak in 1996 with 31.1 million Euro. However, the export performance was dull in 1997, only reaching 17.8 million Euro. (See Fig. 7.11)

All the EU countries[1] imported denim apparel from China since 1994, except for Greece which started importation in 1996, and Portugal, who only imported from China in 1996. Germany was the most important EU importer, capturing 47% of total EU denim apparel imports from China in 1994. Though its market share reduced to 31.7% in 1997, it was still the leading EU importer. Imports to Italy and Denmark rose very quickly, enlarging their shares from 1.7% and 1.1% in 1994 to 10.5% and 13.3% respectively in 1997. However, those to France and the United Kingdom dropped sharply, from 15.7% and 21.2% in 1994 to 4.8% and 10.7% in 1997 respectively. Sweden also imported less denim apparel from China after it joined the EU, from 15.5% in 1995 to 6.7% in 1997.

The average export price increased a little from 4.9 Euro per unit in 1994 to 6.7 Euro per unit in 1997. Those to Italy were imported at the highest unit price at 7.8 Euro; those to France, Belgium, Ireland, and Spain were sold below the average price; and those to Germany, the UK, Netherlands, Denmark, Sweden and Australia were at or near the average price.

REFERENCES

1. Jurg Rupp, A.B., Denim: jack in a box of the fashion industry. ITB International Textile Bulletin, 2000?p.10.
2. Wang, J., Changing fashion world. JTN Monthly, 1999?p.60-61.
3. Anon, Hong Kong: amber lights on apparel exports to Europe and US. JTN Monthly, 1999.
4. Lei Yao, Yi Li., Denim in China, Textile Asia, March 2000

[1] Because Sweden, Finland and Austria didn't join the EU until 1995, their 1994 figures with China were not available.

8 Strategic Implications for Denim Manufacturers and Traders

8.1 DENIM PRODUCT DEVELOPMENT TREND

8.1.1 Development history

8.1.1.1 Physical breakthrough[1]

Denim is one of the oldest fabrics in the world. Its origin can be traced back to the 16th century when a tough cotton fabric was used in Nîmes to make strong trousers for sailors. A Bavarian-born man named Levi Strauss successfully launched denim into a historic legend during the Gold Rush period in San Francisco, though his first products were brown canvas pants instead of indigo blue jeans in 1873. During the 1890s, Levi Strauss manufactured the first jeans under the designation "501 indigo" [2]. However, it was until the 1950s that denim gained worldwide attention and jeans became a favorite among the youngsters.

Basic denim fabric refers to a cotton twill fabric featuring a 3/1 warp faced weave, traditionally made with indigo-dyed yarn for the warp and natural yarn for the weft with the standard weight of 14.5 oz. per square yard. For many years, around 75% of all jeans have been made out of this fabric [3].

With the entry of more denim manufacturers around the world and the sharper market competition, denim has underwent drastic changes in almost every aspects from material selection to final finishes. The denim apparel family has also been expanded to cover all the essential items from tops to bottoms.

A. Material

The traditional denim fabric is made of 100% pure cotton. Today, though cotton is still the main yarn type, there are other materials used to make denim fabrics, including ramie, silk, and cotton blends such as 50% cotton with 50% polyester. The wool denim is a high value added material using wool and cashmere, which was developed by Brad Mill, the largest Australian denim manufacturer, in May 1997. Tencel/Modal (made of blended or mix-woven denims of cotton and rayon) and stretch denim adopting Lycra are also widely accepted nowadays. Levi's "Engineered Jeans" launched at Interjeans in Germany and at 40° in the UK use fabrics made of Tencel/cotton to integrate cotton's toughness and ruggedness with Tencel's comfort and strength [4]. As Caroline Parent, "Levi's" Design Chief said, "it is the fabric that cinches the deal" [4].

B. Spinning, Dyeing & Weaving

The yarn used to be exclusively ring-spun. With the introduction of OE machines, the open-end spun yarn occupied an important place during the spinning process. Both methods have advantages and disadvantages. For example, ring-spun yarn has a softer hand and a natural uneven surface appearance, though the process is slower and

[1] This section is mainly based on the following references: [1];[55]; [56];

more labor-intensive. OE yarn is cheaper and quicker to produce but the hand is not as good as the ring-spun ones. Thanks to the technological improvement, ring optics gives OE yarn the visual characteristics of ring-spun yarn through an irregular spinning process.

Indigo is still the most important dyestuff. There are two basic dyeing methods: indigo dyed and sulphur dyed. Retro dyeing is a new one combining dyeing and surface effect processing in one process, which is energy efficient and more environmentally friendly.

As regards weaving, the varieties are rich as well, ranging from traditional 3/1 weaves to 2/1 and 1/1 ones, from checks and plaids to diamond weaves and jacquards. In the US, weft-elastic fabrics dominate the scene while in the EU, warp and bi-elastic denims are woven in addition to the weft-elastic ones.

C. Color & Weight

Technological advances in various processes greatly enrich the color and weight availability. With the indigo blue as the main stay, there are various colors and color shades, such as black, ecru, and color denim. Denim and jeans has now tended to develop towards two extremes: either extra heavy & thick (over 14 oz.) or ultra light & thin (under 6.5 oz.). The weight range is from 4 oz. to 17 oz.

D. Finishing

Before the introduction of stone wash in 1978, almost all denim sold in the market was unwashed or unrinsed with the typical dark blue appearance. Consumers used to buy jeans and wash them at home to achieve their desired effects. In 1974, the first pre-washed jeans made their appearance in the market. However, it was until the 1980s that washing techniques were fully developed to achieve either hand improvement or the desired final looks such as intentionally destroyed appearance. Stone wash signaled the breakthrough in washing techniques. In 1986, chemical wash won the popularity among denim manufacturers and consumers. Enzyme wash entered the picture partly due to the increased environment-consciousness among the consumers. Other washing methods include snow wash, ice wash, granite wash, marble wash, charcoal and surf wash, etc.

E. Final product line

The past half-century has witnessed the significant changes in the denim family, evolving from the basic denim fabric and jeans to the full product lines. The denim apparel is no longer limited to jeans and expanded into denim jacket, shirt, vest, shorts and skirt as well. As regards denim fabrics, they cover a lot of varieties in addition to the basic cotton twill fabrics, including stretch denim, ecru denim, color denim, re-used denim, sand blast/shotgun denim, heavy/light denim and soda pop denim, etc. Cotton Incorporated predicts that the popularity of denim will continue into spring 2001, with new versions such as metallic coated denim and metallic printed jeanswear on the market [5].

8.1.1.2 Concept evolution

Levi Strauss developed jeans because of its durability and easy care. Therefore, the earliest consumers were gold miners in the west. Jeans were perceived as workwear rather than fashion items in the US before the 1950s. After the WWII came to the end, denim jeans were formally introduced into the Europe, Japan and other countries. In these regions, jeans were viewed more as an American symbol than as functional working clothes. That's the major difference between the US market and other markets such as the Europe and Japan. It can be said that it was since 1950s that

Table 8.1 Summary of the important events in denim development history

Denim development history	
Year	Important events
1873	Levi Strauss made the first jeans from a heavy brown canvas
1890s	Levi Strauss produced the first jeans under the designation "501 indigo"
1950	First zipper jeans
1960s	Denim achieved worldwide growth and acceptance
1978	Stone wash
1986	Chemical wash
1987	The first denim fabrics dyed "superblue indigo"
1992	Soft jeans made of blended yarn or mix woven fabrics; reverse denim
1993	Multicolor warp yarn dyeing-sizing range
1994	Soda pop denim
1995	Retro dyeing process
1997	"Denim Wool"; Water free fading of denim with laser by Icon Inc,. USA

(Source: [45])

denim products entered a golden development age and evolved into today's all-purpose garment appealing to all classes and ages.

Denim and its related products have undergone remarkable changes since the 1950s. Table 8.2 illustrates the important events during the past five decades, which greatly affects the concept of jeans among the consumers and helped promote technological innovations and product development of denim fabrics and apparel.

In view of the denim development history of the past 50 years, it clearly shows that it is closely associated with social and cultural changes during each different period. The movie stars' excellent performance in the films with the West and anti-hero themes in the 1950s injected fresh cultural connotations into the jeans, making them appeal to the youngsters who were pursuing freedom and breaking with rigid social mores. The following decade saw the further development of this social movement and jeans were viewed as the symbol of social equalizer. Back to the mainstream in the 1970s marked the debut of designer jeans. Jeans were no longer confined to the youth. Those over 25 years old began to accept them as a way to remain young and energetic. The successful "Back to basic" campaign launched by Levi Strauss & Co. in the 1980s indicated that man began to pursue a simple way of life. However, the popularity of worn or destroyed look of jeans reflected the ultimate anti-fashion view among some consumers. In the 1990s, worldwide concerns about the environmental pollution finally arouse consumers' attention and green consumption becomes a no strange name. Associated with this environment-consciousness is the appearance of "Eco-denim". Furthermore, as people are looking for a simpler and more casual life, jeans, with its unique cultural connotations accumulated for so many years, are the ideal non-class leisure wear among the consumers. Levi's "Engineered Jeans" well reflect this trend. It is an ergonomic update on the classic five pocket style. Moreover, the production process for "Tencel" has a minimal impact upon the environment and the cellulose used to manufacture the fiber is sourced from managed forests [4].

Though the concept of denim and jeans has evolved progressively, its core concept remains unchanged. Today, denim is well established as street fashion fabric and denim apparel is closely connected with youth, freedom and vitality. The well known "five-nons" embodied in denim, which is non-age, non-sex, non-season, non-border, and non-class, clearly explain its wide acceptance and popularity for so many years.

8.1.2 Future trend

8.1.2.1 Hotter market competition

Since the appearance of the first jeans in San Francisco in 1873, American brands, such as Levi's, Wrangler and Lee, used to be the only players in the denim world until the 1950s. Competition was just concentrated upon the American domestic market during this period.

The European consumers developed an interest in jeans left by American GIs after the end of WWII. With the availability of denim imports from the US, the denim production and consumption in the European markets generated healthy development. European brands emerged to fight for the market share with established American brands and in the 1970s, these labels started their internationalization process in the world market and achieved positive results [46].

Because of relatively easier entry of the textile and apparel sector, many developing countries selected this sector as the engine for their economic growth. Their comparative advantages in the labor cost and land boosted their quick emergence in the world textile and apparel market. The most remarkable example is China, which is now the world No. 1 textile and apparel exporter. With the increased new entrants in the denim sector, the market competition became much sharper and hotter in the 1990s than one hundred years ago.

Table 8.2 Important events related to denim concept development

	Events	Concept development
1950s	*Movie Stars* John Wayne (the west films reflecting cowboy's life James Dean (Rebel without a cause) *Music: Rock 'n' Roll*	Jeans were associated with anti-establishment and a break with convention. They were also the symbol of cowboys who are brave and cool.
1960s	Anti-war demonstration Hippy civel rights movement	Jeans were viewed as social equalizer and wore as uniform of the social revolution.
1970s	The anti-hero "revolution" was integrated with the mainstream of the society Utopian ideals of the world	Jeans were accepted by the adults; Designer jeans were in the picture
1980s	Back to wealth and status Yuppies Levi's "back to basic" campaign	Jeans were developed in two extremes: very fashionable and elegant; and very anti-fashion such as destroyed jeans by sand blast and shotgun
1990s	Trend towards "green" consumption Casual on Friday Pursue family life and simpler life	"Environmental-friend" products such as Wrangler's "Earth Wash jeans" Jeans were perceived as the ideal non-class leisure fashion Soft denim and jeans and the basics boom

(Source: [46])

Table 8.3 Global denim production situation (Source: [6, p35])

Global Denim Production Capacity

Unit: million linear meters

Region	Countries	1992	1997	Over-capacity
Australia	Australia	15	15	2
Western Europe	Italy, Spain, Belgium, Greece, France, Germany, Ireland, Portugal	195	190	52
Eastern Europe	Czech Rep., Poland, CIS, Russia	20	15	4
Middle East	Turkey, Israel, Syria, Dubai, Saudi Arabia	110	160	45
North Africa	Marocco, Egypt	55	45	12
Southern Africa	South Africa, Ivory Coast	45	50	15
North America	USA, Canada	600	630	30
South America	Mexico, Brazil, Argentina, Colombia, Peru, Chile	245	675	190
Southern Asia	India, Pakistan, Sri Lanka, Bangladesh	60	340	100
South East Asia	Indonesia, Philippines, Thailand, Vietnam, Malaysia	50	200	70
Far East	China, HK, Japan, Taiwan, Korea	1025	1700	500
World		2420	4020	1020

Table 8.3 shows that the world denim production capacity increased from 2.4 billion linear meters in 1992 to 4.0 billion in 1997, up by 66.7% within just five years. However, there are signs that the global denim consumption has slowed its growth pace. For example, the annual growth rate in the US market has been down from 9% to 2-3% in recent years. As a result, too rapid production increase undoubtedly resulted in worldwide over-capacity, which was estimated to be 1.0 billion linear meters if global consumption was set to be 3.0 billion in 1996. Asia countries experienced the most severe capacity surplus in recent years due to blind expansion and lack of systematic view for future development. The production capacity in this region was almost doubled from 1.1 billion linear meters in 1992 to 2.2 billion in 1997. Take India and China for example, the former suffered about 25% over-capacity out of its current 200 million meters [7] while the latter experienced about one third glut out of its 1040 million meters output [8]. If the over-capacity problem cannot be solved within the next several years, denim manufacturers will have a very hard time, especially for those engaging in commodity item production which price is the single most important factor.

Denim consumption nowadays can be divided into two broad categories: one is pursuit of new fashion items with high quality and high added value; the other is commodity items appealing to middle and low end markets. Since most developing countries are competing in the latter category, market competition is really cutthroat. The situation will not change much in the future with more and more developing countries entering the picture. The complete quota elimination by the end of 2004 is another positive sign for both current and potential players. Therefore, it can be expected that the market order will be re-organized with new faces and new brands.

8.1.2.2 Focus upon innovation and concept crystallization

When the denim market was dominated by several American brands before the 1970s, little technological breakthrough was achieved. However, when more non-US players joined the game since the late 1970s, innovation in the washing and finishing techniques as well as fabrication and color varieties came in streams. Innovation and product development have become important company strategies to survive the cruel global competition.

Aggressive competition is the major driven force for the drastic changes during the last twenty years. Stable quality is no longer the single determinant in the market competition. Only those who pay more attention on technological innovation and continuous product development can ultimately win the game. Japanese denim industry is one of the best examples for its continuous innovation efforts during its short history. Kurabo boasted that they were the first in the world to develop uneven denim. Nisshinbo developed liquid ammonia treatment and its "Ekian" products enjoy good reputation both home and abroad. It was also Japanese jeans manufacturers that suggested stone-washed jeans which opened a new era for jeans development. Moreover, the extremely demanding consumers in Japan push the manufacturers to spare no efforts in delivering high quality and high value-added products. As a result, the Japanese denim manufacturers successfully build their upmarket image based on these technological breakthrough and continuous improvement.

There is no denying the fact that Levi's is still the world leading brand. But it has faced more challenges in recent years from competitors all over the world. Its sales dropped by US$ 1.1 billion over 1997 in 1998 and its standing in the Fairchild 100 list of fashion's most recognizable brands have slipped from the 5th in 1997 to 23rd in 1999 [9]. With more brands and entrants in the market, it is natural to focus attention on factors other than quality and cost control. Therefore, though denim has a history of more than 100 years, its rapid evolution just started from 20 years ago when more players fight for the market share and brand building. Efforts to find new materials, new processing techniques and new fashion styles will be further enhanced by every ambitious company to secure a favorable market position. Product differentiation is of much more importance especially when global sourcing nowadays has already made traditional comparative advantages, such as low labor cost and abundant raw material supply, available to all the global players.

Denim wonder in the fashion circle during the past century is partly due to its successful connotation and continuous association with the social and cultural trend. Therefore, in addition to the heavy R & D investment in technical innovations, it is of the same importance to develop and realize an appropriate concept for jeans and other denim apparel, which is in line with current hot issues and social trend. To some extent, this may be harder than actual technical breakthrough because it requires a better understanding of consumers' psychology, social and cultural context as well as the product itself. The emergence of "environmental-friend" denim products indicates the "green consumption" trend in the 1990s and company's shift priority in marketing and promotion campaign.

8.1.2.3 Subject to shorter fashion cycle and contemporary life style

Denim fabric was chosen to make jeans because of its fastness and sturdiness. And Jeans, for a long time, were perceived more as workwear than fashion items in America. Therefore, the popularity of jeans before the 1950s was mainly attributed to its practical functions, less subject to fashion cycles.

With jeans and denim apparel more closely associated with fashion trend and life style, the life cycle of denim products promoted every season is much shorter than previous years. In other words, the best seller this year may fail to attract consumers next year. The rich varieties available in the market also give consumers wide choices. Fabric, finish, color and style all have strong influences upon consumer's selection of a denim item. And these four aspects are subject to other known and unknown factors as well, further complicating the issue. Take the European market for example. The emergence of a new fashion culture from the music and dance triggered new interpretations about fashion items. As a result, the market is becoming highly fragmented with no single brand occupying the dominant position as Levi Strauss' 501 campaign did in the 1980s. Brand competition further segments the market into various niches, each appealing to different target consumers. The popularity of destroyed jeans among some consumers in the 1980s is a good example for this. In a word, denim products in the future will be prone to shorter life cycle and be more closely connected with consumer's life style.

8.2 IMPLICATIONS FOR CHINA'S DENIM SECTOR AND THE WHOLE TEXTILE INDUSTRY

Business environment is becoming more and more dynamic, complex, diverse and hostile [10]. Both the external and internal environments are significantly different from those in the 1950s or 1960s. As the world No. 1 textile and apparel exporter and one of the world most important denim manufacturers, it is crucial to depict the overall environment in which China's textile and apparel enterprises are competing for survival and to identify some possible measures they can take to secure a healthy growth.

8.2.1 General external environment

8.2.1.1 Political and legal dimension

A. China's WTO entry

This issue is of political significance as well as economic implications for China. Its impact upon China's textile and apparel industry has been explained in detail in Section 2. However, its effects upon China are far beyond this scope. China's entry into WTO means that its economy will be truly integrated with the world economy and act within the WTO-based framework. As mentioned earlier, to individual industries, the changes of game rules due to WTO entry are of consequence, including liberalization of foreign trade regime and domestic distribution channels, elimination of NTBs and reduction of tariff rates, and commitment to improving market access conditions of professional services and the telecommunication sector, etc. These are no longer piecemeal changes without a definite time frame for implementation. Rather, they are regulated within the whole package of WTO-related agreements. Significant changes can be expected in the next five years if China successfully enters WTO by the end of 2000.

B. SOE reform

SOE reform is one of the most important aspects in China's economic restructuring from a planned economy to a market one. During the past 2 years, reform of SOEs has achieved some positive results, especially in the textile and apparel sector where some SOEs turns loss into profits this year.

Current SOE reform places much emphasis upon corporatization, that is, transforming traditional state-owned enterprises into corporations. The state will not interfere with enterprise's' daily operation and ultimately withdraw the state assets in industries without strategic importance to the nation, such as the textile and apparel industry. The major aim is to pull SOEs out of the current dilemma on a market-oriented basis. Other measures include mergers, bankruptcy, reorganization of group companies, debt-to-equity swaps, and so on (See Section 2.1.2.5). Whether these steps will achieve the desirable results is an issue requiring detailed research and analysis, which is beyond this study's scope. But one thing is certain: the final result of the reform will influence the future direction of the overall economic restructuring path in China.

C. Changes in legal system

The rapid economic development needs solid support of a well-developed legal framework. However, the legal system in China lags behind the economic development, failing to provide a firm foundation for future sustainable economic growth. Having realized this problem, China is now trying to work out new laws and regulations compatible with current economic situations. Enforcement of the new contract law and the drafting of a competition law have sent a positive signal, indicating that domestic enterprises will face a better regulated legal environment with more clearly defined rights and obligations. However, since China's economic reform is unique in the world without a specific theory or pattern to model after, it will take a very long time for China to work out a legal system solidly supporting its economic development.

8.2.1.2 Social and cultural dimension

A. New social security system

In the past, the employers particularly those SOEs, were responsible for the employees' welfare benefits, including pension and medical fees, which was a very heavy burden for the enterprises' healthy growth. To guarantee a sound economic development and solve this byproduct out of a planned economy, a new social security system has been established in China since the early 1990s, aiming to shift the welfare responsibility for employees from enterprises to this new system. It consists of basic living costs for laid-off workers, pensions for the retired workers and a minimum living standard for urban residents, together with pension, unemployment and medical insurance [11]. After several years' enforcement, there have been some positive achievements. For example, about 120 million people have joined the insurance-based retirement program accounting for 90% of employees in urban areas. By the end of September 1999, 668 cities and 1638 towns had participated in the system and 2.8 million received minimum living allowances [12]. As regards laid-off workers, they receive a monthly allowance ranging from RMB 140 to 380 after enrolling at enterprise-based re-employment service centers [11]. Specific vocational training programs have been developed in these re-employment centers to enhance their skills and help them find new jobs. The Regulations on Subsistence Security for Urban Citizens went into force on October 1, 1999 [13], establishing a platform for the whole aid package. The government also allocated RMB 400 million to help improve living standards of people in the Central and Western China [12].

The social security system is still at its early stage and leaves much room for further improvement. For example, though about RMB 50 billion has been collected in pension schemes nationwide, many cities and provinces have already run up a deficit [11]. According to Premier Zhu, the coverage of social welfare insurance programs should be expanded in accordance with law and social welfare insurance must be made mandatory in urban areas. New funding channels should be sought to supplement social security funds, together with a well-developed labor market [14].

Though it is still a long way to go for China to establish an effective social security system, this move has already caused profound social changes. First, when it is the society at large rather than the enterprises that are responsible for the employees' welfare, the enterprises can concentrate upon more business-oriented activities. Second, state-owned enterprises are no longer the "apple of the eye" when people decides their career path since the past "security" is not there any more. Third, people began to realize that their future welfare is mostly dependent upon their own efforts. This change of attitude is, to some extent, more important.

B. Departure from lifelong employment and non-performance-based promotion

This point has close relationship with the previous one. The "iron rice bowl" has been broken, putting a period to the lifelong employment and egalitarian income distribution system. Employees are paid and promoted according to their actual working performance and ability. Those who are not suitable or who do a bad job may be fired at last. Therefore, to work in a state-owned enterprise is no longer desirable since it has lost most privileges it once enjoyed. More and more people are now working in non-SOE sector. For instance: employees in FIEs accounted for 0.83% of the total employment in 1997, compared with 0.01% in 1985 [15].

Since people know that they are responsible for their own behavior and their efforts will be rewarded, they tend to work harder and be more initiative. The productivity and efficiency in enterprises are boosted and employees see a clearer future than before when one's performance had nothing to do with one's final payment or promotion.

C. Increased exposure to world fashion culture

Before the 1980s when China was relatively isolated from the outside world, people wore almost the same style clothes with dull color. Few people paid attention to fashion items or fashion trend, and even if some did, they didn't have alternatives to change since the offered products were either identical or similar. Department stores were the only place where people purchased textiles and apparel. There were almost no fashion magazines, foreign brands or fashion exhibitions and shows. It could be called "fashion desert" where people had no sense and knowledge of fashion and textiles and apparel were merely functional products.

The past two decades have witnessed great changes in China, particularly since the 1990s. Many domestic as well as foreign brands have made their appearance in various forms of retailing outlets. People are more exposed to world fashion culture and have more dispensable income to spend on fashion goods. Textiles and apparel are not viewed just as necessities, instead, they are associated with one's status, individuality and image. People, especially urban youngsters, are quick to accept different cultures and customs and hence, develop a taste which is closer to world fashion development. Consumers become more demanding and sophisticated than before.

D. More informal and leisure life style

Since the mid-1990s, China began to enforce a new working schedule, featuring five working days. People have more leisure time to relax and enjoy life. What's more, the quick development of the entertainment sector has led to increased emphasis upon life quality. The trend towards casual life in the world also influences China.

Another thing worth noting is that there tend to be longer holidays around some major festivals, such as the New Year and National Day, which often last about one week. More and more people choose to travel around or shop during the holiday, which creates a very good business opportunity.

8.2.1.3 Technological dimension

A. Technological advances in textiles and apparel

In Section 8.1.1.1, the physical development of denim has been discussed. As regards textiles, various technical advances have been achieved, such as the synthetic fiber, spandex, non-woven fabrics and non-water printing. Cost reduction, efficiency improvement and niche positioning are the major driving forces behind these breakthroughs. Increased environmental consciousness among the consumers and industry has encouraged use of "renewable" or "sustainable" resources in textile manufacturing [16].

Technological advance in the apparel sector is much slower because of its labor-intensive nature, especially during the 1960s and 1970s when labor productivity was mainly the result out of incremental improvements rather than out of the major technological leap experienced by the textile sector such as the development of open-end spinning and shuttleless looms [47]. The application of micro-processor in apparel production in the late 1970s, such as compute-aided design (CAD) and computer-controlled-cutting (CCC), makes flexible manufacturing economically viable and possibly find a way out of the dilemma between flexibility and lower unit costs. Other benefits include productivity boost, fabric and labor saving and better quality control. For example, by using CAD in the apparel sector, labor savings may reach as high as 20:1 or even 100:1 [47]. Computer controlled cutting is a related micro-processor innovation. The result is that one lower paid, unskilled operator can do the equivalent work of three of four highly paid, skilled cutters in approximately one tenth of the time [17]. As a result, traditional manual skill requirements have been changed to machine operation due to this deskilling vs. enskilling process [48].

B. Information technology (IT) in the supply chain

The application of IT in the textile and apparel industry has triggered a re-engineering of business process and a re-configuration of business networks within the supply chain [10]. The emergence of e-commerce, non-store retailing, EDI and Internet has exercised great influence upon present ways of thinking. For example, traditional retailing forms are challenged by e-retail where it costs only US$ 1000 to set up shop [18]. In the US, the proportion of on-line shoppers for apparel products has increased from 10% in 1997 to 25% in 1999 [19]. The overwhelming reasons to build EDI links with suppliers are the expected enhanced accuracy of information flows, reduced lead time and inventory, and strengthened customer service.

To put it simple, IT has resulted in a redefinition of business scope and opportunities, forcing the involved companies to compete and cooperate in a drastically changed environment. In this sense, the textile and apparel sector may undergo some significant changes brought about by the IT revolution rather than by pure technological advances, especially in the organizational structure and competitive strategies. There has emerged "virtual corporation" which uses IT to co-ordinate its supply and marketing activities without owning any of them [10].

Information technology in China has developed quickly during the past decade. According to the China National Network Information Center, there were 1.2 million Internet users by June 1998, or one in a thousand now has access to the Internet [20, p118] However, compared with those developed countries, the IT development still lags far behind the international level.

8.2.1.4 Demographic dimension

A. Nuclear family with only one child

Chinese families once featured bigness with several generations living under the same roof. Now, there have been some remarkable changes in the family structure. Families made up of one couple and one child became prominent, especially in the urban areas.

The family planning policy after more than 20 years' enforcement in China has formed some Chinese-unique features. The single child has become the focus of the family, receiving sole attention from six adults: two parents and four grand parents. These children can have whatever they want and parents hope to give them the best: good education, extensive hobby development and branded apparel such as Nike shoes, Levi's jeans and Crocodile T-shirt. The market potential for children and youngster are indeed very large. Since the family planning policy will continue into the future, such nuclear family will dominate the scene, posing a new force to the already dynamic business environment.

B. Emergence of a middle-income class and a consumer market

The broken of "iron rice bowl" and the entry of MNCs have greatly enlarged the income gap between the rich and the poor. At the same time, there emerged a middle-income class. As for textile and apparel products, the two extremes are not the desirable consumers since the markets are relatively small. It is the middle-income class that has large market potential for textiles and apparel. With the economic development, this class will grow as well, with young and middle-aged people constituting the major part.

Though China has a population of more than 1.2 billion, the real consumers are far less than the figure. According to James Glasse, consumer market in China now numbers some 150 million people, a massive rise compared with the 1980s when there was virtually no consumer market at all [20, p17-18].

C. Less willing to work in textile enterprises

Today, less and less people are willing to work in a textile or apparel enterprise, particularly those with higher educational background. (See Section 2.1.2.1) They are attracted by industries offering good salary and a foreseeable brighter future, such as hi-tech or service sectors. Working in textile and apparel enterprises is considered as "no future" and "non-competent", which is now a prevalent opinion held by most youngsters and their parents. As a result, the average age in the textile and apparel industry becomes older in recent years and the enterprises find it harder to recruit talented people to join them.

8.2.1.5 Macroeconomic dimension

Though current economic growth in China is not as rapid as that in the early 1990s, it is still much faster than the average world growth level. In 1999, the GDP growth in China reached 7.1% [21].

Investment is mainly concentrated upon infrastructure, most of which has been directed toward the central and west regions. Between 1998 and 1999, about two-thirds of infrastructure investment is channeled to the central and west [49]. In the first seven month this year, the growth rate of infrastructure investment in the central and west regions reached 19.4% and 14.4% respectively, compared with 5% in the east region [22], indicating the government's determination to develop these regions through policy encouragement and orientation.

Foreign trade registered a remarkable growth thanks to the increase of export rebate rate and heavy attack upon smuggling. In the first eight months of this year, the total import and export value reached US$ 301.7 billion, of which the export value was US$ 159.3 billion, 34.6% up over the same period last year [23]. Growth of foreign direct investment (FDI) was stalled in 1999, fallen by 20.9% from the 1998 level [24]. However, it recovered quickly this year. In the first seven month of this year, FDI reached US$ 27.6 billion, increased by 24% over the same period last year [25].

Domestic market demand was sluggish for a very long time. The retail price of commodities in 1998 declined by 2.6 percent over 1997 and further dropped by 3% in 1999 [26]. The government adopted various measures to stimulate demand including consecutive interest rate cuts and the imposition of 20% interest tax upon the bank deposit. Now, there are some signs showing a little recovery from the deflation situation started since April 1998 [27]. In February the consumer price index increased by 0.7% on a year-on-year basis [27] and in August, another 0.3% over last August [50]. The total retailing volume for consumer goods increased by 10% in the first seven months of this year [28]. Nevertheless, it is hard to say now that the consumer market will gain momentum. Chinese people expect that they will spend more on housing, medical care, pension insurance and education of their next generation. Many Chinese consumers may further close their pockets in the future or become more discreet in money spending.

As domestic economy becomes more integrated with the world, especially after China enters WTO, the macro-economic condition will undergo some fundamental changes. For example, according to Mr. Dai Xianglong, the president of Bank of China, the interest rate will be liberalized to be subject to market forces within three years as the first step to lay a foundation for further reform on exchange rate and free convertibility of RMB [29].

8.2.1.6 Global dimension

A. Globalization vs. regionalism

The economy is global, but the world is organized politically into nation-states [30, p23]. As a result, both globalization and regionalism are parallel on the horizon, characterized by the establishment of WTO and various regional preferential agreements.

The world economy is integrated more than ever before. Companies with a global vision adopt various strategies to make their presence in other countries' markets. However, no matter what strategies they pursue, their consideration is constrained by many political factors, and the rising trend of regionalism is one of them. Regional trade arrangement peaked in the 1950s and 1960s and again in the late 1980s and 1990s with different features. There are 107 free trade agreements (including Customs Union) went into effect around the world as June 1999 [31].

Today's regionalism is smaller countries actively approaching larger countries which are usually their most important trading partners, aiming to securing the necessary market access key to their economic development. Mexico's eagerness to join NAFTA and other countries' expressed interest in acceding, such as Chili, all reflect this security-based consideration. At the same time, larger countries want to use such negotiation as a leverage to handle non-trade issues with the smaller country and to increase their bargaining power with other large countries [51]. Whether regionalism can boost the welfare of member countries and whether it can ultimately promote or frustrate the multilateral trading system are still open questions despite numerous academic discussion such as Viner's "trade creation and diversion" argument and Bhagwati's "stumbling and building blocks" argument [52]. But one thing is certain at present: Regionalism is developing faster than globalization and its scope has been extended to cover countries located in different continents. For example, in March 2000, Mexico and the EU signed NAFTA-like trade agreement, which has passed in Mexico and waited for review in the EU. It aims to eliminate 95% of trade barriers between the two countries by the year of 2007. About 82% of goods exported from Mexico to the EU will be duty-free immediately, with the rest 18% be freed progressively before 2003 [33].

Despite the political factors in those regional trade arrangements, trade creation within member countries and diversion from non-members concerning textile and apparel products is obvious, such as Mexico with the US, and Turkey with the EU. The international apparel/textile production division and QR requirements in the textile and apparel trade foster an ideal environment for regional trade development and these preferential agreements become the final stimulus accountable for the rapidly increased textile and apparel trading volume among member countries.

B. Managed trade regime

This point has been explained in Section 1.3.4.3. Today's trade regime is more complicated than fifty years ago. Tariff is no longer the most commonly used weapon to manipulate trade after several rounds of multi-lateral talks under the GATT. Instead, non-tariff measures, such as quota, technical standards and anti-dumping investigation, are extensively employed to prevent imports and protect domestic industries.

In addition, developed countries want to include some non-trade issues in new multi-lateral round of talk, such as labor standards, environmental protection and competition policy. In the US, there is a proposal under which US companies would voluntarily adopt a "NO SWEAT" label showing that apparel and other products are not manufactured in "sweatshops" and agree to be inspected [34]. Though nothing significant was achieved during the last Seattle Round, these issues may become the major topics in the future.

8.2.2 Competitive internal environment

8.2.2.1 Threat of new entry

Entry barrier in the textile and apparel sector is much lower compared with those hi-tech industries. Moreover, most denim enterprises in China are engaging in production of basic items featuring low value-added and labor-intensive nature. Product differentiation in most enterprises is not obvious, making them more vulnerable to competition. As mentioned earlier, China's WTO entry will encourage foreign investment and more foreign brands will seek to have a direct presence in this huge domestic market. Therefore, threat of new entry, both from home and abroad will be much more intense in the future.

8.2.2.2 Threat of substitutes

As regards denim apparel family, they face challenges from other causal wear, such as khaki pants. (See Section 1.3.4.6) Though efforts have been made to enlarge denim apparel's age appealing, such as "gentlemen's jeans" and "designer jeans", it cannot be denied that denim apparel has its own limits and restraints. It is the most popular item among those aged below 35 and is worn in very casual occasion. The khaki pants, on the other hand, appeal to a wider consumer group and can be worn in more occasions. The emergence of other casual items, such as "casual suit", is another threat to denim apparel. In this sense, though denim apparel still enjoys dominance in the casual wear category at present, it is facing hotter competition from other substitutes.

8.2.2.3 Power of suppliers

In China, most production enterprises in the textile and apparel supply chain are small in size and have no obvious competitive advantage over each other. The same case in the denim sector. Therefore, bargaining power of suppliers is not strong enough to exert significant influence upon an enterprise' s daily operation and decision-making. For example, according to the survey conducted in 1997 and 1998 on 27 denim manufacturers, 84.8% focus upon standard products [53]. This indicates that downstream enterprises can easily change their purchase orders from A supplier to B supplier without encountering too many difficulties or increasing transaction costs. As a result, the relationship with suppliers is very loose and unstable.

8.2.2.4 Power of buyers

In developed countries, textile and apparel production is more demand-pull rather than supply-push. The emphasis is shifted from "cost-plus" methods of pricing towards "value-minus" methods, that is, to satisfy customer's needs at the lowest cost with the maximum added value [35]. The establishment of the buyer's market in China regarding denim products indicates that the bargaining power has switched from suppliers to retailers/wholesalers and from retailers to ultimate consumers. After more than twenty years' reform and open-up, both the distribution channel and consumers have undergone remarkable changes and their bargaining power has increased a lot. Enterprises who are still not market-oriented and who still neglect consumers' needs are more likely to lose the competition. Today's consumers want value for money, including price, quality, service (product-related and corresponding available speed), and product features. They are more sensible and discreet in current "supply-surplus" situation. The reform in the distribution channel and entry of foreign entities, such as Wal-Mart, Carrefour and Metro, all contribute to the strengthening role of wholesalers and retailers. According to Taplin, large retailers exercise oligopsony power upon manufacturers for cost reduction and more responsive to retailer needs [17]. That's one of the major reasons that domestic enterprises are now paying much attention to the market needs instead of "closing-door-manufacturing". However, the loose relationship within the supply chain makes domestic enterprises still slow and inefficient to meet consumers' needs, a big problem for future competition.

8.2.2.5 Rivalry

Section 2.1.2.3 (B) and 2.4.2.3 have explained in detail the rivalry condition in domestic textile and apparel market, both at present and after China's WTO entry. In a word, competition in China is still at its preliminary stage, concentrated more on simple price cut than long-term planning and creative efforts. When foreign competitors turn to global sourcing and outward processing, move upmarket with high value-added activities, or pursue globalization to survive the competition, many domestic enterprises are still obsessed with quick returns in the short run through simple imitation or large discounts. It's no denying that market competition in China has been boosted to a new level with participation of more enterprises in a more open environment. But it still takes time to form a more order and fair competition, which is crucial to the healthy economic growth.

8.2.3 Some suggestions to domestic textile and apparel (denim) enterprises

Based on the above analysis, it is clear that various internal and external factors are interacting together to form a more dynamic and complicated environment in which enterprises have to make corresponding changes to fit. Moreover, the final success is dependent not only upon individual companies' efforts, but also upon strategic cooperation among companies and the appropriate measures taken by the government.

8.2.3.1 Supply chain reform

To some extent, this reform is of parallel importance to the current SOE reform. The highly fragmented supply chain is one of the most serious bottlenecks for development in the textile and apparel sector. There are various definitions about SCM in the literature. To sum up, supply chain management refers to an integrated management system based on well-defined relationship with suppliers and customers and aimed to facilitate and control the flow of physical goods and associated information throughout the distribution channel in order to create the greatest comparative value for the customer [36]. Since it is an issue far too complicated to discuss in detail here, only some points are raised to shed some light upon the possible reform directions.

A. Quick response (QR)

The key element of fashion is time [54]. The most important aim of supply chain reform for textile and apparel products is to realize quick response, that is, developing the right product and delivering it to the right place at the right time through well-designed coordination of the retailers, suppliers and manufacturers [54, p16-18].

Today, QR is playing a more and more important role in sourcing decisions. The emergence of Mexico and Turkey as the major textile suppliers to America and EU can be partly attributed to their geographic proximity which makes QR easier to realize in addition to the relevant preferential arrangements mentioned above. Labor cost is not as determinative as before, even in some basic items. The increased bargaining power of buyers, especially those large retailers, puts much pressure upon the manufactures to shorten lead time. A survey conducted by the US shows that apparel enterprises adopting QR enjoys 15.4% return on sales, compared with 1.3% without QR and 6.5% if importing from the Far East [37, p43].

Enterprises in China had been rigidly separated horizontally and vertically under the central-planning system. Though the market economy has been established for some time, many enterprises still see competition as vicious and daily operation independent of cooperation with others. As a result, there is little coordination between downstream and upstream enterprises and between manufactures and retailers or wholesalers. Such separatism places the enterprises at a very disadvantageous place when quick response has become a part of strategy in international competition. Table 8.4 shows that lead time in apparel production in

China is longer than other Far East countries and much longer than East European countries and countries such as Algeria and Morocco.

To facilitate production and make the end products more appealing in the market, parties involved in the chain should, first of all, share the information in a true two-way traffic. Application of hard wares such as Bar-coding, Point of Sales scanning and Cross Docking can be solved with enough capital and expertise support. The most difficult part is the actual enforcement requiring trust, frankness and understanding, a kind of philosophy still strange to most textile and apparel enterprises. Moreover, it should be admitted that management level in Chinese enterprises is still low, a problem greatly hindering the enforcement of QR strategy. At present, several things should be done first to pave the way for the QR strategy:

- Changing the old attitude towards competition. Management in the enterprises should realize that strategic cooperation with other chain members and even potential competitors is sometimes necessary and important.
- Sending people to some benchmarked companies to learn their QR practice and gain experience from them. For example, the QR system in Levi Strauss is composed of five parts, including bar-code, retail-related EDI, efficient physical flow, manufacturer-related EDI and flexible manufacturing system [38, p159].[2]
- Conducting experiments in some product lines on a small scale. For example, one manufacturer with one supplier and one retailer concerning one or two product lines. These enterprises should be capital-strong and open-minded since QR strategy requires relatively heavy investment and good management expertise. After such network is successful, then it can be carefully extended to include more participants.

B. Strategic alliance rather than vertical integration

There is now a trend in China to form large company groups through backward or forward vertical integration. This strategy has advantages such as greater control over the costs and supply and higher fixed overhead costs [39], however, it may fail to be flexible in view of quick response requirements because of big size and multi-layered management functions. Moreover, most Chinese enterprises lack necessary capital and management expertise to run the integrated company. Consequently, it may be better to form strategic alliance among chain members rather than vertical integration. The Italian apparel company Benetton here provides an example of de-integration for Chinese textile and apprel enterprises. Benetton produced end products based on orders received. In addition to the IT-based information system and CAD/CAM

Table 8.4[3] Lead times in apparel production in some countries and regions

Average international lead times in apparel production		
Supplier	First order	Repeat order
China	6-7 months	3 months[a]
Bangladesh	6-7 months	not possible
Other Far East	4-5 months	3-5 months[b]
East Europe, Algeria, Morocco, Tunisia, Turkey, Greece	3[c]-6 weeks	3[c]-6 weeks
Other EC	2[c]-4 weeks	2[c]-4 weeks

a: Repeat orders possible but facility rarely used

b: Minimum 7 weeks if goods are shipped by air

c: If fabric is readily available at the factory

(Source: EIU cited in JSDC Vol.110, May/June 1994)

[2] This book is translated by Yan Xiong from Japanese to Chinese

[3] From "Supply Chain Issues in the Textile and Clothing Industry", Graham et al. p36

systems, its supply base is vertically de-integrated, composed of major suppliers with whom Benetton has majority shareholdings and small subcontractors who agree to work exclusively for Benetton in return for stable demand and a 10% profit margin on their sales. Activities that are high value-added or high technology-/capital-intensive are conducted centrally through the major suppliers, such as fabric knitting and weaving, CAD/CAM design, computerized cutting, and dyeing. On the other hand, some low value-added work is decentralized to the subcontractors, including piece work, pressing and garment finishing when economies of scale cannot be achieved and where the quality of customer service will not be hampered. Based upon such arrangement, the company can take advantage of economies of scale where they exist and reduce the company's costs and financial exposure where economies of scale are low. Analysts estimate that the company's production can be compatible with that of the producers in Asia thanks to high technology employed and this de-integrated supply system [54, p26-32]

Strategic alliances are of more flexibility and less financial and risk exposure than actual merger and acquisition, which may be more suitable for present conditions in Chinese textile and apparel enterprises. In addition, the textile and apparel production process can be more easily divided up among individual companies compared with other industries. Enterprises with strength should learn to effectively cooperate with other suppliers, trying to form stable long-term rather than random short-term relationship with other members in the supply chain. At the same time, efforts should be made to improve current poor commercial credit situation by first solving the serious "triangular debt" among enterprises in order to enhance mutual trust and create a friendly business environment.

C. Emphasis upon on-site training

Reform of the supply chain, in the final analysis, is dependent upon competition over human resources. Both information management and supply chain coordination require talented personnel. Moreover, the success of supply chain needs whole-hearted participation of every individual no matter what position he or she is in the company. On-site training gives employees a good opportunity to continue study in today's information-explosive era, which is important to boost morale in addition to material incentives.

Though education in China is now receiving more and more attention, the gap with the developed countries is hard to be narrowed in the short run. According to the survey conducted in 27 denim fabric manufacturers in China between 1997 and 1998, only about 12% of the management received higher education [53]. Lack of qualified technicians and managers may turn out to be the most serious bottleneck in winning the competition for Chinese textile and apparel enterprises. Though government is responsible for developing an educational system meeting the new economic conditions, enterprises should do their share in providing on-site training for their employees, including skill enhancing and horizon widening. It may increase cost in the short run, but it will finally benefit the enterprise in view of management improvement and productivity boost. Employees will have more incentives to work since they gain a lot at the same time. Even when they leave the enterprise, the knowledge they learned will benefit the new one as well, which is ultimately good to the whole society. It is the largest plus out of on-site training.

D. Macro-regulation from the government

As already mentioned in Section 2.4.4.2 (B), government role in current economic restructuring should be shifted from micro-levl to macro-one. As regards supply chain reform, the government support is indispensable, especially concerning the reform in the cotton distribution system.

A relatively stable cotton price and supply is key to Chinese textile industry, particularly to the denim manufacturers who are very heavy cotton users. The reform was carried out since last September by introducing market mechanism to this highly regulated sector with the cotton price set according to the market condition rather than direct administrative order. The cotton price dropped at first. However, when domestic demand rose with the improved situation in the textile industry and high-quality new cotton was in short supply, the cotton price went up again this year. For example, the price of Grade 229 rose from RMB 9,300 per ton in October 1999 to RMB 11,000 in January 2000 [40]. The price of Grade 327 is now up to RMB 12,800 per ton, 24% up within just two months and domestic price for long-staple cotton is RMB 8000 higher than that in the international market [41]. The National Cotton Auction Market began to sell the nation's cotton reserves since December 3,1999 [40] in the hope of alleviating the supply shortage and curbing the price increase. Whether the reform will achieve the desired result is hard to ascertain now, but at least the first step has been taken to liberalize the raw material distribution system, which is good news to the whole supply chain construction.

Another urgent task for the Chinese government is to construct a legal system conducive to the sustainable economic development. The deteriorating commercial credit and serious corruption scandals indicate that current laws and regulations are very inadequate to guarantee an open, fair and transparent competitive environment. If these problems cannot be solved, it will greatly frustrate the supply chair reform whose foundation is built upon a clearly defined legal framework.

8.2.3.2 Focus upon core activities concerning individual companies

This issue is somewhat related to Section 8.2.3.1 (B). Constant changes of consumer tastes and fashion styles as well as shorter life cycle make product-related investment decision more risky and difficult. Moreover, most Chinese textile and apparel enterprises are scarce in capital and expertise. Therefore, it will be more reasonable for them to work out a strategy concentrated upon specialization rather than diversification. However, most enterprises often make decisions regardless of its own strength and weakness as well as demand/supply conditions, leading to either unpractical ambition or shortsighted behavior.

To maintain or capture a larger share in the low-/middle-end markets featuring commodity products, cost may be the most decisive factor in the competition, partly offsetting the long geographical distance with the major importing markets. The most obvious advantage Chinese enterprises enjoy now is the labor and land cost and availability of relatively skilled workers compared with other developing countries such as Mexico. Therefore, how to effectively control the production cost, including the material, labor, capital and management cost, is very important for China's textile and apparel industry to be competitive in these market segments. Cost leadership could be the core strategy for most export-oriented enterprises serving the mass markets.

For those who want to build up their own brands and corporation image, cost is not the single consideration. What matters more is their design and marketing strength, which are the two most important value-added activities in the whole value chain. Marketing is particularly important for a fashion enterprise aiming to build brands and expand market share. Mass customization, which is now valued by some world famous enterprises such as Levi's, is still new in China. It not only needs computer-based technology mentioned earlier, but also requires customer-oriented attitude. To some extent, the latter may be harder to achieve since it is rooted in daily operation and ways of thinking. With Chinese consumers being widely exposed to world fashions either through traditional mass media or through Internet, they want products that are more individual and fashion-oriented. Therefore, enterprises, such as denim manufacturers, should work out marketing strategy based on this new consumer trend to upgrade their service and increase intangible value. A very hot issue these days is e-commerce. However, Chinese textile and apparel enterprises should be careful in taking this approach. Levi Strauss is withdrawing from its direct e-commerce sales, focusing more on its consumers and retail partnership. According to Jeff Beckman, a Levi's spokesman, "running a world class e-commerce site had become unaffordable", while at the same time, upsetting their traditional retail outlets [42]. In view of China's current hard and soft environment, it is still not mature for Chinese enterprises to carry out direct on-line sales as its competitive strategy. It will be more cost efficient for them to concentrate upon traditional marketing approaches since they are more suitable for current economic conditions. This doesn't mean that e-commerce is not viable in China, but blind adoption regardless of reality will just push the enterprises into a more serious dilemma.

In a word, enterprises with vision should focus upon these core activities based on their strength and environment, and at the same time, establish close and long-term relationship with some key suppliers for product manufacturing and with retailers for information gathering and collaboratively product development. Such specialization could more effectively raise the entry barrier and form a competitive edge hard to be imitated or acquired.

8.2.3.3 Think global

It is the time for Chinese enterprises to be global-minded if they really want to survive the international competition. There are two reasons. First, as said before, the global trade is now characterized by regionalism and various complicated non-tariff barriers. One possible way to overcome these obstacles is to get into the regional bloc by directly establishing plants or factories in the target region, such as Mexico or Turkey. Second, compared with some later movers, usually countries in the Latin America or Africa, Chinese textiles and apparel have benefited from their learning curve and enjoyed better value for money. Based on these two reasons, Chinese textile and apparel enterprises with strength can move their production facilities to some developing countries in order to be an insider rather than an outsider under current regionalism move.

Instead of investing individually, it will be much better for several enterprises to act together and coordinatively. This can not only reduce risks and resource inputs for individual company, but also enhance the overall strength and competitiveness. Enterprises should be well informed of the foreign market before they really take steps. Relevant government agencies should play a role here. It can help gather information about the target market, conduct study on its general business environment and practice, and contact foreign association and government.

8.2.3.4 Well prepared for the network economy

The world today is very different from the one twenty years ago. The explosion of information technology, together with significant organizational changes at the global level has ushered in a "new world information economy", featuring deep ubiquitous networking [43, 44]. Textile and apparel sector is an old and traditional industry benefited a lot from the industrial revolution. Though technology in its production process, especially in the apparel sector, evolved slowly during the past 50 years, the network economy will definitely re-shape traditional organizational structure and the way to view and do business.

"The central economic imperative of the network economy is to amplify relationships." [44, p118] Current competition is not just concentrated upon cost or prompt delivery, rather, it is the result of various interactions. It is now necessary to re-think the relationship between the mainland and HK textile and apparel enterprises. Section 2.5.3.5 has discussed the possible impact upon the old "front-store-back-factory" relationship after China enters WTO, with analysis focus upon the WTO-related environmental changes. Under the network economy, fundamental changes are necessary and inevitable. If the two sides fail to change correspondingly, it may turn out to be fatal to the final result.

Both the mainland and HK textile and apparel enterprises are facing big challenges. For instance, though China enjoys low labor cost, it is always easy to find other suppliers with lower labor cost; HK is a fashion center in the Asia, but it still cannot compete with France or Italy concerning fashion design. Moreover, no matter how efficient the infrastructure is, both sides are at a disadvantage over those periphery countries around the US and EU due to long geographic distances. Therefore, enterprises in both sides should build relationships based on "network consideration" rather than "pure labor division", a new cooperation pattern carefully designed to generate synergy, making 1 plus 1 larger than 2. The following "4R" strategy is just a preliminary discussion, hoping to provide some suggestions for the enterprises both in the mainland and in HK.

First, rationalize the value chain by eliminating non value-added processes. For example: in current cooperation pattern, transportation of the final products first from the mainland to HK and then from HK to the final importer is actually a step adding no value but delaying the responsive time. Infrastructure in some coastal cities such as Shanghai and Shenzhen has improved a lot. Possible low efficiency there can be partly attributed to lack of experienced personnel. Therefore, HK trading companies can directly establish offices in some key cities to handle freight forwarding or help train local employees.

Second, realize instant information exchange by installing EDI among the cooperative enterprises. This measure can not only save time and labor in handling paper work, but also greatly facilitate information flow and promote better communication. Speed is key to fashion business. The more time it saves, the more value it may create.

Third, recover new competitive niches by integrating design with hi-tech contents, such as new-type fabric or sophisticated manufacturing process. Fashion design is hard to be protected by patents since a small change can successfully evade the piracy charge. Moreover, both HK and the mainland designers are still not in a position to compete with those in France and Italy. Though it is possible for them to catch up in the future, it is better to establish their own unique competitive advantages by finding new niches now.

The ability to meet consumers' demand has become a basic requirement. To create and change what the customers want beyond today's capabilities is more crucial for the success [36, 44]. As mentioned in Section 2.5.3.5, if the mainland and HK can

8.1 Strategies corresponding to current dynamic external and internal environment

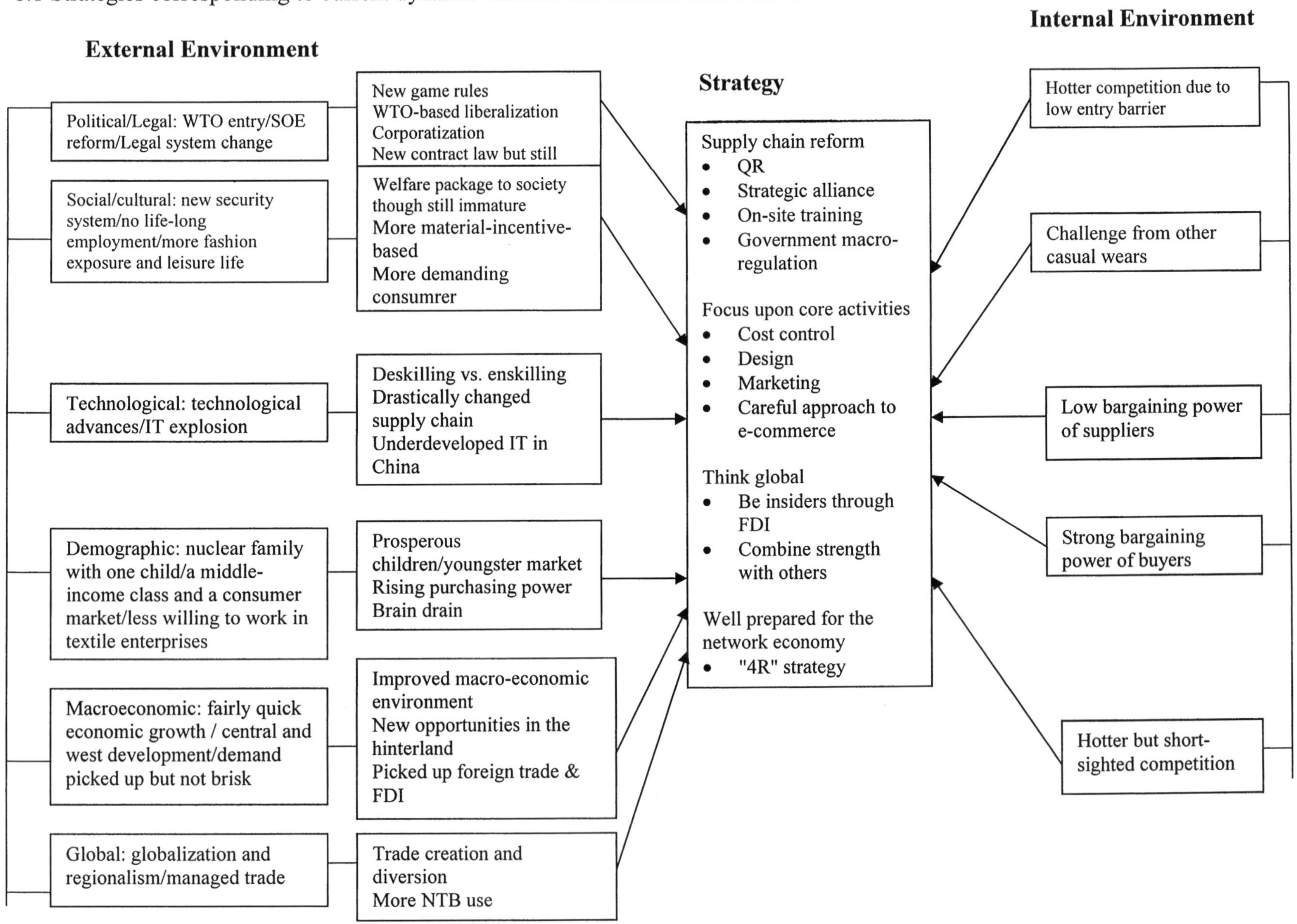

collaborate on R & D, based on the former's strength in basic scientific researches and the latter's good market sense and strong capital support, they have the chance to move upmarket and secure a position relatively difficult to be attacked. The success largely depends on the close cooperation between the two sides in identifying the market niche and analyzing its feasibility and future prospect.

Fourth, re-design current network by adding new members into the net. "As the number of nodes in a network increases arithmetically, the value of the network increases exponentially." [44, p23] HK investment is currently focused upon the Pearl River Delta. In view of China's huge domestic market which is actually composed of several different sub-markets, there is a large space for the current network to be expanded to other regions, such as those second-tier cities in the central and west of China. The "product life cycle" theory used by some MNCs to make investment decision can be applied to the mainland market as well. The product lines can be moved progressively from the east to the west and from the coast to the hinterland based on the product life cycle, which can not only prolong the product's life, but also maximize profit. This again requires members in the net to act together in market research and eliminate obstacles in the distributive channel.

8.2.4 Summary

The above analysis draws a picture for current external and internal environment in which domestic textile and apparel enterprises are competing. Figure 8.1 is the summary of the above-discussed issues. No enterprise can win the game without paying full attention to the drastically changed environment. Only those who clearly understand today's market can find a way out of the hyper-competition.

REFERENCES

1. Fong, W.S., Denim developments. Textile Asia, 1987 p.186-189.
2. Downey, L., History of Denim.2000 Levi Strauss & Co.
3. Duetsch, L.L., Industry Studies.1993 Englewood Cliffs, N.J: Prentice Hall.182-206.
4. Anon, Levis Launch Engineered Jeans at 40. JTN Monthly, 2000 p.50.
5. Carpio, L., Denim is back. Textile Asia, 2000 p.93.
6. Zimmermann, R., Denim: what might the future hold for the world's hottest textile sector ITS Textile Leader, 1998 p.34-48.
7. Modi, S.K., India--The emerging denim giant. JTN Monthly, 1997 p.78.
8. Lu, Q., Chinese jeans makers face crisis. JTN Monthly, 1999 p.41.
9. Graham, D., Denim: a century's working fabric, in The Toronto Star.1999.
10. Kilduff, P., Textiles in a New Environment. Textile Asia, 2000 p.44-48.
11. Anon, Strengthening social security.1999 China Daily.
12. Anon, Medical Net Cast Wider.2000 China Daily.
13. Anon, State plans to expand social net.2000 China Daily.
14. Anon, Nation moves boldly forward.2000 China Daily.
15. Gao, Z., Review of the competitive mechanism in the Chinese mainland's labor market (in Chinese). www.macrochina.com.cn.
16. Lennox-Kerr, P., Innovations in Fibers, Textiles and Machinery. Textile Outlook International, 1999 p.129-157.

17. Taplin, I.M., Backwards into the future: new technologies and old work organization in the US clothing industry, in Rethinking global production: a comparative analysis of restructuring in the clothing industry J.W. Ian M Taplin Editor.1997 Ashgate Aldershot, Brookfield USA, Singapore, Sydney.p.50.
18. Hamilton, P., The changing textile world. Textile Asia, 2000 p.58-60.
19. Anon, Development prospect of HK apparel industry (in Chinese).2000 www.ctei.gov.cn.
20. Glasse, J., Supply chain management in China: pitfalls and opportunities to 2005.1999 London: Financial Times Retail & Consumer.
21. Anon, Economy mixed for 2000.2000 China Daily.
22. Anon, Government policy plays a role: the macro economic situation in July 2000 (in Chinese).2000 The State Information Center, together with provincial and municipal information centers.
23. Anon, Import and Export in August 2000 (in Chinese).2000 MOFTEC.
24. Anon, FDI in China in 1999 (in Chinese).2000 China Ministry of Foreign Trade and Economic Cooperation.
25. Anon, FDI in China between January and July 2000 (in Chinese).2000 China Ministry of Foreign Trade and Economic Cooperation.
26. Yuan, Z., The review and forecast for the China's apparel industry (in Chinese). China Textile & Apparel, 2000 p.86.
27. Anon, Index: deflation easing.2000 China Daily.
28. Anon, Retailing volume for consumer goods between January and July in 2000, (in Chinese).2000 The State Information Center, together with provincial and municipal information centers.
29. Anon, Daixianglong: the overall plan for liberalization of interest rate within three years (in Chinese).2000 www.macrochina.com.cn.
30. Lawrence, R.Z., Regionalism, multilateralism, and deeper integration: changing paradigms for developing countries, in Trade rules in the making: challenges in regional and multilateral negotiations P.L. Miguel R. Mendoza, and Barbara Kotschwar Editor.1999 Organization of American States, Brookings Institution Press Washington, D. C.p.23.
31. Anon, Why a free trade agreement (FTA) now .2000 www.jetro.go.jp (Japan External Trade Organization).
32. Viner, J., The custom union issue, in Trading blocs: alternative approaches to analyzing preferential trade agreements P.K. Jagdish Bhagwati, Arvind Panagariya Editor.1999 MIT press Cambridge, Mass.p.105-119.
33. Anon, Mexico-EU bilateral trade agreement to benefit garment and knitting industries (in Chinese). China Textile & Apparel, 2000 p.38.
34. VanGrasstek, C., "Labor Rights", by, p509, in the book, in Trade rules in the making: challenges in regional and multilateral negotiations P.L. Miguel R. Mendoza, and Barbara Kotschwar Editor.1999 Organization of American States, Brookings Institution Press Washington DC.p.509.
35. Rigby, D., The Future of European Textiles--Mills or Heritage Center JTN Monthly, 1997 p.15-19.
36. Ross, D.F., Competing through supply chain management: creating market-winning strategies through supply chain partnership.1997 Boston Kluwer Academic Publichers.
37. Graham, J.a.K., Supply Chain Issues in the Textile & Clothing Industry.1995 Textile Intelligence Center.

38. MITI, J., The Future of Japanese Textile Industry.1998 Shanghai China Textile University Press.
39. Miller, A., Strategic Management.1998 Boston Irwin/McGraw-Hill.68-310.
40. Lu, Q., Outlook on Cotton Market. JTN Monthly, 2000 p.46-47.
41. Anon, An overview of China's textile and apparel industry (in Chinese). CHINA TEXTILE & APPAERL, 2000 p.78.
42. Anon, Levi Strauss Stepping Away from Direct E-Commerce Sales.2000.
43. Martin Carnoy, M.C., Stephen S. Cohen, Fernando Henrique Cardoso, The New Global Economy in the Information Age.1993 Pennsylvania The Pennsylvania State University Press.6.
44. Kelly, K., New Rules for the New Economy: 10 Ways the Network Economy is Changing Everything.1998 London Fourth Estate Limited.
45. Pushpa Bajaj, Rohit Agarwal., Innovations in denim production. American Dyestuff Reporter. p 26-38
46. The jeans encyclopedia of legend, past and present, 1992 Sportswear International
47. Kurt Hoffman, Howard Rush., Microelectronics and clothing, the impact of technical change on a global industry, 1988, New York: Praeger, p11
48. Jonathan Winterton, Ian M Taplin., Making sense of strategies for survival: clothing in high wage economies. Ian M Taplin, Jonathan Winterton edited, Rethinking global production: a comparative analysis of restructuring in the clothing industry. Aldershot, Ashgate.
49. Hong Kong TDC., Economic development in the Northwest of China: business opportunities for HK (in Chinese). Hong Kong Trade Development Council, 2000, pi
50. Anon., The State Information Center, together with provincial and municipal information centers, 2000, September 14, http://dailynews.cei.gov.cn/t091414.htm
51. John Whalley, Colleen Hamilton., The Trading System after the Uruguay Round, Washington, DC, Institute for International Economics, p 103-122
52. Jagdish Bhagwati., Regionalism and multilateralism, Jagdish Bhagwati, etc. edited, Trading blocs: alternative approaches to analyzing preferential trade agreements, Cambridge, Mass. MIT press, p 3-33
53. Zhiming Zhang, Chen Gong, Gary Ching., Analysis of the China denim wear industry (in Chinese), China Textile & Apparel, Aug/Sept.2000, p 54-55.
54. Graham, JF and Kilduf., Supply Chain Issues in the Textile & Clothing Industry, Textile Intelligence Center
55. Pushpa Bajaj, Rohit Agarwal., Innovations in denim production,. American Dyestuff Reporter, p 26-38
56. Roshan Paul, Sandeep R. Naik, Jegadeesh Thampi., Denim: the evergreen favorite, Textile Dyer & Printer, August 14, 1996, p 15-1

9 Conclusion

The past century has not washed the charm of denim, rather, it has been welcomed by more and more people in different countries. The US, as the home of denim, remains as the leading denim consumer in the world. With the establishment of NAFTA in 1994, trade with Mexico has prospered, providing about 52.3% of the denim products to the world between January and November 1998. EU is the second largest market for denim products, who is a net denim fabric exporter and jeans importer. About 53% of EU denim product imports and 79.4% of EU denim product exports were conducted within EU member in 1997, underlying the more and more importance of regional trading. Though Japan was a late mover in denim manufacturing, it quickly gained wide recognition for its ability to supply quality and technologically innovative products. Its trade focus is upon the Asia, especially the East Asian countries, despite lack of any form of FTAs.

Since the three countries consume about three quarters of world denim [Zimmermann, 1998 #2], their trade policy and priority exert much influence upon the overall international trading pattern. In view of current textile and apparel trade in the world, several distinctive features can be identified, including international division of labor, increased share of developing countries at the expense of developed ones, rapid trade creation within member nations and frequent use of non-tariff barriers to manage trade. Analysis of denim trading in these markets further strengthens these arguments.

Though HK is still a very important textile and apparel supplier, it has been overtaken by the Chinese mainland in recent years. However, thanks to its well-developed infrastructure and trade expertise, its entrepot trade volume is now much larger than normal one. China's WTO entry is both an opportunity and a threat to HK's textile and apparel industry. The final result is largely dependent upon the industry's own efforts and determination to meet the possible challenges, such as hotter competition in the major export markets and further facilities re-location.

Textile and apparel industry in China is now undergoing restructuring process with SOE reform as the most priority task. The entry into WTO puts another drastic change to the external environment. Though most public opinion thinks that this sector will benefit a lot from WTO entry, analysis of the diamond structure shows that it cannot be overoptimistic about the sector's future, especially before the year 2005. Moreover, if the diamond fails to upgrade to a higher level, the sector will face a more difficult condition when more players enter the picture and when competition is conducted in a quota-free environment. Lack of a textile cluster may prove to be fatal to its future development and move upmarket.

Today's business environment is very different from that of 50 years ago. Tariff protection is replaced by various complicated non-tariff ones, which are hard to be compromised through multilateral talk. Regionalism is on the rise, with the formation of NAFTA and envisioned EU enlargement as typical examples. Proliferation of information technology makes QR technologically feasible, partly leading to the rapid changes in the whole supply chain. Consumers want more value for money and in some developed countries, "green" consumption has been a trend now. All of these political, economic and social changes are interacting together, pressurizing textile and apparel enterprises to make strategically adjustment in the new network economy. Within the network, individual companies cannot be as independent as before. They may have to be closely involved in a certain form of alliance with others to survive the more adverse and dynamic environment. No matter what long-term vision they hold, both the business scope and strategy should be re-defined to meet the new economy. Some may argue that the textile and apparel industry is labor-intensive in nature, therefore, it is less subject to impacts encountered by those hi-tech or service industries. It is somewhat true if the industry is viewed in an isolated situation. However, today's world is interwoven together by various kinds of relationships and multi-layered sophisticated information technology. No single industry can be left intact under such a drastically changed competitive environment. Moreover, with more entrants and relatively stable demand, the competition frontier has been moved to more abstract areas, such as QR, green products, more segmented market niches and new marketing approaches. In this sense, current information era will definitely re-shape the industrial structure as well as individual company's ways of thinking. Therefore, Chinese textile and apparel enterprises should be well prepared for the coming hotter competition in a more extensive frontier and in a more dynamic environment.

Printed in Poland
by Amazon Fulfillment
Poland Sp. z o.o., Wrocław

36399849R00148